Systems of Evolution Equations
with Periodic and Quasiperiodic Coefficients

Mathematics and Its Applications (*Soviet Series*)

Volume 87

Systems of
Evolution Equations
with Periodic and
Quasiperiodic Coefficients

by

Yu. A. Mitropolsky,

A. M. Samoilenko

and

D. I. Martinyuk
Institute of Mathematics,
Kiev, Ukraine, C.I.S.

SPRINGER-SCIENCE+BUSINESS MEDIA, B.V.

Library of Congress Cataloging-in-Publication Data

Mitropol'skiĭ, ĨU. A. (ĨUriĭ Alekseevich), 1917-
Systems of evolution equations with periodic and quasiperiodic
coefficients / by Yu. A. Mitropolsky, A.M. Samoilenko, and
D.I. Martinyuk.
      p.   cm. -- (Mathematics and its applications. Soviet series ;
87)
   Includes index.
   ISBN 978-94-010-5210- 8   ISBN 978-94-011-2728-8 (eBook)
   DOI 10.1007/978-94-011-2728-8
   1. Evolution equations.   I. Samoĭlenko, A. M. (Anatoliĭ
Mikhaĭlovich) II. Martinyuk, D. I. III. Title. IV. Series:
Mathematics and its applications (Kluwer Academic Publishers).
Soviet series ; 87.
QA371.M625 1992
515'.353--dc20   92-37365

   ISBN 978-94-010-5210-8

This is a translation of the original work
Systems of Evolution Equations with
Periodic and Conditionally Periodic Coefficients
Published by Nauka, Moscow, © 1984

SERIES EDITOR'S PREFACE

'Et moi, ..., si j'avait su comment en revenir, je n'y serais point allé.'

Jules Verne

The series is divergent; therefore we may be able to do something with it.

O. Heaviside

One service mathematics has rendered the human race. It has put common sense back where it belongs, on the topmost shelf next to the dusty canister labelled 'discarded nonsense'.

Eric T. Bell

Mathematics is a tool for thought. A highly necessary tool in a world where both feedback and non-linearities abound. Similarly, all kinds of parts of mathematics serve as tools for other parts and for other sciences.

Applying a simple rewriting rule to the quote on the right above one finds such statements as: 'One service topology has rendered mathematical physics ...'; 'One service logic has rendered computer science ...'; 'One service category theory has rendered mathematics ...'. All arguably true. And all statements obtainable this way form part of the raison d'être of this series.

This series, *Mathematics and Its Applications*, started in 1977. Now that over one hundred volumes have appeared it seems opportune to reexamine its scope. At the time I wrote

"Growing specialization and diversification have brought a host of monographs and textbooks on increasingly specialized topics. However, the 'tree' of knowledge of mathematics and related fields does not grow only by putting forth new branches. It also happens, quite often in fact, that branches which were thought to be completely disparate are suddenly seen to be related. Further, the kind and level of sophistication of mathematics applied in various sciences has changed drastically in recent years: measure theory is used (non-trivially) in regional and theoretical economics; algebraic geometry interacts with physics; the Minkowsky lemma, coding theory and the structure of water meet one another in packing and covering theory; quantum fields, crystal defects and mathematical programming profit from homotopy theory; Lie algebras are relevant to filtering; and prediction and electrical engineering can use Stein spaces. And in addition to this there are such new emerging subdisciplines as 'experimental mathematics', 'CFD', 'completely integrable systems', 'chaos, synergetics and large-scale order', which are almost impossible to fit into the existing classification schemes. They draw upon widely different sections of mathematics."

By and large, all this still applies today. It is still true that at first sight mathematics seems rather fragmented and that to find, see, and exploit the deeper underlying interrelations more effort is needed and so are books that can help mathematicians and scientists do so. Accordingly MIA will continue to try to make such books available.

If anything, the description I gave in 1977 is now an understatement. To the examples of interaction areas one should add string theory where Riemann surfaces, algebraic geometry, modular functions, knots, quantum field theory, Kac-Moody algebras, monstrous moonshine (and more) all come together. And to the examples of things which can be usefully applied let me add the topic 'finite geometry'; a combination of words which sounds like it might not even exist, let alone be applicable. And yet it is being applied: to statistics via designs, to radar/sonar detection arrays (via finite projective planes), and to bus connections of VLSI chips (via difference sets). There seems to be no part of (so-called pure) mathematics that is not in immediate danger of being applied. And, accordingly, the applied mathematician needs to be aware of much more. Besides analysis and numerics, the traditional workhorses, he may need all kinds of combinatorics, algebra, probability, and so on.

In addition, the applied scientist needs to cope increasingly with the nonlinear world and the extra mathematical sophistication that this requires. For that is where the rewards are. Linear models are honest and a bit sad and depressing: proportional efforts and results. It is in the non-linear world that infinitesimal inputs may result in macroscopic outputs (or vice versa). To appreciate what I am hinting at: if electronics were linear we would have no fun with transistors and computers; we would have no TV; in fact you would not be reading these lines.

There is also no safety in ignoring such outlandish things as nonstandard analysis, superspace and anticommuting integration, p-adic and ultrametric space. All three have applications in both electrical engineering and physics. Once, complex numbers were equally outlandish, but they frequently proved the shortest path between 'real' results. Similarly, the first two topics named have already provided a number of 'wormhole' paths. There is no telling where all this is leading - fortunately.

Thus the original scope of the series, which for various (sound) reasons now comprises five sub-series: white (Japan), yellow (China), red (USSR), blue (Eastern Europe), and green (everything else), still applies. It has been enlarged a bit to include books treating of the tools from one subdiscipline which are used in others. Thus the series still aims at books dealing with:

-   a central concept which plays an important role in several different mathematical and/or scientific specialization areas;
-   new applications of the results and ideas from one area of scientific endeavour into another;
-   influences which the results, problems and concepts of one field of enquiry have, and have had, on the development of another.

The shortest path between two truths in the real domain passes through the complex domain.

J. Hadamard

La physique ne nous donne pas seulement l'occasion de résoudre des problèmes ... elle nous fait pressentir la solution.

H. Poincaré

Never lend books, for no one ever returns them; the only books I have in my library are books that other folk have lent me.

Anatole France

The function of an expert is not to be more right than other people, but to be wrong for more sophisticated reasons.

David Butler

Bussum

Michiel Hazewinkel

CONTENTS

vii

PREFACE

The investigation of a great number of problems in celestial mechanics, physics and engineering involves the study of oscillating systems governed by systems of nonlinear ordinary differential equations or by systems of partial differential equations. The methods of investigation of periodic and quasiperiodic solutions for these systems are highly developed and presented in many fundamental monographs.

The recent growth of interest in the theory of differential equations with deviating argument and, especially, in the theory of differential equations with lag (and equations reducible to these) has been caused by practical needs and, in the first place, by the rapid development of the theory of automatic regulation systems. Equations with retarded argument have a wide spectrum of applications to the problems arising in automation, telemechanics, radiolocation, electrical and radio communication, rocket engineering, shipbuilding, theoretical cybernetics, medicine, biology, etc. (Zverkin, Kamensky, Norkin, and Elsgoltz, 1962, 1963; Krasovsky, 1959; Martinyuk, 1971; Myshkis, 1972; Pelyukh and Sharkovsky, 1974; Pinney, 1961).

Advances in technical sciences also contributed to the growth of interest in the difference equations which proved to be a convenient model for the description of pulse and discrete dynamical systems and systems which include digital computing devices as a component (Boltyansky, 1973; Halanay and Veksler, 1971). Moreover, difference equations are encountered when various classes of differential equations are solved numerically by the finite differences method.

The problems concerning the existence of periodic solutions and construction of algorithms for calculating these solutions are among the most important aspects of the investigation of nonlinear systems of differential equations with lag and various classes of integro-differential and difference equations. Many well-known methods give good solutions for weakly nonlinear systems with lag (i.e., for systems with a small parameter). Among these one should mention Bogolyubov-Krylov's asymptotical method (Bogolyubov and Mitropolsky, 1963; Volosov and Morgunov, 1971; Grebennikov, 1968; Grebennikov and Ryabov, 1978, 1979; Krylov and Bogolyubov, 1937), Poincaré's method of a small parameter, Shimanov's method of auxiliary systems, and Lyapunov's method of functional equations which was developed for weakly nonlinear systems by Yu.A.Ryabov (see, e.g., (Bogolyubov, 1945; Bogolyubov, Mitropolsky, and Samoilenko, 1969; Martinyuk and Kozubovskaya, 1968; Martinyuk and Kolomiets, 1968; Martinyuk, Mironov, and Kharabovskaya, 1971; Martinyuk and Fodchuk, 1963, 1966; Mitropolsky and Martinyuk, 1969, 1979; Mitropolsky, 1964; Mitropolsky and Moseenkov, 1976; Rubanik, 1969; Rozhkov, 1968, 1970; Ryabov, 1960a,b, 1961, 1962, 1964; Hale, 1966; Shimanov, 1957, 1959, 1960, 1965, 1970)).

However, the application of these already classical methods to the investigation of periodic solutions of strongly nonlinear systems with lag (i.e., systems without a small parameter) is possible not always. The investigation of periodic solutions for these systems can be carried out efficiently by employing the topological methods (Borisovich, 1963, 1967; Borisovich and Subbotin, 1967; Krasnoselsky, Lifshitz, and Strygin, 1967; Krasnoselsky, 1963, 1965, 1966). But these methods cannot give an algorithm for calculating periodic solutions, and therefore, the creation of methods, which can give both the proof of the theorems on the existence of periodic solutions and the algorithm for calculating these solutions, is a current problem.

Bubnov-Galerkin's method is one of these. For nonlinear systems of ordinary differential equations the fairly complete justification of this method was presented in the works by M.Urabe (1965, 1966) and L.Cessari (1963). However, when justifying the applicability of Bubnov-Galerkin's method to systems with lag, one should overcome the fundamental difficulties. The idea how to do this was suggested in the work by Samoilenko and Nurzhanov (1979) devoted to the investigation of a certain class of integro-differential equations.

The numerical-analytic method for the investigation of periodic solutions of nonlinear systems of ordinary differential equations has been proposed by one of the authors. This method enables us to construct periodic solutions for these systems in the form of uniformly convergent sequences of periodic functions. In the monograph, we justify the applicability of the numerical-analytic method to the investigation of periodic solutions for various classes of nonlinear systems with heredity. We also clarify the influence of a lag on the existence of these solutions.

In many problems of celestial mechanics, physics, and engineering we encounter processes with time dependence which is not periodic but can be expressed in the form of trigonometric sums. This was the reason which gave rise to the investigation of almost periodic solutions of differential equations, and of differential equations with almost periodic coefficients. An important subclass of almost periodic functions should be particularly mentioned. They have the following characteristic peculiarity: Their spectrum is everywhere dense on the real axis (Bol, 1961). The problems concerning the existence of quasiperiodic solutions were studied for various classes of ordinary differential equations by N.Bogolyubov (1945), Yu.Mozer (1968), M.Urabe (1972, 1973), A.Samoilenko and I.Parasyuk (Samoilenko and Parasyuk, 1977; Parasyuk, 1978), and others. In the problems of nonlinear mechanics, the construction of quasiperiodic solutions by computer was realized by Yu.Ryabov and I.Tolmachev (1969).

At the same time, in many cases quasiperiodic solutions are not preserved under perturbations whereas isolated invariant toroidal manifolds are more stable. The perturbation theory for invariant toroidal manifolds of systems of ordinary differential equations was developed by Yu.Mozer (1968), A.Samoilenko (1970a,b), R.Sacker (1965, 1969), Yu.Neimark (Neimark, 1972, 1978; Gurtovnik, Kogan and Neimark, 1975), and others. The problems concerning the existence of compact closed invariant manifolds (not necessary toroidal) were studied by N.Bogolyubov, O.Lykova, Ya.Kurzweil, Yu.Neimark, V.Pliss, S.Smale, Ch.Pew, M.Peysoto, I.Kupka (see, e.g., Mitropolsky and Lykova,

1974; Pliss, 1964)).

For systems with lag, these problems have been investigated less satisfactorily. Indeed, though a series of results has been obtained up to now in the theory of local invariant manifolds for systems with lag by V.Fodchuk (1965, 1970), A.Halanay (1965), J.Hale (1966), Yu.Neimark (1975), A.Zverkin (1970), and others, there exists a comparatively small number of works devoted to the study of invariant toroidal manifolds for these systems (Martinyuk and Samoilenko, 1974a, 1976; Ordynskaya, 1976, 1977, 1979).

After the works by A.Kolmogorov and V.Arnold (see, e.g., (Arnold, 1961, 1963a,b; Grebennikov, and Ryabov, 1971)), N.Bogolyubov (1964) developed the modernized method of successive changes of variables which guarantees the rapid convergence of the corresponding expansions. By using this method, Yu.Mitropolsky (1968) studied the behavior of solutions of a nonlinear system of differential equations in the vicinity of a quasiperiodic solution. The problems concerning the reducibility of the nonlinear systems to linear systems with constant coefficients were also investigated in this work. The behavior of trajectories of the vector field of finite smoothness on the m-dimensional torus and reducibility of linear systems with quasiperiodic coefficients were examined in the works (Mitropolsky and Samoilenko, 1964, 1965, 1972a,b, 1976a,b; Mitropolsky, Samoilenko and Perestyuk, 1977; Samoilenko, 1964, 1968).

The important results concerning the location of integral curves of systems of nonlinear equations in the vicinities of smooth toroidal invariant manifold and compact invariant manifold were also obtained by use of the method of rapid convergence (Bogolyubov, Mitropolsky, and Samoilenko, 1969; Samoilenko, 1966a,b).

When the method of rapid convergence had been created, there appeared the possibility of the fairly complete investigation of the reducibility of linear systems of difference-differential (Bortei and Fodchuk, 1976, 1979) and difference equations with quasiperiodic coefficients, and of nonlinear systems of difference equations on toroidal sets and in their vicinities. Note that include all these types of equations in a single class and call them evolution equations, because they can describe real systems whose rate of changes at given time depends on the state of the system at this and preceding time moments.

Hence, the construction and investigation of periodic and quasiperiodic solutions for various classes of differential equations with deviating argument, for integro-differential and difference equations, and for other types of equations are current and important problems. The solutions of many of these problems are presented in the monograph.

In Chapter 1, we study periodic solutions for nonlinear systems of evolution equations including differential equations with lag, systems of neutral type, various classes of nonlinear systems of integro-differential equations, etc. The numerical-analytic method for the investigation of periodic solutions of these evolution equations is presented. According to this method, we seek periodic solutions as a limit of uniformly convergent sequence of periodic functions. These sequences can be constructed with the help of the algorithm suggested in this chapter; moreover, by using these sequences one can prove the existence of the corresponding exact periodic solutions of the evolution systems. The scheme of calculation (which enables us to construct periodic solutions approximately) and the estimate of accuracy of this scheme are also given in this chapter. This scheme is rather convenient in practice, since it reduces the construction of a uniformly convergent sequence to the calculation of integrals of trigonometric polynomials.

In Chapter 2 and Chapter 3, the problems concerning the existence of periodic and quasiperiodic solutions for systems with lag are examined. For a nonlinear system with quasiperiodic coefficients and lag, we prove the theorem, which establishes the conditions under which quasiperiodic solutions exist. When proving this theorem, we point out the method for calculating these solutions. According to this method, the determination of quasiperiodic solutions is reduced to finding periodic solutions of the special system of partial differential equations. In order to calculate periodic solutions, we use Bubnov-Galerkin's method. We present the justification of applicability of this method and prove the important theorems on existence of Bubnov-Galerkin's approximations and convergence of these approximations to the exact periodic solution. For this purpose, we exploit a new (unknown before) idea: to employ Green's functions for the problem of periodic solutions when constructing these solutions.

Chapter 4 is devoted to the study of invariant toroidal manifolds for various classes of systems of differential equations with quasiperiodic coefficients. We prove theorems on the existence of invariant toroidal manifolds for various classes of systems with lag and point out the conditions under which these manifolds exist. We also present the method, with the help of which these manifolds can be constructed, and investigate the behavior of trajectories of nonlinear systems with quasiperiodic coefficients and lag in the vicinities of exponentially stable toroidal manifolds.

In Chapter 5, we examine the problem concerning the reducibility of a linear system of difference equations with quasiperiodic coefficients to a linear system of difference equations with constant coefficients. We consider the case when the right-hand side of the equation is not analytic but has a finitely many derivatives.

Chapter 6 contains the investigation of invariant toroidal sets for the systems of difference equations with quasiperiodic coefficients. Here, as we see this, the interesting result is obtained, namely, the theorem on existence of the continuous torus is proved for the discrete dynamical system. In this chapter, we also present a method which enables us to construct the toroidal set in the form of a uniformly convergent sequence of toroidal sets and study the behavior of trajectories of a nonlinear system of difference equations on the torus and in its vicinity.

The Authors

1. NUMERICAL-ANALYTIC METHOD OF INVESTIGATION OF PERIODIC SOLUTIONS FOR SYSTEMS WITH AFTEREFFECT

One of the most important problems in the investigation of nonlinear systems of differential equations with aftereffect is the study of their periodic solutions and the construction of an algorithm for calculating these. Different methods are employed for the investigation of periodic solutions; they vary from graphic to topological ones. Most of these methods give good results for weakly nonlinear systems with aftereffect (systems that contain a so-called small parameter). The asymptotic method, Poincaré's method of a small parameter, Shimanov's method of auxiliary systems, etc., are all methods of this sort.

However, the application of these already-classical methods to the study of periodic solutions of systems with aftereffect which are essentially nonlinear (i.e., systems without a small parameter) is sometimes impossible. Therefore, the selection of certain classes of nonlinear systems of differential equations and the creation of methods which can always be applied to these classes are current problems.

In this chapter, we consider the numerical-analytic method for the investigation of periodic solutions to various classes of nonlinear systems with aftereffect. This method was suggested by Samoilenko (1965, 1966a) for the investigation of periodic solutions to nonlinear systems of ordinary differential equations.

Investigations of the nonlinear systems with lag (Martinyuk, Samoilenko, 1967a, b; 1970), neutral type systems (Samoilenko and Ronto, 1976; Samoilenko, Martinyuk, and Perestyuk, 1973; Samoilenko and Nurzhanov, 1979; Tkach, 1969), countable systems with lag (Martinyuk, 1967, 1968a, b, c), different classes of nonlinear systems of integro-differential equations (Vakhabov, 1968, 1969; Vuitovich, 1982; Vuitovich and Nurzhanov, 1982; Martinyuk and Kharabovskaya, 1970), etc., have favored the development of this method.

Using the numerical-analytic method, we can find periodic solutions of systems with aftereffect in the form of uniformly convergent sequences of periodic functions. Moreover, by employing the functions of this sequence we can determine whether these solutions exist. We first introduce some notations and prove some auxiliary statements, and then proceed to the presentation of the method itself.

1

§1. Notations and Basic Ideas. Auxiliary Lemmas

Let $x = (x_1, x_2, ..., x_n), y = (y_1, y_2, ..., y_n)$ and $f = (f_1, f_2, ..., f_n)$ be points of an n-dimensional Euclidean space E_n, let D be a closed bounded region of E_n, and let Γ_D be its boundary. For $x \in E_n$, we denote by $|x|$ a vector $|x| = (|x_1|, |x_2|, ..., |x_n|)$ and by $\|x\|$ its octahedral norm $\|x\| = \sum_{i=1}^{n} |x_i|$. M denotes a vector with real valued nonnegative components $M = (M_1, M_2, ..., M_n)$ and $K_1 = \{k'_{ij}\}$ and $K_2 = \{k''_{ij}\}$ denote $(n \times n)$-dimensional matrices with nonnegative elements k'_{ij} and k''_{ij}, in the ith row and jth column, respectively. For the vector function $f(t, x, y) = (f_1(t, x, y), f_2(t, x, y), ..., f_n(t, x, y))$, we denote the following vector

$$|f(t, x, y)|_0 = (\max_t |f_1(t, x, y)|, \max_t |f_2(t, x, y)|, , \max_t |f_n(t, x, y)|),$$

and scalar

$$\|f(t, x, y)\|_0 = \max_i \sum_{i=1}^{n} |f_i(t, x, y)|$$

by $|f(t, x, y)|_0$ and $\|f(t, x, y)\|_0$, respectively.

The integral time average of a vector function periodic in t with period ω

$$\overline{f(t, x, y)} = \frac{1}{\omega} \int_0^\omega f(t, x, y)\, dt$$

will be denoted by $\overline{f(t, x, y)}$.

$D - M\omega/2$ denotes the set of points from D, belonging to D with their $M\omega/2$ vicinity (an $M\omega/2$ vicinity of a point $\bar{x} \in E_n$ is a set of points $x \in E_n$ satisfying the inequalities $|x_i - \bar{x}_i| \leq M_i \omega/2$ $(i = 1, 2, ..., n)$.

Consider the system of different equations

$$\frac{dx(t)}{dt} = f(t, x(t), x(t - \Delta)), \qquad (1.1)$$

where Δ is a constant that characterizes a lag in the system; $f(t, x, y)$ is a vector function periodic in t with period ω defined for all $-\infty < t < \infty, x \in D$ and $y = x(t - \Delta) \in D$.

Assume that $f(t, x, y)$ is a continuous function with respect to its variables $t, x,$ and y such that

$$|f(t, x, y)| \le M \tag{1.2}$$

and the Lipschitz condition with matrices K_1 and K_2

$$|f(t, x_1, y_1) - f(t, x_2, y_2)| \le K_1 |x_1 - x_2| + K_2 |y_1 - y_2| \tag{1.3}$$

holds for all $-\infty < t < \infty, x, x_1, x_2, y, y_1, y_2 \in D$. The inequalities (1.2) and (1.3) should be understood as follows

$$|f_i(t, x, y)| \le M_i;$$

$$|f_i(t, x_1, y_1) - f_i(t, x_2, y_2)| \le \sum_{j=1}^{n} k'_{ij} |x_{1j} - x_{2j}| + \sum_{j=1}^{n} k''_{ij} |y_{1j} - y_{2j}|.$$

Further, we shall consider only those systems (1.1) for which the region D, vector M, matrices K_1, K_2, lag Δ and period ω satisfy the following conditions:

(i) the set $D - M\omega/2$ is nonempty;
(ii) the largest eigenvalue λ_{max} of the matrix

$$Q = (K_1 + K_2) \left(\frac{\omega}{3} + \frac{3}{2\omega} \Delta^2 \left(1 - \frac{\Delta}{\omega} \right)^2 \right)$$

is less that unity ($\lambda_{max} < 1$).

As follows from Perron's theorem (Bellman, Kalaba, 1968), the largest eigenvalue of the matrix Q is real and nonnegative (since its elements are nonnegative). It can be estimated from above by

$$\left(\frac{\omega}{3} + \frac{3}{2\omega} \Delta^2 \left(1 - \frac{\Delta}{\omega} \right)^2 \right) \min \left\{ \max_i \sum_{j=1}^{n} (k'_{ij} + k''_{ij}), \max_j \sum_{i=1}^{n} (k'_{ij} + k''_{ij}) \right\}. \tag{1.4}$$

If (1.3) is given in terms of the vectors K_1 and K_2 as

$$|f(t, x_1, y_1) - f(t, x_2, y_2)| \le K_1 |x_1 - x_2| + K_2 |y_1 - y_2| \tag{1.5}$$

or in terms of scalars k_1 and k_2 as

$$|f_i(t, x_1, y_1) - f_i(t, x_2, y_2)| \le k_1 \| x_1 - x_2 \| + k_2 \| y_1 - y_2 \|, \tag{1.6}$$

then, by virtue of (1.4), the condition $\lambda_{\max} < 1$ is satisfied for

$$q_1 = \left(\|K_1\|+\|K_2\|\right)\left(\frac{\omega}{3}+\frac{3}{2\omega}\Delta^2\left(1-\frac{\Delta}{\omega}\right)^2\right) < 1; \tag{1.7}$$

$$q_2 = \left(k_1+k_2\right)n\left(\frac{\omega}{3}+\frac{3}{2\omega}\Delta^2\left(1-\frac{\Delta}{\omega}\right)^2\right) < 1, \tag{1.8}$$

respectively.

Note also that, for solutions of (1.1) periodic in t with period ω, we can assume without loss of generality that $0 \le \Delta \le \omega$. In fact, Δ can otherwise be represented as $\Delta = m\omega + \Delta_1$, where m is an integer, $m \ge 1$, and Δ_1 satisfies the inequality $0 \le \Delta_1 \le \omega$. Since the solution of (1.1) is periodic in t with period ω, the ω-periodic solutions of the equation

$$\frac{dx(t)}{dt} = f(t, x(t), x(t-\Delta_1))$$

and the periodic solution of (1.1) clearly coincide.

The next statements (Samoilenko, 1965; Elsgolts and Norkin, 1971) are of great importance for the study of periodic solutions.

Lemma 1.1. *Let f(t) be a continuous function on the segment $0 \le t \le \omega$. Then for all t from this segment the inequality*

$$\left|\int_0^t \left[f(t)-\overline{f(t)}\right]dt\right| \le 2|f(t)|_0\, t\left(1-\frac{t}{\omega}\right)$$

holds.

Proof. Let us rewrite the left-hand side of this inequality as follows

$$\int_0^t\left[f(t)-\frac{1}{\omega}\int_0^\omega f(t)dt\right]dt = \int_0^t f(t)dt -\frac{t}{\omega}\int_0^\omega f(t)dt$$

$$= \int_0^t f(t)dt -\frac{t}{\omega}\int_0^t f(t)dt -\frac{t}{\omega}\int_t^\omega f(t)dt$$

$$= \left(1-\frac{t}{\omega}\right)\int_0^t f(t)dt -\frac{t}{\omega}\int_t^\omega f(t)dt.$$

Hence,

$$\left| \int_0^t [f(t) - \overline{f(t)}] \, dt \right| \le \left(1 - \frac{t}{\omega}\right) \int_0^t |f(t)|_0 \, dt + \frac{t}{\omega} \int_t^\omega |f(t)|_0 \, dt$$

$$= \left(1 - \frac{t}{\omega}\right) t \, |f(t)|_0 + \frac{t}{\omega} (\omega - t) |f(t)|_0 = 2 \, |f(t)|_0 \, t \left(1 - \frac{t}{\omega}\right).$$

Lemma 1.2. *Let* $\psi(t)$ *be an initial function given on the initial set* $t_0 - \Delta \le t \le t_0$, *and let a continuous function* $f(t, x, y)$ *and a function* $\Delta(t)$ *be periodic in t with period* ω *(it is possible that* $\Delta = \text{const}$*). Then a solution* $x_\psi(t) = \varphi(t)$ *of the system*

$$\frac{dx(t)}{dt} = f(t, x(t), x(t - \Delta(t))),$$

periodic in t with period ω *is determined by the initial vector function* $\psi(t)$ *which is a periodic continuation of the solution* $\varphi(t)$ *onto the initial set.*

Proof. Let us choose an integer m such that the inequality

$$\max_{t_0 \le t \le t_0 + m\omega} \Delta(t) < m\omega$$

holds.

The solution $\varphi(t)$ is defined for $t \ge t_0 + m\omega$ by the initial vector function $\varphi(t)$ given on the initial set $E_{t_0 + m\omega}$ (all points of this set satisfy the inequality $t \ge t_0$). Consequently, for $t \ge t_0 + m\omega$, we have

$$\frac{d\varphi(t)}{dt} \equiv f(t, \varphi(t), \varphi(t - \Delta(t))).$$

By virtue of the periodicity of both its sides, this identity remains valid if t is replaced by $t - m\omega$, provided that the vector function $\varphi(t)$ on the right-hand side is assumed to be periodically extended onto E_{t_0}.

In what follows (unless otherwise stated), we study only those periodic solutions (determined by initial vector functions) which are periodic extensions of periodic solutions onto initial sets.

§2. Algorithm for Finding Periodic Solutions of Nonlinear Systems with Lag

Suppose that (1.1) has a periodic solution with period ω, and the point $x_0 \in D - M\omega/2$ through which this solution passes at time $t = t_0 = 0$ is known. Then the algorithm for finding this solution is given by

Theorem 1.1 (Martinyuk, Samoilenko, 1967a). *Suppose that the system* (1.1) *satisfies conditions (i) and (ii) of §1, and let* $x(t) = \varphi(t)$ *be a solution of* (1.1) *periodic in t with period* ω. *Then*

$$\varphi(t) = \lim_{m \to \infty} x_m(t, x_0) \tag{2.1}$$

in the sense of uniform convergence with respect to $-\infty < t < \infty$, $x_0 \in D - M\omega/2$, *and*

$$|\varphi(t) - x_m(t, x_0)| \leq Q^m (E - Q)^{-1} \frac{M\omega}{2} \tag{2.2}$$

for all $m = 0, 1, 2, \ldots$, *where* $x_m(t, x_0)$ $(x_0(t, x_0) = x_0)$ *are periodic functions with period* ω *determined by*

$$x_m(t, x_0) = x_0 + \int_0^t [f(t, x_{m-1}(t, x_0), x_{m-1}(t - \Delta, x_0))$$

$$- \overline{f(t, x_{m-1}(t, x_0), x_{m-1}(t - \Delta, x_0))}] \, dt. \tag{2.3}$$

Proof. Consider the sequence of functions

$$x_1(t, x_0) = x_0 + \int_0^t [f(t, x_0, x_0) - \overline{f(t, x_0, x_0)}] \, dt \,;$$

$$x_2(t, x_0) = x_0 + \int_0^t [f(t, x_1(t, x_0), x_1(t - \Delta, x_0))$$

$$- \overline{f(t, x_1(t, x_0), x_1(t - \Delta, x_0))}] \, dt \,;$$

. . . .

$$x_m(t, x_0) = x_0 + \int_0^t [f(t, x_{m-1}(t, x_0), x_{m-1}(t - \Delta, x_0))$$

$$- \overline{f(t, x_{m-1}(t, x_0), x_{m-1}(t - \Delta, x_0))}] \, dt.$$

Each is periodic in t with period ω. Moreover, by virtue of lemma 1.1, we have, for $x_0 \in D$ and $0 \le t \le \omega$,

$$|x_m(t, x_0) - x_0| \le 2 M t (1 - t/\omega) = \alpha_1(t) M, \qquad (2.4)$$

where $\alpha_1(t) = 2t (1 - t/\omega)$. For $0 \le t \le \omega$, we have $\alpha_1(t) \le \omega/2$; therefore, (2.4) yields

$$|x_m(t, x_0) - x_0| \le 2 M \omega/2,$$

i.e., $x_m(t, x_0) \in D$ for $x_0 \in D - M\omega/2$ and $0 \le t \le \omega$.

By induction we find that for all $m = 0, 1, 2, \ldots$ and each $x_0 \in D - M\omega/2$ the functions $x_m(t, x_0)$ exist, are periodic in t with period ω and belong to D.

Employng (1.2), (1.3) and conditions (i), (ii), we can prove the convergence of the sequence $\{x_m(t, x_0)\}$. With a view to doing this, we represent the difference $x_2(t, x_0) - x_1(t, x_0)$ as follows

$$x_2(t, x_0) - x_1(t, x_0) = \left(1 - \frac{t}{\omega}\right) \int_0^t [f(t, x_1(t, x_0), x_1(t - \Delta, x_0))$$

$$(2.5)$$

$$-f(t, x_0, x_0)] \, dt - \frac{t}{\omega} \int_0^\omega [f(t, x_1(t, x_0), x_1(t - \Delta, x_0)) - f(t, x_0, x_0)] \, dt.$$

If we estimate (2.5), taking into account (2.4), then we get

$$|x_2(t, x_0) - x_1(t, x_0)| \le \left(1 - \frac{t}{\omega}\right) \int_0^t [K_1 |x_1(t, x_0) - x_0|$$

$$+ K_2 |x_1(t - \Delta, x_0) - x_0|] \, dt + \frac{t}{\omega} \int_t^\omega [K_1 |x_1(t, x_0) - x_0|$$

$$+ K_2 |x_1(t - \Delta, x_0) - x_0|] \, dt = \left[\left(1 - \frac{t}{\omega}\right) \int_0^t \alpha_1(t) \, dt + \frac{t}{\omega} \int_t^\omega \alpha_1(t) \, dt\right] K_1 M$$

$$+K_2\left[\left(1-\frac{t}{\omega}\right)\int\limits_{-\Delta}^{\omega-\Delta}|x_1(t_1,x_0)-x_0|\,dt_1+\frac{t_1}{\omega}\int\limits_{t-\Delta}^{\omega-\Delta}|x_1(t_1,x_0)-x_0|\,dt_1\right] \tag{2.6}$$

for all $0\le t\le\omega$. Since the difference $x_1(t_1,x_0)-x_0$ is periodic in t_1, it follows from the inequality (2.4) that

$$|x_1(t_1,x_0)-x_0|\le\begin{cases}\alpha_1(t_1+\omega)M,&-\Delta\le t_1\le0,\\\alpha_1(t_1)M,&0\le t_1\le\omega-\Delta.\end{cases} \tag{2.7}$$

Here, we assume that $\alpha_1(t)$ can be periodically extended to $t\in(-\infty,\infty)$. Taking (2.7) into account, we can rewrite the inequality (2.6) as follows

$$|x_2(t,x_0)-x_1(t,x_0)|\le\alpha_2(t)K_1M+\beta_2(t,\Delta)K_2M, \tag{2.8}$$

where

$$\alpha_2(t)=\left(1-\frac{t}{\omega}\right)\int\limits_0^t\alpha_1(t)\,dt+\frac{t}{\omega}\int\limits_t^\omega\alpha_1(t)\,dt=\frac{\omega}{6}\alpha_1(t)+\frac{\alpha_1^2(t)}{3}; \tag{2.9}$$

$$\beta_2(t,\Delta)=\begin{cases}\beta_2'(t,\Delta),&0\le t\le\Delta,\\\beta_2''(t,\Delta),&\Delta\le t\le\omega;\end{cases} \tag{2.10}$$

$$\beta_2'(t,\Delta)=\left(1-\frac{t}{\omega}\right)\int\limits_{-\Delta}^{\omega-\Delta}\alpha_1(t_1+\omega)\,dt_1+\frac{t}{\omega}\left[\int\limits_{t-\Delta}^0\alpha_1(t_1+\omega)\,dt_1+\int\limits_0^{\omega-\Delta}\alpha_1(t_1)\,dt_1\right];$$

$$\beta_2''(t,\Delta)=\left(1-\frac{t}{\omega}\right)\left[\int\limits_{-\Delta}^0\alpha_1(t_1+\omega)\,dt_1+\int\limits_0^{t-\Delta}\alpha_1(t_1)\,dt_1\right]+\frac{t}{\omega}\int\limits_{t-\Delta}^{\omega-\Delta}\alpha_1(t_1)\,dt_1.$$

For $\beta_2'(t,\Delta)$ and $\beta_2''(t,\Delta)$, we find

$$\beta_2'(t,\Delta)=\alpha_2(t)+t(1-t/\Delta)(1-2t/\omega)\alpha_1(\Delta);$$

$$\beta_2''(t,\Delta)=\alpha_2(t)-\Delta(1-\Delta/t)(1-2t/\omega)\alpha_1(t). \tag{2.11}$$

Estimating $\beta_2'(t,\Delta)$ for $0\le t\le\Delta$, we obtain

$$\beta_2'(t,\Delta)\le\alpha_2(t)\le\alpha_1(t)\frac{\omega}{3},\qquad t\ge\frac{\omega}{2}; \tag{2.12}$$

$$\beta_2'(t, \Delta) \le \alpha_2(t) + \alpha_1(t)\left(1 - \frac{2t}{\omega}\right)\frac{\alpha_1(\Delta)}{2}$$

$$= \alpha_1(t)\left[\frac{\omega}{6} + \frac{\alpha_1(t)}{3} + \left(1 - \frac{2t}{\omega}\right)\frac{\alpha_1(\Delta)}{2}\right] \le \alpha_1(t)\left[\frac{\omega}{3} + \frac{3\alpha_1^2(\Delta)}{8\omega}\right], \quad t \le \frac{\omega}{2};$$

(2.13)

the inequalities (2.12) and (2.13) together yield

$$\beta_2'(t, \Delta) \le \alpha_1(t)\left[\frac{\omega}{3} + \frac{3\alpha_1^2(\Delta)}{8\omega}\right], \quad 0 \le t \le \Delta. \tag{2.14}$$

Estimating $\beta_2''(t, \Delta)$ for $\Delta \le t \le \omega$, we find

$$\beta_2''(t, \Delta) \le \alpha_2(t) \le \alpha_1(t)\frac{\omega}{3}, \quad t \le \frac{\omega}{2}; \tag{2.15}$$

$$\beta_2''(t, \Delta) \le \alpha_2(t) + \frac{\alpha_1(\Delta)}{2}\alpha_1(t)\left(\frac{2t}{\omega} - 1\right)$$

$$\le \alpha_1(t)\left[\frac{\omega}{6} + \frac{\alpha_1(t)}{3} + \left(\frac{2t}{\omega} - 1\right)\frac{\alpha_1(\Delta)}{2}\right] \le \alpha_1(t)\left[\frac{\omega}{3} + \frac{3\alpha_1^2(\Delta)}{8\omega}\right], \quad t \ge \frac{\omega}{2}.$$

(2.16)

Combining (2.15) with (2.16), we obtain

$$\beta_2''(t, \Delta) \le \alpha_1(t)\left[\frac{\omega}{3} + \frac{3\alpha_1^2(\Delta)}{8\omega}\right], \quad 0 \le t \le \omega. \tag{2.17}$$

The inequalities (2.14) and (2.17) yield the following estimate

$$\beta_2(t, \Delta) \le \alpha_1(t)\left[\frac{\omega}{3} + \frac{3\alpha_1^2(\Delta)}{8\omega}\right], \quad 0 \le t \le \omega. \tag{2.18}$$

By using (2.9) and (2.18), we can rewrite the inequality (2.8) as follows

$$|x_2(t, x_0) - x_1(t, x_0)| \le \alpha_1(t)\left[\frac{\omega}{3} + \frac{3\alpha_1^2(\Delta)}{8\omega}\right](K_1 + K_2)M, \quad 0 \le t \le \omega. \tag{2.19}$$

Assume now that $x_m(t, x_0) - x_{m-1}(t, x_0)$ satisfies the inequality

$$| x_m(t, x_0) - x_{m-1}(t, x_0) | \leq \alpha_1(t) \left[\left(\frac{\omega}{3} + \frac{3\alpha_1^2(\Delta)}{8\omega} \right) (K_1 + K_2) \right]^{m-1} M = \alpha_1(t) M_m, \quad (2.20)$$

for all t from the segment $[0, \omega]$ and fixed $m \geq 2$. Let us show that $| x_{m+1}(t, x_0) - x_m(t,x_0) |$ satisfies the inequality

$$| x_{m+1}(t, x_0) - x_m(t, x_0) | \leq \alpha_1(t) \left[\left(\frac{\omega}{3} + \frac{3\alpha_1^2(\Delta)}{8\omega} \right)^m (K_1 + K_2) \right]^m M = \alpha_1(t) M_{m+1}. \quad (2.21)$$

If we represent the difference $x_{m+1}(t, x_0) - x_m(t, x_0)$ as follows

$$x_{m+1}(t, x_0) - x_m(t, x_0) = \left(1 - \frac{t}{\omega} \right) \int_0^t [f(t, x_m(t, x_0), x_m(t - \Delta, x_0))$$

$$- f(t, x_{m-1}(t, x_0), x_{m-1}(t - \Delta, x_0))] dt - \frac{t}{\omega} \int_t^\omega [f(t, x_m(t, x_0), x_{m1}(t - \Delta, x_0))$$

$$- f(t, x_{m-1}(t, x_0), x_{m-1}(t - \Delta, x_0))] dt$$

and estimate it taking into account (2.20), then we get

$$| x_{m+1}(t, x_0) - x_m(t, x_0) | \leq \alpha_1(t) K_1 M_m + \beta_1(t, \Delta) K_1 M_m$$

$$\leq \alpha_2(t) \left[\frac{\omega}{3} + \frac{3\alpha_1^2(\Delta)}{8\omega} \right] (K_1 + K_2) M_m \leq \alpha_1(t) \left[\left(\frac{\omega}{3} + \frac{3\alpha_1^2(\Delta)}{8\omega} \right) (K_1 + K_2) \right]^m M ,$$

i.e., (2.21). By induction, we find that

$$| x_{m+1}(t, x_0) - x_m(t, x_0) | \leq \alpha_1(t) \left[\left(\frac{\omega}{3} + \frac{3\alpha_1^2(\Delta)}{8\omega} \right) (K_1 + K_2) \right]^m M \qquad (2.22)$$

for all $0 \leq t \leq \omega$ and all $m = 1, 2, \dots$.

Since $\alpha_1(t) \leq \omega/2$ for $0 \leq t \leq \omega$, and $x_m(t, x_0)$ is a function periodic in t with period ω, the inequality (2.22) implies

$$| x_{m+1}(t, x_0) - x_m(t, x_0) | \leq \frac{\omega}{2} \left[\left(\frac{\omega}{3} + \frac{3\alpha_1^2(\Delta)}{8\omega} \right) (K_1 + K_2) \right]^m M \qquad (2.23)$$

for all $-\infty < t < \infty$ and all $m = 1, 2, \ldots$ This inequality together with condition (ii) proves the uniform convergence of the sequence (2.3) with respect to $(t, x_0) \in (-\infty, \infty) \times (D - M \omega/2)$.

Denoting the limiting function of the sequence $\{x_m(t, x_0)\}$ by $x_\infty(t, x_0)$ and proceeding to the limit in (2.3) as $m \rightarrow \infty$, we find that $x_\infty(t, x_0)$ is a periodic solution of the equation

$$x(t, x_0) = x_0 + \int_0^t [f(t, x(t, x_0), x(t - \Delta, x_0))$$

$$- \overline{f(t, x(t, x_0), x(t - \Delta, x_0))}] \, dt. \tag{2.24}$$

Moreover, the following estimates are valid for $x_\infty(t, x_0)$: According to Lemma 1.1, we have

$$|x_\infty(t, x_0) - x_0| \leq M \omega/2 \tag{2.25}$$

for all $-\infty < t < \infty$ and $x_0 \in D - M \omega/2$. Taking into account condition (ii) and the inequality

$$|x_{m+k}(t, x_0) - x_m(t, x_0)| \leq \alpha_1(t) \sum_{i=0}^{k-1} \left[\left(\frac{\omega}{3} + \frac{3\alpha_1^2(\Delta)}{8\omega} \right)(K_1 + K_2) \right]^{m+i} M, \tag{2.26}$$

which can easily be derived from (2.23), we find

$$|x_\infty(t, x_0) - x_m(t, x_0)| \leq Q^m (E - Q)^{-1}(M \omega/2), \tag{2.27}$$

where E is the unit matrix.

The function $x(t) = \varphi(t)$ is a periodic solution of the equation (1.1) by condition of the theorem. This means that it satisfies the equation

$$x(t) = x_0 + \int_0^t f[t, x(t), x(t - \Delta)] \, dt \tag{2.28}$$

and possesses the property

$$\overline{f(t, \varphi(t), \varphi(t - \Delta))} = 0. \tag{2.29}$$

It follows from (2.24), (2.28), and (2.29) that $\varphi(t)$ is a periodic solution of eqn.(2.24) (just as $x_\infty(t, x_0)$). Therefore, to complete the proof of the theorem, it suffices to show that (2.24) cannot have two different solutions. We prove this by contradiction.

Let $x(t, x_0)$ and $y(t, x_0)$ be two different solutions of (2.24). Then we can estimate the difference $x(t, x_0) - y(t, x_0)$, for $0 \leq t \leq \omega$, as follows

$$| x(t, x_0) - y(t, x_0) | \leq \left(1 - \frac{t}{\omega} \right) \int_0^t K_1 | x(t, x_0) - y(t, x_0) | \, dt$$

$$+ \frac{t}{\omega} \int_t^\omega K_1 | x(t, x_0) - y(t, x_0) | \, dt + \left(1 - \frac{t}{\omega} \right) \int_0^t K_2 | x(t - \Delta, x_0) - y(t - \Delta, x_0) | \, dt$$

$$+ \frac{t}{\omega} \int_t^\omega K_2 | x(t - \Delta, x_0) - y(t - \Delta, x_0) | \, dt . \qquad (2.30)$$

We set $| x(t, x_0) - y(t, x_0) | = r(t)$ and rewrite (2.30) as

$$r(t) \leq (K_1 + K_2) \left[\left(1 - \frac{t}{\omega} \right) \int_0^t |r(t)|_0 \, dt + \frac{t}{\omega} \int_t^\omega |r(t)|_0 \, dt \right], \qquad (2.31)$$

hence,

$$r(t) \leq (K_1 + K_2) \, \alpha_1(t) \, |r(t)|_0, \quad 0 \leq t \leq \omega. \qquad (2.32)$$

Substitution of the right-hand side of (2.32) for $r(t)$ (on the right-hand side of (2.30)) yields

$$r(t) \leq \left(1 - \frac{t}{\omega} \right) \int_0^t K_1 \alpha_1(t) (K_1 + K_2) \, |r(t)|_0 \, dt + \frac{t}{\omega} \int_t^\omega K_1 \alpha_1(t) (K_1 + K_2) \, |r(t)|_0 \, dt$$

$$+ \left(1 - \frac{t}{\omega} \right) \int_0^t K_2 \, |r(t - \Delta)| \, dt + \frac{t}{\omega} \int_t^\omega K_2 |r(t - \Delta)| \, dt$$

$$\leq K_1 (K_1 + K_2) \alpha_2(t) |r(t)|_0 + K_2 (K_1 + K_2) \beta_2(t, \Delta) |r(t)|_0$$

$$\leq \alpha_1(t) (K_1 + K_2)^2 \left(\frac{\omega}{3} + \frac{3\alpha_1^2(\Delta)}{8\omega} \right) |r(t)|_0. \qquad (2.33)$$

If we now substitute the right-hand side of (2.33) for $r(t)$ on the right-hand side of (2.30) and iterate this procedure again and again, then after the nth iteration we get

$$r(t) \le \alpha_1(t) \, (K_1 + K_2) \, Q^{n-1} \, |r(t)|_0, \tag{2.34}$$

(the matrix Q was defined in §1). It follows from (2.34) that

$$|r(t)|_0 \le \frac{\omega}{2}(K_1 + K_2) \, Q^{n-1} \, |r(t)|_0. \tag{2.35}$$

Proceeding to the limit in (2.35) as $n \to \infty$ and taking condition (ii) into account, we obtain

$$|r(t)|_0 \le 0,$$

i.e., $r(t) \equiv 0$. This completes the proof of Theorem 1.1.

§3. Existence of Periodic Solutions

It follows from Theorem 1.1 that the process of finding a periodic solution of the system (1.10) can be reduced to the computation of the functions $x_m(t, x_0)$ provided that this solution exists and the point x_0 through which it passes at $t = 0$, is known.

If $x_m(t, x_0)$ are known, the problem of the existence of periodic solutions can be solved as follows.

Denote

$$T(x_0) = \frac{1}{\omega} \int_0^\omega f(t, x_\infty(t, x_0), x_\infty(t - \Delta, x_0)) \, dt, \tag{3.1}$$

where $x_\infty(t, x_0)$ is the limit of the sequence of periodic functions with period ω

$$x_m(t, x_0) = x_0 + \int_0^t [f(t, x_{m-1}(t, x_0), x_{m-1}(t - \Delta, x_0))$$

$$- \overline{f(t, x_{m-1}(t, x_0), x_{m-1}(t - \Delta, x_0))}] \, dt. \tag{3.2}$$

Since $x_\infty(t, x_0)$ is a solution of the equation

$$x_\infty(t, x_0) = x_0 + \int_0^t [f(t, x_\infty(t, x_0), x_\infty(t - \Delta, x_0))$$

$$-\overline{f(t, x_\infty(t, x_0), x_\infty(t-\Delta, x_0))}]\,dt, \tag{3.3}$$

it is a periodic solution of (1.1) provided that $T(x_0) = 0$.

So, the problem of the existence of periodic solutions for (1.1) is connected with the problem of the existence of zeros of the function $T(x_0)$. The points x_0 for which $T(x_0) = 0$ are singular points of the mapping

$$T: D - M\,\omega/2 \to E_n, \quad T(x_0) = \overline{f(t, x_\infty(t, x_0), x_\infty(t-\Delta, x_0))}. \tag{3.4}$$

We can find the mapping (3.4) only approximately, e.g., by calculating the functions

$$T^m(x_0) = \frac{1}{\omega}\int_0^\omega f(t, x_m(t, x_0), x_m(t-\Delta, x_0))\,dt. \tag{3.5}$$

Therefore, we encounter the following problem: To solve the problem of zeros of the mapping (3.4), and consequently, the problem of periodic solutions for (1.1), beginning with the mapping (3.5). The solution of this problem is given by the following theorem

Theorem 1.2. (Martinyuk, Samoilenko, 1967a). *Suppose that the system*

$$\frac{dx(t)}{dt} = f(t, x(t), x(t-\Delta)), \tag{3.6}$$

given in some region D of the space E_n, satisfies conditions (i) and (ii) and the follo-wing conditions

(a) for some integer m, the function $T^m(x_0)$ has an isolated singular point

$$T^m(x_0^0) = 0;$$

(b)   the index of this point is nonzero;

(c)   there exists a closed convex region D_1 that belongs to $D - M\,\omega/2$ and has the unique singular point x_0^0 such that the inequality

$$\inf_{x\in\Gamma_{D_1}} \sum_{i=1}^n |T_i^m(x)| \geq \sum_{i=1}^n \left\{ Q^m(E-Q)^{-1}(K_1+K_2)\frac{M\omega}{3} \right\}_i \tag{3.7}$$

holds on its boundary.

Then (3.6) has a periodic solution x(t) for which $x(0) \in D_1$.

Proof. The index of an isolated singular point x^0 of the continuous mapping

$T^m(x_0)$ is equal, by definition, to the characteristic of the vector field (generated by T^m) on a sufficiently small sphere S^n centered at x^0. The region D_1 contains no singular points other than x^0 and is homeomorphic to the unit sphere in E_n (see, e.g., Krasnoselsky, Perov, Povolotsky, and Zabreiko, 1963); therefore, the characteristic of the vector field T^m on the sphere S^n is equal to the characteristic of the same field on Γ_{D_1}.

The fields T^m and T are homotopic on Γ_{D_1}.

The last statement follows from the fact that the family of vector fields (which are continuous everywhere on Γ_{D_1})

$$V(\Theta, x_0) = T^m(x_0) + \Theta(T(x_0) - T^m(x_0)), \tag{3.8}$$

continuously depending on the parameter Θ ($0 \le \Theta \le 1$) and connecting the fields $V(0,x_0) = T^m(x_0)$ and $V(1, x_0) = T(x_0)$, is nonzero everywhere on Γ_{D_1}.

Indeed, by using the relations (2.27), (3.1), and (3.5), we find

$$| T(x_0) - T^m(x_0) | \le \frac{1}{\omega} \int_0^\omega [K_1 |x_m(t, x_0) - x_\infty(t, x_0)| + K_2 | x_m(\Delta, x_0)$$

$$- x_\infty(t - \Delta, x_0)|] \, dt \le Q^m (E - Q)^{-1} \left\{ \frac{K_1}{\omega} \int_0^\omega \alpha_1(t) dt \right.$$

$$\left. + \frac{K_2}{\omega} \left[\int_{-\Delta}^0 \alpha_1(t_1 + \omega) dt_1 + \int_0^{\omega - \Delta} \alpha_1(t_1) dt_1 \right] \right\} M$$

$$= Q^m (E - Q)^{-1} (K_1 + K_2) \frac{M\omega}{3} \le Q^m (E - Q)^{-1}; \tag{3.9}$$

hence, the inequality

$$\| V(\Theta, x_0) \| \ge \| T^m(x_0) \| - \| T(x_0) - T^m(x_0) \| > 0 \tag{3.10}$$

holds on Γ_{D_1}.

Since the characteristics of the fields homotopic on compact sets are equal (Krasnoselsky, Burd, and Kolesov, 1970; Krasnoselsky, Perov, Povolotsky, and Zabreiko, 1963), the characteristic of the field T on Γ_{D_1} is equal to the index of the singular point x^0 of the field T^m and is thus nonzero. By virtue of Theorem 5.12 in (Alexandrov, 1947), this is sufficient in order that the vector field T possess a singular point in D_1, i.e., the point x_0^0 such that

$$T(x_0^0) = 0. \tag{3.11}$$

Hence, the point x_0^0 for which $T(x_0^0) = 0$ exists, and therefore, the periodic solution of (3.6) exists, as well.

In applications of this theorem to all concrete systems of equations, it is necessary to find indices of singular points and determine the region D_1 such that (3.7) holds on its boundary.

These indices can always be found in the case of a plane, i.e., when $E_n = E_2$ (see, e.g., (Krasnoselsky, Perov, Povolotsky, and Zabreiko, 1963)). For the space with dimensionality more than two, the calculation of the index is more difficult. However, in this case, we have a number of criteria which allow us to determine whether the index is non-zero. Thus, in particular, if the right-hand side of (3.6) is differentiable in the vicinity of the point x^0 and $\det \left| \left| \dfrac{dT^m(x_0)}{dx} \right| \right| \neq 0$, then the index of the point x^0 is nonzero. Moreover, it is also nonzero in the case when $T^m(x_0)$ maps the neighborhood of x^0 onto its image continuously and in a one-to-one manner. For the continuously differentiable vector field, the index of a singular point can be also found in the n-dimensional case (see, e.g., (Petrovsky, 1964)).

The choice of the region D_1 such that (3.7) holds on its boundary can be made somewhat arbitrarily. In particular, for systems periodic in time which have the standard form

$$\frac{dx(t)}{dt} = \varepsilon X(t, x(t), x(t - \Delta)), \qquad \Delta > 0, \tag{3.12}$$

where ε is a small positive parameter, any sufficiently small sphere centered at the singular point can be taken as the region D_1 provided that this singular point is isolated.

As follows from the above discussion, the problem of existence of periodic solutions for the system (3.12) is solved by the following statement.

Theorem 1.3. (Martinyuk and Samoilenko, 1967a). *Assume that the right-hand side of* (3.12) *is defined in the region*

$$-\infty < t < \infty, \qquad (x, y) \in D \times D \tag{3.13}$$

is periodic in t *with period* ω, *and continuous with respect to its variables* t, x, *and* y. *Suppose that it satisfies the inequalities*

$$|X(t, x, y)| \leq M,$$

$$\tag{3.14}$$

$$|X(t, x_1, y_1) - X(t, x_2, y_2)| \leq K_1 |x_1 - x_2| + K_2 |y_1 - y_2|.$$

Assume also that the averaged system

$$\frac{d\xi(t)}{dt} = \varepsilon X_0(\xi(t), \xi(t-\Delta)),$$

$$\left(X_0(x, y) = \frac{1}{\omega}\int_0^\omega X(t, x, y)\,dt\right) \tag{3.15}$$

has the isolated equilibrium point $\xi = \xi_0$:

$$X_0(\xi_0, \xi_0) = 0, \tag{3.16}$$

and the index of the point ξ_0 *of the mapping* $X(\xi_0, \xi_0)$ *is nonzero.*

Then the system (3.12) *has a periodic solution with period* ω *for sufficiently small* ε.

Theorem 1.3 justifies the averaging principle for the equations with the retarded argument of the standard form.

For the case when $E_n = E_1$, i.e., when x is a scalar, Theorem 1.2 can be strengthened, since we need not demand that the singular point must be isolated.

Theorem 1.4. (Martinyuk and Samoilenko, 1967a). *Let the right-hand side of the equation*

$$\frac{dx(t)}{dt} = f(t, x(t), x(t-\Delta)) \tag{3.17}$$

be periodic in t *with period* ω, *continuous with respect to its variables* t, x, *and* y, *and defined in the region* $-\infty < t < \infty$, $a \le x \le b$, *and* $a \le y \le b$. *Suppose that the following inequalities hold for it*

$$|f(t, x, y)| \le M,$$

$$|f(t, x_1, y_1) - f(t, x_2, y_2)| \le k_1|x_1 - x_2| + k_2|y_1 - y_2|,$$

$$q = (k_1 + k_2)\frac{\omega}{3} + \frac{3}{2\omega}\Delta^2\left(1 - \frac{\Delta}{\omega}\right)^2 < 1.$$

Suppose also that the function (3.5) *satisfies, for some* $m \ge 1$, *the inequalities*

$$\min_{a+\frac{\omega}{2}M \le x \le b-\frac{\omega}{2}M} T^m(x) \le -\frac{q^m}{1-q}(k_1+k_2)\frac{M\omega}{3},$$

(3.18)

$$\max_{a+\frac{\omega}{2}M \le x \le b-\frac{\omega}{2}M} T^m(x) \ge \frac{q^m}{1-q}(k_1+k_2)\frac{M\omega}{3}.$$

Then (3.17) *has a periodic solution* $x = x(t)$ *such that* $a + M\,\omega/2 \le x(0) \le b - M\omega/2$.

Proof. Let x_1 and x_2 be points of the segment $\left[a+\frac{\omega}{2}M, b-\frac{\omega}{2}M\right]$ such that

$$T^m(x_1) = \min_{a+\frac{\omega}{2}M \le x \le b-\frac{\omega}{2}M} \{T^m(x)\}, \quad T^m(x_2) = \max_{a+\frac{\omega}{2}M \le x \le b-\frac{\omega}{2}M} \{T^m(x)\} .$$

Taking into account the inequalities (3.9) and (3.18), we get

$$T(x_1) = T^m(x_1) + (T(x_1) - T^m(x_1)) \le 0 ,$$

(3.19)

$$T(x_2) = T^m(x_2) + (T(x_2) - T^m(x_2)) \ge 0 .$$

By virtue of the continuity of $T(x)$, the relations (3.19) imply that there exists a point x^0, $x^0 \in [x_1, x_2]$ such that $T(x_0) = 0$. This means that the equation (3.17) has a solution periodic with period ω.

§4. Construction of an Approximate Periodic Solution

Suppose that the system (1.1) possesses a periodic solution. Then, in order to construct this solution, one should compute the sequence of functions $x_m(t, x_0)$ given by (2.3) and find the point x_0 through which this solution passes at the initial time $t = t_0 = 0$. When finding $x_m(t, x_0)$, one must integrate different periodic functions, and this can be done by employing the following computation scheme.

We approximate $f(t, x, y)$ by some periodic function $P(t, x, y)$ polynomial with respect to $x = (x_1, x_2, ..., x_n)$ and $y = (y_1, y_2, ..., y_n)$:

$$P(t, x, y) = \sum_{i=1}^{l} \alpha_i(t) P_i(x, y),$$ (4.1)

where $P_i(x, y)$ are polynomials with respect to x and y, and $\alpha_i(t)$ are periodic functions. Suppose that the deviation of $f(t, x, y)$ from $P(t, x, y)$ does not exceed $a = (a_1, a_2, \ldots, a_n)$, i.e.,

$$|f(t, x, y) - P(t, x, y)| \le a.$$ (4.2)

Having calculated the integrals

$$\bar{a}_0^i = \frac{1}{\omega} \int_0^\omega \alpha_i(t) \, dt,$$

$$\bar{a}_k^i = \frac{2}{\omega} \int_0^\omega k^2 \alpha_i(t) \cos k\omega_1 t \, dt,$$ (4.3)

$$\bar{b}_k^i = \frac{2}{\omega} \int_0^\omega k^2 \alpha_i(t) \sin k\omega_1 t \, dt, \qquad \omega_1 = \frac{2\pi}{\omega},$$

for $k = 1, 2, \ldots, N_i$, we can construct the expression

$$P_1(t, x, y) = \sum_{i=1}^{l} \left\{ \bar{a}_0^i + \sum_{k=1}^{N_i} \left(\frac{\bar{a}_k^i}{k^2} \cos k\omega_1 t + \frac{\bar{b}_k^i}{k^2} \sin k\omega_1 t \right) \right\} P_i(x, y).$$ (4.4)

The sequence of functions $x_m(t, x_0)$ defined by (2.3) can be now determined by the formula

$$x_m(t, x_0) = x_m(t) \approx x_m^1(t) = x_0 + \int_0^l [P_1(t, x_{m-1}^1(t), x_m^1(t - \Delta))$$

$$- \overline{P_1(t, x_{m-1}^1(t), x_{m-1}^1(t - \Delta))}] \, dt.$$ (4.5)

These scheme is quite convenient for calculation, because for all x_0, only trigonometric polynomials are to be integrated in order to find $x_m^1(t)$.

Let us now estimate the accuracy of the approximation of $x_m(t)$ given by (4.5). Suppose that the functions $\alpha_i(t)$ $(i = 1, 2, \ldots, l)$ are twice piecewise differentiable, and moreover

$$\left| \frac{d^2 \alpha_1(t)}{dt^2} \right| \le L_i, \qquad L_i = (L_{i1}, L_{i2}, ..., L_{in}); \qquad (4.6)$$

$$b = \max_{t,\, x \in D,\, y \in D} |P(t, x) - P_1(t, x)|$$

$$< \sum_{i=1}^{l} \left[\delta_i \left(1 + \frac{\pi^3}{3} \right) + \frac{L_i(2N+1)}{\omega_1^2 N_i(N_i+1)} \right] P_i, \quad b = (b_1, b_2, ..., b_n), \qquad (4.7)$$

where $P_i = \max\limits_{x \in D,\, y \in D} |P_i(t, x)|$; δ_i are errors of the calculation of the integrals (4.3). The inequality (4.7) follows from the expansion

$$P(t, x) - \overline{P(t, x)} = \sum_{i=1}^{l} \left\{ \sum_{k=1}^{\infty} \left(\frac{a_k^i}{k^2} \cos k\omega_1 t - \frac{b_k^i}{k^2} \sin k\omega_1 t \right) \right\} P_i(x, y)$$

and the estimate

$$|a_k^i| + |b_k^i| \le \frac{2L_i}{\omega_1^2}.$$

Taking the relations (2.3), (4.2), (4.5), and (4.7) into account, we obtain

$$|x_m(t) - x_m^1(t)| \le \alpha_1(t)\,(a+b) + \left(1 - \frac{t}{\omega} \right) \int_0^t K_1 |x_m(t) - x_{m-1}^1(t)|\, dt$$

$$+ \frac{t}{\omega} \int_t^{\omega} [K_1|x_{m-1}(t-\Delta) - x_{m-1}^1(t-\Delta)| + \left(1 - \frac{t}{\omega} \right) \int_0^t K_2 |x_{m-1}(t-\Delta)$$

$$- x_{m-1}^1(t-\Delta)|\, dt \; + \; \frac{t}{\omega} \int_0^{\omega} K_2 |x_{m-1}(t-\Delta) - x_{m-1}^1(t-\Delta)| \qquad (4.8)$$

for all $m \ge 1$ provided that $x_m^1(t) \in D,\, 0 \le t \le \omega$, and $x_0 \in D - \dfrac{M+a+b}{2}\omega$. It follows from (4.8) that

$$|x_m(t) - x_m^1(t)| \le \left\{ \alpha_1(t) + \alpha_1(t) \left[(K_1 + K_2)\left(\frac{\omega}{3} + \frac{3}{2\omega}\Delta^2 \left(1 - \frac{\Delta}{\omega} \right)^2 \right) \right] + ... \right.$$

$$+ \alpha_1(t) \left[(K_1 + K_2) \left(\frac{\omega}{3} + \frac{3}{2\omega} \Delta^2 \left(1 - \frac{\Delta}{\omega} \right)^2 \right) \right]^{m-1} \right\} (a+b) \qquad (4.9)$$

and therefore,

$$| x_m(t) - x_m^1(t) | \leq \left\{ \sum_{i=1}^{m-1} \left[(K_1 + K_2) \left(\frac{\omega}{3} + \frac{3}{2\omega} \Delta^2 \left(1 - \frac{\Delta}{\omega} \right)^2 \right) \right]^i \right\} (a+b) \frac{\omega}{2}. \qquad (4.10)$$

We assume that the function $x_m^1(t)$ is the mth approximation (calculated according to the above-mentioned scheme) to the periodic solution $\varphi(t)$ of the system (1.1) which passes through the point x_0. By virtue of (2.27), we have the following bound for the deviation of $x = \varphi(t)$ from this solution $x_m^1(t)$:

$$| \varphi(t) - x_m^1(t) | \leq | \varphi(t) - x_m(t, x_0) | + | x_m(t, x_0) - x_m^1(t, x_0)|$$

$$\leq Q^m (E - Q)^{-1} \frac{M\omega}{2} + \sum_{i=1}^{m-1} Q^i (a+b) \frac{\omega}{2}. \qquad (4.11)$$

The determination of the point x_0, through which the periodic solution passes at $t = t_0 = 0$, is a quite complicated problem. However, in some special cases this problem can be solved easily. One of these cases is given by the following theorem.

Theorem 1.5. (Martinyuk and Samoilenko, 1967a). *Suppose that the right-hand side of a system*

$$\frac{dx(t)}{dt} = f(t, x(t), x(t - \frac{\omega}{2})) \qquad (4.12)$$

is defined in the region

$$-\infty < t < \infty, \quad x \in D, \quad y \in D \qquad (4.13)$$

and satisfies the following conditions:

(i)   *the function $f(t, x, y)$ is periodic in t with period ω; it is bounded and satisfies the Lipschitz condition with respect to x and y with the matrices K_1 and K_2, i.e.,*

$$|f(t, x, y)| \leq M,$$

$$(4.14)$$

$$|f(t, x_1, y_1) - f(t, x_2, y_2)| \leq K_1 |x_1 - x_2| + K_2 |y_1 - y_2|;$$

(ii) for all $(x, y) \in D \times D$ *and* $-\infty < t < \infty$, *we have*

$$f(t, x, y) = -f(-t, x, y). \qquad (4.15)$$

Then, for an arbitrarily chosen point $x_0 \in D - M \, \omega/2$, *one can always find the* ω– *periodic solution* $x = x(t, x_0)$ *of the system (4.12) which passes through this point at t = 0.*

Proof. Assume that $x_0 \in D - M \, \omega/2$. Then the first approximation to the periodic solution of (4.12) is given by the formula

$$x_1(t) = x_0 + \int_0^t f(t, x_0, x_0) \, dt = x_0 + \int_0^t [f(t, x_0, x_0)$$

$$- \overline{f(t, x_0, x_0)}] \, dt = x_1(t + \omega), \qquad (4.16)$$

since $f(t, x_0, x_0)$ is an odd function, and hence, $\overline{f(t, x_0, x_0)} = 0$. Moreover, $x_1(t) = x_1(-t)$ as the integral of an odd function and, by virtue of Lemma 1.1, we have $| x_1(t) - x_0 | \le M\omega/2$, i.e., $x_1(t) \in D$.

Let us prove that the function $f\left(t, x_1(t), x_1\left(t - \dfrac{\omega}{2}\right)\right)$ is odd. This follows from the relation

$$f\left(-t, x_1(-t), x_1\left(-t - \frac{\omega}{2}\right)\right) = f\left(-t, x_1(t), x_1\left(t + \frac{\omega}{2}\right)\right)$$

$$= f\left(-t, x_1(t), x_1\left(t - \frac{\omega}{2}\right)\right) = -f\left(t, x_1(t), x_1\left(t - \frac{\omega}{2}\right)\right). \qquad (4.17)$$

Since the function $f\left(t, x_1(t), x_1\left(t - \dfrac{\omega}{2}\right)\right)$ is periodic and odd, we get $\overline{f\left(t, x_1(t), x_1\left(t - \dfrac{\omega}{2}\right)\right)} = 0$, and thus,

$$x_2(t) = x_0 + \int_0^t f\left(t, x_1(t), x_1\left(t - \frac{\omega}{2}\right)\right) dt = x_0 + \int_0^t \left[f\left(t, x_1(t), x_1\left(t - \frac{\omega}{2}\right)\right) \right.$$

$$\left. - \overline{f\left(t, x_1(t), x_1\left(t - \frac{\omega}{2}\right)\right)} \right] dt = x_2(t + \omega),$$

$$x_2(t) = x_2(-t), \quad |x_2(t) - x_0| \le M\omega/2.$$

One can easily conclude by induction that, for all $m \ge 1$, the functions $x_m(t)$ given by

$$x_m(t) = x_0 + \int_0^t f\left(t, x_{m-1}(t), x_{m-1}\left(t - \frac{\omega}{2}\right)\right) dt \qquad (4.18)$$

are periodic in t with period ω and satisfy the relation $x_m(t) = x_m(-t)$ and the inequality

$$|x_m(t) - x_0| \le M\omega/2. \qquad (4.19)$$

The uniform boundedness and equicontinuity of the family $x_m(t)$ follow from (4.19) and (4.18), respectively.

By virtue of the Arzelà theorem, one can select from the sequence $\{x_m(t)\}$ a uniformly convergent subsequence

$$\{x_{m_k}(t)\}: x_{m_k} \underset{k \to \infty}{\to} x(t). \qquad (4.20)$$

By proceeding to the limit in (4.18), we find that the limiting function $x(t)$ of the subsequence (4.20) is a periodic solution of the system (4.12) such that $x(0) = x_0$. Consequently, Theorem 1.5 states that the region $D - M\omega/2$ consists of the initial values of periodic solutions.

In general, the initial values of periodic solutions to the system (1.1) should be found numerically. In this case the following theorem (Martinyuk and Samoilenko, 1967a) proves to be important.

Theorem 1.6. *Suppose that*

$$T(x_0) = \frac{1}{\omega} \int_0^\omega f(t, x_\infty(t, x_0), x_\infty(t - \Delta, x_0)) \, dt, \qquad (4.21)$$

where $x_\infty(t, x_0)$ is the limit of the sequences (2.3). Then the inequalities

$$|T(x_0)| \le M,$$

$$|T(x_0) - T(x_0')| \le (K_1 + K_2)\left[E + \frac{(K_1 + K_2)\omega}{3}\right.$$

$$\left. + \frac{(K_1 + K_2)\omega}{3} Q(E - Q)^{-1}\right]|x_0 - x_0'| \qquad (4.22)$$

hold for all $-\infty < t < \infty$, $x_0 \in D - M\omega/2$, *and* $x_0' \in D - M\omega/2$.

Proof. By virtue of the properties of the function $x_\infty(t, x_0)$, which have been established by Theorem 1.1, the function $T(x_0)$ is continuous, periodic, and bounded for $x_0 \in D - M\omega/2$. It follows from (4.21) that

$$
|T(x_0) - T(x_0')| \leq \frac{K_1}{\omega} \int_0^\omega |x_\infty(t, x_0) - x_\infty(t, x_0')| dt
$$

$$
+ \frac{K_2}{\omega} \int_0^\omega |x_\infty(t - \Delta, x_0) - x_\infty(t - \Delta, x_0')| dt. \tag{4.23}
$$

Taking into account the fact that $x_\infty(t, x_0)$ satisfies eqn.(3.3), we find

$$
|x_\infty(t, x_0) - x_\infty(t, x_0')| \leq |x_0 - x_0'| + K_1 \left[\left(1 - \frac{t}{\omega}\right) \int_0^t |x_\infty(t, x_0) - x_\infty(t, x_0')| dt \right.
$$

$$
\left. + \frac{t}{\omega} \int_t^\omega |x_\infty(t, x_0) - x_\infty(t, x_0')| dt \right] + K_2 \left[\left(1 - \frac{t}{\omega}\right) \int_0^t |x_\infty(t - \Delta, x_0) \right.
$$

$$
\left. - x_\infty(t - \Delta, x_0')| dt + \frac{t}{\omega} \int_t^\omega |x_\infty(t - \Delta, x_0) - x_\infty(t - \Delta, x_0')| dt \right]. \tag{4.24}
$$

We set $|x_\infty(t, x_0) - x_\infty(t, x_0')| = r(t)$. Solving the inequality (4.24), we get

$$
r(t) \leq [E + (K_1 + K_2)\alpha_1(t) + (K_1 + K_2)Q\alpha_1(t) + \ldots
$$

$$
+ (K_1 + K_2)Q^{n-1}\alpha_1(t)] |x_0 - x_0'| + (K_1 + K_2)Q^n\alpha_1(t)| r(t)|_0. \tag{4.25}
$$

By virtue of the inequality (4.25) and condition (ii) (see §1), we obtain

$$
r(t) \leq \left[E + \sum_{m=1}^\infty Q^{m-1}(K_1 + K_2)\alpha_1(t) \right] |x_0 - x_0'|. \tag{4.26}
$$

Since $\int_0^\omega \alpha_1(t) dt = \omega^2/3$, the inequalities (4.25) and (4.26) yield

$$| T(x_0) - T(x_0') | \le (K_1 + K_2) \left[E + \frac{(K_1 + K_2)\omega}{3} \right.$$

$$\left. + \frac{(K_1 + K_2)\omega}{3} Q(E - Q)^{-1} \right] | x_0 - x_0' |. \qquad (4.27)$$

The next lemma follows from the relations (2.27) and (3.5), Theorem 1.6, and the assumption that $T(x) = 0$ at the point $x \in D_1$.

Lemma 1.3. (Martinyuk and Samoilenko, 1967a). *Suppose that the system* (1.1) *is given in the region D and that D_1 is a subset of $D - M\omega/2$. Then, in order that there exist a point $x_0 \in D_1$ such that $T(x_0) = 0$, it is necessary that the inequality*

$$| T^m(x_1) | \le \sup_{x \in D_1} (K_1 + K_2) \left[E + \frac{(K_1 + K_2)\omega}{3} \right.$$

$$\left. + \frac{(K_1 + K_2)\omega}{3} Q(E - Q)^{-1} \right] | x - x_1 | + \frac{\omega}{3} (K_1 + K_2) Q^m (E - Q)^{-1} M \qquad (4.28)$$

should hold for $t = 0$, all integer m, and arbitrary $x_1 \in D_1$.

We now use Lemma 1.3 to find the initial values of periodic solutions. Let us decompose the set $D - M\omega/2$ into a finite number of subsets D_i. On each D_i, we choose a single point $x = x^i$ and find $T^m(x^i)$ for some m. Comparing $T^m(x^i)$ with the right-hand side of (4.28), we reject those sets D_i for which this inequality is violated. These sets do not contain the points through which periodic solutions pass at $t = 0$. The sets D_i which remain form the set $\mathcal{M}_m^{(i)}$ whose points are the only ones through which periodic solutions of the system (1.1) may pass.

As $i \to \infty$ and $m \to \infty$, the set $\mathcal{M}_m^{(i)}$ tends to the set $\mathcal{M}$ of the initial values of periodic solutions. Therefore, each point $x_1 \in \mathcal{M}_m^{(i)}$ can be chosen as the mth approximation to the initial value x_0 of the periodic solution, and the accuracy of this approximation is given by the inequality

$$| x_1 - x_0 | < \sup_{x \in \mathcal{M}_m^{(i)}} | x_1 - x | \qquad (4.29)$$

§5. Periodic Problem of Control for Systems with Lag

Let us illustrate the application of the results obtained above by the following example. Consider a mechanical system whose motion is described by the system of equations

$$\frac{dx(t)}{dt} = f(t, x(t), x(t - \Delta)) - u \qquad (5.1)$$

with the right-hand side periodic in t with period ω and with the constant control $u = (u_1, u_2, ..., u_n)$. The problem is to choose the control u which would guarantee that the trajectory which passes through the given point x_0 at given time $t = t_0 = 0$ is ω-periodic.

If the function $f(t, x, y)$ satisfies condition of Theorem 1.1 in the region $-\infty < t < \infty$, $x \in D$, $y \in D$ and if $x_0 \in D - M\omega/2$, then this control exists and is unique. It is given by

$$u = \overline{f(t, x_\infty(t, x_0), x_\infty(t - \Delta, x_0))} \qquad (5.2)$$

or, approximately, by

$$u_m = \overline{f(t, x_m(t, x_0), x_m(t - \Delta, x_0))}, \qquad (5.3)$$

where $x_m(t, x_0)$ are determined by the recursion relation (2.3), and $x_\infty(t, x_0)$ is the limiting function of the sequence $\{x_m(t, x_0)\}$.

We now prove the uniqueness of the control u. Assume the opposite, i.e., that there exist controls $u^{(1)}$ and $u^{(2)}$ ($u^{(1)} \neq u^{(2)}$) such that solutions $x(t, x_0, u)$ of the system (5.1) which pass through x_0 at $t = t_0 = 0$, are ω-periodic for $u = u^{(1)}$ and $u = u^{(2)}$. Then

$$| x(t, x_0, u^{(1)}) - x(t, x_0, u^{(2)}) |$$

$$\leq K_1 \left[(1 - t/\omega) \int_0^t | x(t, x_0, u^{(1)}) - x(t, x_0, u^{(2)}) | \, dt \right.$$

$$\left. + (t/\omega) \int_t^\omega | x(t, x_0, u^{(1)}) - x(t, x_0, u^{(2)}) | \, dt \right]$$

$$+ K_2 \left[(1 - t/\omega) \int_0^t | x(t - \Delta, x_0, u^{(1)}) - x(t - \Delta, x_0, u^{(2)}) | \, dt \right.$$

$$+ (t/\omega) \int_t^\omega | x(t - \Delta, x_0, u^{(1)}) - x(t - \Delta, x_0, u^{(2)}) | \, dt \left. \vphantom{\int_0^t} \right]. \qquad (5.4)$$

Setting here $| x(t, x_0, u^{(1)}) - x(t, x_0, u^{(2)}) | = r(t)$, we find

$$r(t) \leq \alpha_1(t) \, (K_1 + K_2) \, Q^{n-1} \, | r(t) |_0. \qquad (5.5)$$

This yields

$$| r(t) |_0 \leq \frac{\omega}{2} \, (K_1 + K_2) \, Q^{n-1} \, | r(t) |_0. \qquad (5.6)$$

By virtue of condition (ii) (see §1), the inequality (5.6) implies that $| r(t) |_0 = 0$ and, hence, that $u^{(1)} = u^{(2)}$. It follows from Theorem 1.6 that this control is stable. This means that a small change of the initial point x_0 involves a small change of the control u. The deviation of the control u from its mth approximation u_m can be estimated as follows

$$| u - u_m | \leq (K_1 + K_2) \, Q^m \, (E - Q)^{-1} \, (M\omega/3). \qquad (5.7)$$

§6. The Concrete Example of Application of the Numerical-Analytic Method

We now illustrate the general aspects of the method given above by the following example. Assume that the right-hand side of the equation

$$\frac{dx(t)}{dt} = \varepsilon[x^2(t) - \sin t - x(t - \Delta)\cos t + 0.001], \quad \varepsilon \geq 0, \ 0 \leq \Delta \leq 2\pi, \qquad (6.1)$$

i.e., the function

$$f(t, x, y) = \varepsilon(x^2 - \sin t - y \cos t + 0.001),$$

satisfies the inequalities

$$|f(t, x, y)| \leq 1.32 \, \varepsilon,$$

$$|f(t, x_1, y_1) - f(t, x_2, y_2)| \leq \varepsilon \, (0.5 \, |x_1 - x_2| + |y_1 - y_2|) \tag{6.2}$$

in the region

$$|x| \leq 0.25, \quad |x| \leq 0.25. \tag{6.3}$$

In this region the equation (6.1) satisfies conditions of Theorem 1.2 for all ε such that $0 \leq \varepsilon \leq 0.06$.

Every 2π-periodic solution of eqn.(6.1) which passes at $t = 0$ through the point x_0 belonging to the region

$$|x_0| \leq 0.25 \, (1 - 5.28 \, \pi \, \varepsilon), \tag{6.4}$$

is a limit of a uniformly convergent sequence of functions

$$x_m(t) = x_0 + \varepsilon (\cos t - 1) + \varepsilon \int_0^t \{[x_{m-1}^2(t) - \cos t \, x_{m-1}(t - \Delta)]$$

$$- [\overline{x}_{m-1}^2(t) - \overline{\cos t \, x_{m-1}(t - \Delta)}]\} \, dt, \quad m = 1, 2, ... \tag{6.5}$$

We have

$$x_1(t) = x_0 + \varepsilon \, (\cos t - 1 - \sin t \, x_0), \tag{6.6}$$

in the first approximation, and

$$x_2(t) = x_0 + \varepsilon \, (\cos t - 1 - \sin t \, x_0) + \varepsilon^2 (2x_0 \sin t - 2x_0 + \frac{\sin 2t}{4}$$

$$- \frac{\sin(2t - \Delta)}{4} - \frac{\sin \Delta}{4} + \sin t - \frac{x_0}{4} \cos(2t - \Delta) + \frac{x_0}{4} \cos \Delta)$$

$$+ \varepsilon^3 \left(-2\sin t - \frac{x_0^2}{4} \sin 2t - \frac{x_0}{2} \cos 2t - \frac{3}{2} x_0 - 2x_0 \cos t \right), \tag{6.7}$$

in the second approximation.

In order to solve the problem of existence of 2π-periodic solutions of eqn.(6.1), we now calculate $T^m(x_0)$:

$$T^{(0)}(x_0) = \varepsilon(x^2 + 0.001),$$

(6.8)

$$T^{(1)}(x_0) = \varepsilon\left[\left(1 + \frac{\varepsilon^2}{2}\right)x_0^2 - \varepsilon\left(2 + \frac{\sin\Delta}{2}\right)x_0 + 0.001 + \varepsilon\,\frac{3\varepsilon - \cos\Delta}{2}\right].$$

The relation (6.8) implies that $T^{(0)}(x_0) \neq 0$ and that $T^{(1)}(x_0) = 0$ for

$$x_0^{(1,2)} = \frac{(4 + \sin\Delta)\varepsilon \pm \sqrt{(4 + \sin\Delta)^2\varepsilon^2 - 4(\varepsilon^2 + 2)(0.002 + 3\varepsilon^2 - \varepsilon\cos\Delta)}}{2 + \varepsilon^2}.$$

(6.9)

For sufficiently small ε such that

$$(4 + \sin\Delta)^2\varepsilon^2 < 4(\varepsilon^2 + 2)(0.002 + 3\varepsilon^2 - \varepsilon\cos\Delta),$$

(6.10)

the equation (6.1) has no 2π-periodic solutions of for any $0 \leq \Delta \leq 2\pi$. If $\cos\Delta > 0.002/\varepsilon + 3\varepsilon$, then the equation $T^{(1)}(x_0) = 0$ always possesses real solutions $x_0^{(1)}$ and $x_0^{(2)}$ which have opposite signs.

We set $\varepsilon = 0.001$ and $\cos\Delta > 0.23$. Then the positive root $x_0^{(1)}$ of the equation $T^{(1)}(x_0) = 0$ lies in the region

$$|x_0^{(1)}| < 0.25\,(1 - 5.28 \cdot 0.01\pi);$$

(6.11)

it is isolated and its index is nonzero.

Let us take the interval

$$0 \leq x_0 \leq 0.2$$

(6.12)

to be the region D_1. Then on its boundary we have

$$\inf_{x \in \Gamma_{D_1}} |T^{(1)}(x)| = \inf\{|T^{(1)}(0),\ T^{(1)}(0.2)|\} = 0.01\,|0.00115 - 0.05\cos\Delta|. \quad (6.13)$$

The right-hand side of the inequality (3.7) is bounded from above by 0.01×0.002, i.e.,

$$Q\,(E - Q)^{-1}(K_1 + K_2)\,(M\omega/3) \leq 0.01 \cdot 0.002.$$

Conditions of Theorem 1.2 are satisfied as soon as

$$0.01\,|0.00115 - 0.05\cos\Delta| \geq 0.01 \cdot 0.002, \quad \text{i. e. } \Delta < 5\pi/18.$$

Therefore, for $\varepsilon = 0.01$ and for all Δ from the intervals $0 \leq \Delta \leq 5\pi/18$ and $31\pi/18 \leq \Delta$ $\leq 2\pi$, there exists a 2π-periodic solution of the equation (6.1) which passes through the point x_0 (which is close to $x_0^{(1)}(\Delta)$) at $t = 0$. The first and the second approximations of this periodic solution are given by (6.6) and (6.7) respectively (one should only replace x_0 by the solution $x_0 = x_0^{(1)}(\Delta)$ which is defined by (6.9)).

§7. Periodic Solutions of Differential Equations of the Second Order with Retarded Argument

Consider a differential equation of the form

$$\frac{d^2 x(t)}{dt^2} = f\left(t, x(t), x(t-\Delta), \frac{dx(t)}{dt}, \frac{dx(t-\Delta)}{dt}\right). \tag{7.1}$$

We can always reduce this equation to the system of differential equations with retarded argument and then apply the theory formulated above to the investigation of the periodic solutions of this system. However, it is more convenient to apply the results obtained above directly to the equation (7.1).

Assume that the right-hand side of eqn.(7.1) is defined in the region

$$-\infty < t < \infty, \quad a \leq x(t) \leq b, \quad a \leq x(t-\Delta) = x_\Delta \leq b,$$

$$c \leq y = \frac{dx(t)}{dt} \leq d, \quad c \leq \frac{dx(t-\Delta)}{dt} = y_\Delta \leq d, \tag{7.2}$$

and is periodic in t with period ω and continuous with respect to all the variables $t, x,$ $x_\Delta, y,$ and y_Δ. Suppose also that it satisfies the inequalities

$$|f(t, x, x_\Delta, y, y_\Delta)| \leq M, \tag{7.3}$$

$$|f(t, x_1, x_{1\Delta}, y_1, y_{1\Delta}) - f(t, x_2, x_{2\Delta}, y_2, y_{2\Delta})| \leq K_1 |x_1 - x_2|$$

$$+ K_2 |x_{1\Delta} - x_{2\Delta}| + K_3 |y_1 - y_2| + K_4 |y_{1\Delta} - y_{2\Delta}|, \tag{7.4}$$

where $K_1, K_2, K_3, K_4,$ and M are positive constants. Further, let the constants $a, b, c, d,$ $M, K_1, K_2, K_3,$ and K_4 satisfy the inequalities

$$b - a \leq M \, \omega^2/4, \quad c \leq -5M \, \omega/6 \leq 5M \, \omega/6 \leq d; \tag{7.5}$$

$$q = (\omega^2/4) \, (K_1 + K_2) + (5\omega/6) \, (K_3 + K_4) < 1. \tag{7.6}$$

We introduce an operator L, according to the formula

$$L f(t) = \int_0^t [f(t) - \overline{f(t)}] \, dt, \tag{7.7}$$

where $f(t)$ is a function periodic in t with period ω. Then

$$L^2 f(t) = L(L f(t)) = \int_0^t \left\{ \int_0^t [f(t) - \overline{f(t)}] \, dt - \overline{\int_0^t [f(t) - \overline{f(t)}] \, dt} \right\} dt. \tag{7.8}$$

Clearly, since $f(t)$ is ω-periodic, the functions $Lf(t)$ and $L^2 f(t)$ are also ω-periodic and for all $t \in [0, \omega]$ we have

$$| Lf(t) | \leq \alpha_1(t) \, |f(t)|_0; \tag{7.9}$$

$$|L^2 f(t)| \leq \alpha_1(t) \, |Lf(t)|_0 \leq \frac{\omega}{2} \, \alpha_1(t) \, |f(t)|_0 \leq \frac{\omega^2}{4} |f(t)|_0, \tag{7.10}$$

by virtue of Lemma 1.1. The algorithm for construction of the periodic solution of the equation (7.1) is established by the following statements.

Theorem 1.7. (Martinyuk, 1967). *Suppose that the function $f(t, x, x_\Delta, y, y_\Delta)$ defined in the region (7.2) is periodic in t with period ω and continuous with respect to its variables $t, x, x_\Delta, y,$ and y_Δ. Suppose also that it satisfies the inequalities (7.3) and (7.4) and conditions (7.5) and (7.6). Then the sequence of functions periodic in t with period ω*

$$x_{m+1}(t, x_0) = x_0 + L^2 f(t, x_m(t, x_0), x_m(t - \Delta, x_0), \dot{x}_m(t, x_0), \dot{x}_m(t - \Delta, x_0)) \tag{7.11}$$

converges as $m \to \infty$ uniformly in

$$-\infty < t < \infty, \quad a + M \, \omega^2/4 \leq x_0 \leq b - M \, \omega^2/4. \tag{7.12}$$

The limiting function $x_\infty(t, x_0)$ is defined in the region (7.12); it is periodic in t with period ω and, furthermore, it is the unique solution of the equation

$$x(t, x_0) = x_0 + L^2 f(t, x(t, x_0), x(t-\Delta, x_0), \dot{x}(t, x_0), \dot{x}(t-\Delta, x_0)). \qquad (7.13)$$

Proof. In (7.11), we set $m = 0$. By virtue of (7.9), we obtain

$$|x_1(t, x_0) - x_0| \le \alpha_1(t) |Lf(t, x_0, x_0, 0, 0)| \le \alpha_1(t) M \omega/2 \le M \omega^2/4. \qquad (7.14)$$

Differentiating the relation (7.11) with $m = 0$, we find

$$\dot{x}_1(t, x_0) = Lf(t, x_0, x_0, 0, 0) - \overline{Lf(t, x_0, x_0, 0, 0)}. \qquad (7.15)$$

By majorizing the right-hand side of this equality, we get

$$|\dot{x}_1(t, x_0)| \le \alpha_1(t) M + \frac{1}{\omega} \int_0^\omega \alpha_1(t) M \, dt \le \left(\alpha_1(t) + \frac{\omega}{3} \right) M \le \frac{5M\omega}{6}. \qquad (7.16)$$

By virtue of the periodicity of $x_1(t, x_0)$ and $\dot{x}_1(t, x_0)$, we have

$$|x_1(t-\Delta, x_0) - x_0| \le M \omega^2/4, \quad |\dot{x}_1(t-\Delta, x_0)| \le 5M \omega/6. \qquad (7.17)$$

The inequalities (7.5), (7.14), (7.16), and (7.17) together imply that

$$a \le x_1(t, x_0) \le b, \quad a \le x_1(t-\Delta, x_0) \le b,$$

$$c \le \dot{x}_1(t, x_0) \le d, \quad c \le \dot{x}_1(t-\Delta, x_0) \le d, \qquad (7.18)$$

provided that $a + M \omega^2/4 \le x_0 \le b - M \omega^2/4$. One can easily prove by induction that the functions from the sequence (7.11) satisfy the inequalities

$$a \le x_m(t, x_0) \le b, \quad a \le x_1(t-\Delta, x_0) \le b,$$

$$c \le \dot{x}_m(t, x_0) \le d, \quad c \le \dot{x}_1(t-\Delta, x_0) \le d \qquad (7.19)$$

for all $m = 0, 1, \ldots$, $t \in (-\infty, \infty)$ and $a + M \omega^2/4 \le x_0 \le b - M \omega^2/4$.

Let us now prove that the sequence (7.11) is convergent. For this purpose, we estimate the differences

$$x_2(t, x_0) - x_1(t, x_0), \quad \dot{x}_2(t, x_0) - \dot{x}_1(t, x_0).$$

We have

$$|x_2(t, x_0) - x_1(t, x_0)| \leq$$

$$\leq |L^2(f(t, x_1(t, x_0), x_1(t - \Delta, x_0), \dot{x}_1(t, x_0), \dot{x}_1(t - \Delta, x_0)) - f(t, x_0, x_0, 0, 0))|$$

$$\leq \alpha_1(t) \frac{\omega}{2} [K_1 |x_1(t, x_0) - x_0|_0 + K_2 |x_1(t - \Delta, x_0) - x_0|_0$$

$$+ K_3 |\dot{x}_1(t, x_0)| + K_4 |\dot{x}_1(t - \Delta, x_0) - x_0|_0]; \tag{7.20}$$

$$|\dot{x}_2(t, x_0) - \dot{x}_1(t, x_0)|$$

$$\leq |L(f(t, x_1(t, x_0), x_1(t - \Delta, x_0), \dot{x}_1(t, x_0), \dot{x}_1(t - \Delta, x_0)) - f(t, x_0, x_0, 0, 0))|$$

$$- \overline{L(f(t, x_1(t, x_0), x_1(t - \Delta, x_0), \dot{x}_1(t, x_0), \dot{x}_1(t - \Delta, x_0)) - f(t, x_0, x_0, 0, 0))}$$

$$\leq \left(\alpha_1(t) + \frac{\omega}{3}\right)|f(t, x_1(t, x_0), x_1(t - \Delta, x_0), \dot{x}_1(t, x_0), \dot{x}_1(t - \Delta, x_0)) - f(t, x_0, x_0, 0, 0)|_0$$

$$\leq \left(\alpha_1(t) + \frac{\omega}{3}\right) [K_1 |x_1(t, x_0) - x_0|_0 + K_2 |x_1(t - \Delta, x_0) - x_0|_0$$

$$+ K_3 |\dot{x}_1(t, x_0)| + K_4 |\dot{x}_1(t - \Delta, x_0) - x_0|_0]. \tag{7.21}$$

Taking the relations (7.14), (7.16), and (7.17) into account, we can rewrite (7.20) and (7.21) as follows

$$|x_2(t, x_0) - x_1(t, x_0)| \leq \alpha_1(t) \frac{\omega}{2}\left[(K_1 + K_2)\frac{M\omega^2}{4} + (K_3 + K_4)\frac{5M\omega}{6}\right];$$

$$\tag{7.22}$$

$$|\dot{x}_2(t, x_0) - \dot{x}_1(t, x_0)| \leq \left(\alpha_1(t) + \frac{\omega}{3}\right)M\left[(K_1 + K_2)\frac{M\omega^2}{4} + (K_3 + K_4)\frac{5M\omega}{6}\right].$$

One can easily prove by induction that the following inequalities hold

$$|x_{m+1}(t, x_0) - x_m(t, x_0)| \leq \alpha_1(t) \frac{M\omega}{2} Q^m, \tag{7.23}$$

$$|\dot{x}_{m+1}(t, x_0) - \dot{x}_m(t, x_0)| \leq \left(\alpha_1(t) + \frac{\omega}{3}\right)M Q^m, \tag{7.24}$$

for any $m \geq 1$, where

$$Q = (K_1 + K_2)\frac{M\omega^2}{4} + (K_3 + K_4)\frac{5M\omega}{6}.$$

These relations yield

$$|x_{m+1}(t, x_0) - x_m(t, x_0)|_0 \leq \frac{M\omega^2}{4}Q^m,$$

$$|x_{m+1}(t - \Delta, x_0) - x_m(t - \Delta, x_0)|_0 \leq \frac{M\omega^2}{4}Q^m; \qquad (7.25)$$

$$|\dot{x}_{m+1}(t, x_0) - \dot{x}_m(t, x_0)|_0 \leq \frac{5M\omega}{6}Q^m,$$

$$|\dot{x}_{m+1}(t - \Delta, x_0) - \dot{x}_m(t - \Delta, x_0)|_0 \leq \frac{5M\omega}{6}Q^m \qquad (7.26)$$

for all $0 \leq t \leq \omega$ and $m = 0, 1, 2, \ldots$ Since the functions $x_m(t, x_0)$ and $\dot{x}_m(t, x_0)$ are periodic in t with period ω, the inequalities (7.25) and (7.26) are valid for all $-\infty < t < \infty$ and $m = 0, 1, 2, \ldots$

By virtue of the inequalities (7.25), (7.26), and condition (7.6), the sequences of periodic functions $\{x_m(t, x_0)\}$ and $\{\dot{x}_m(t, x_0)\}$ converge uniformly, i.e.,

$$\lim_{m \to \infty} x_m(t, x_0) = x_\infty(t, x_0), \quad \lim_{m \to \infty} \dot{x}_m(t, x_0) = \dot{x}_\infty(t, x_0). \qquad (7.27)$$

Moreover, by using (7.25) and (7.26), we obtain the following estimates for the deviations of $x_m(t, x_0)$ and $\dot{x}_m(t, x_0)$ from $x_\infty(t, x_0)$ and $\dot{x}_\infty(t, x_0)$, respectively:

$$|x_\infty(t, x_0) - x_m(t, x_0)|_0 \leq Q^m(1 - Q)^{-1}\left(\frac{M\omega^2}{4}\right); \qquad (7.28)$$

$$|\dot{x}_\infty(t, x_0) - \dot{x}_m(t, x_0)|_0 \leq Q^m(1 - Q)^{-1}\left(\frac{5M\omega}{6}\right). \qquad (7.29)$$

By proceeding to the limit in (7.11) as $m \to \infty$, we find that $x_\infty(t, x_0)$ is a periodic solution of eqn.(7.13). The uniqueness of the function $x_\infty(t, x_0)$ can be easily proved by contradiction.

We now clarify the relation between the limiting function $x_\infty(t, x_0)$ and the periodic solution $\varphi(t)$ of eqn.(7.1).

Theorem 1.8. (Martinyuk, 1967). *Assume that the right-hand side of eqn.* (1.7) *satisfies conditions of Theorem* 1.7 *and that the equation*(7.1) *has an* ω*–periodic solution* $x = \varphi(t)$ *which passes through the point* $\varphi(0) = x_0$ *of the interval* $a + M\omega^2/4$ $\leq x_0 \leq b - M\omega^2/4$ *at* $t = 0.$ *Then*

$$\varphi(t) = x_\infty(t, x_0) = \lim_{m \to \infty} x_m(t, x_0). \tag{7.30}$$

Proof. Since $\varphi(t)$ is a periodic solution of eqn.(7.1), we have

$$\overline{f(t, \varphi(t), \varphi(t-\Delta), \dot\varphi(t), \dot\varphi(t-\Delta))} = 0,$$

$$\dot\varphi(0) = -\int_0^t \overline{f(t, \varphi(t), \varphi(t-\Delta), \dot\varphi(t), \dot\varphi(t-\Delta))}\, dt\,. \tag{7.31}$$

By virtue of (7.13), the function $\varphi(t)$ satisfies also the following equation

$$\varphi(t) = x_0 \int_0^t \left\{ \int_0^t [f(t, \varphi(t), \varphi(t-\Delta), \dot\varphi(t), \dot\varphi(t-\Delta)) \right.$$

$$- \overline{f(t, \varphi(t), \varphi(t-\Delta), \dot\varphi(t), \dot\varphi(t-\Delta))}]\, dt$$

$$\left. - \int_0^t [f(t, \varphi(t), \varphi(t-\Delta), \dot\varphi(t), \dot\varphi(t-\Delta)) - \overline{f(t, \varphi(t), \varphi(t-\Delta), \dot\varphi(t), \dot\varphi(t-\Delta))}]\, dt \right\} dt$$

$$= x_0 + L^2 f(t, \varphi(t), \varphi(t-\Delta), \dot\varphi(t), \dot\varphi(t-\Delta)).$$

Therefore, $\varphi(t)$ is the periodic solution of the equation (7.13) (along with $x_\infty(t, x_0)$). The relation (7.30) follows thus from the uniqueness of this solution.

Consequently, Theorems 1.7 and 1.8 state that if we consider eqn.(7.1), the right-hand side of which satisfies the conditions (7.3)-(7.6), then every ω-periodic solution of (7.1), which passes through the point $a + M\omega^2/4 = x_0 \leq b - M\omega^2/4$ at $t = 0$, can be represented as the limit of the sequence of periodic functions (7.11); moreover, the difference between the exact periodic solution $\varphi(t)$ and its approximation $x_m(t, x_0)$ can be estimated with the help of the inequalities (7.28) and (7.29).

Consider now the problem of existence of periodic solutions of eqn.(7.1). We denote

$$T(x_0) = \overline{f(t, x_\infty(t, x_0), x_\infty(t-\Delta, x_0), \dot x_\infty(t, x_0), \dot x_\infty(t-\Delta, x_0))} \tag{7.32}$$

where $x_\infty(t, x_0)$ is a limit of the sequence of periodic functions (7.11).

Since the equation (7.13) turns into the equation (7.1) for $T(x_0) = 0$, the investigation of the problem of existence of periodic solutions is reduced to the investigation of the problem of existence of zeros of the function $T(x_0)$. This means that a single ω-periodic solution of the equation (7.1) corresponds to each zero of the function $T(x_0)$ and that the number of periodic solutions of eqn.(7.1) is equal to the number of zeros of $T(x_0)$. Since the function $\Delta(x_0)$ can be determined only approximately, we define, by use of the sequence of functions (7.11),

$$T^m(x_0) = \overline{f(t, x_m(t, x_0), x_m(t - \Delta, x_0), \dot{x}_m(t, x_0), \dot{x}_m(t - \Delta, x_0))}. \qquad (7.33)$$

Each of the functions (7.11) is defined for $x_0 \in [a + M\omega^2/4, b - M\omega^2/4]$; moreover, these functions are continuous with respect to x_0 on this region. In addition, the inequalities (7.28) and (7.29) yield

$$| T(x_0) - T^m(x_0) | \le d_m, \qquad (7.34)$$

where $d_m = M Q^m (1 - Q)^{-1}$. Taking this inequality and the continuity of the functions $T^m(x_0)$ into account, one can easily prove a statement similar to Theorem 1.4.

Theorem 1.9. (Martinyuk, 1967). *Suppose that the right-hand side of the equation (7.1) satisfies conditions of Theorem 1.7 and that the function $T^m(x_0)$ satisfies the inequalities*

$$\min_{a+\frac{M\omega^2}{4}\le x\le b-\frac{M\omega^2}{4}} T^m(x) \le -d_m, \qquad \max_{a+\frac{M\omega^2}{4}\le x\le b-\frac{M\omega^2}{4}} T^m(x) \ge d_m. \qquad (7.35)$$

for some m. Then the equation (7.1) possesses an ω-periodic solution $x = x(t)$ for which

$$a + M\omega^2/4 \le x(0) \le b - M\omega^2/4.$$

In the special case, when the right-hand side of eqn.(7.1) is a polynomial with respect to x, x_Δ, y, and y_Δ, taking into account that $T(x_0)$ is a scalar function of a single argument and using the theorem on zeros of analytic function, one can prove the following statement.

Theorem 1.10. (Martinyuk, 1967). *Let the right-hand side of equation (7.1) be a polynomial with respect to x, x_Δ, y, and y_Δ which satisfies conditions of Theorem 1.7. Suppose that the equation (7.1) has an ω-periodic solution. Then either a single* *periodic solution of this equation corresponds to each value $x(0) = x_0$ from the interval $[a + M \omega^2/4, b - M \omega^2/4]$, or this is true only for finitely many values of this sort.*

Indeed, the function $T(x_0)$ is analytic, since it is a limit of the sequence of poly-nomials uniformly convergent on a layer of the complex plane which contains the inter-val $[a + M \omega^2/4, b - M \omega^2/4]$. Therefore, $T(x_0)$ is either identically zero or has a finite number of zeros on this interval. In the first case, the initial values $x(0) = x_0$, $dx(t)/dt \mid_{t=0}$ $= y_0$ of periodic solutions fill in some straight line on the rectangle

$$D = [a + M \omega^2/4, b - M \omega^2/4] \times [c + 5M \omega/6, d - 5M \omega/6].$$

In the second case, these initial values are isolated points of this rectangle.

As mentioned above, the determination of the initial values x_0 of periodic solutions is equivalent to the determination of zeros of the function $T(x_0)$. The inequality (7.34) implies that, for every zero x^0 of $T(x_0)$, we have

$$|T^m(x_0)| \le d_m, \quad m = 0, 1, \dots. \tag{7.36}$$

Thus, all zeros of the function $T(x_0)$ belong to the set of solutions of the inequality (7.36). If we denote by $\mathfrak{M}_m$ the set of points from the interval $a + M \omega^2/4 \le x_0 \le b - M\omega^2/4$ which satisfy the inequality (7.36), then every point of this set can be zero of the function $T(x_0)$. However, the set $\mathfrak{M}_n$ tends as $m \to \infty$ to the set $\mathfrak{M}$ of zeros of the function $T(x_0)$, and therefore, one can take any point from $\mathfrak{M}_m$ as the m-th approxi-mation to the initial value of the periodic solution. Hence, the determination of the initial values of the periodic solutions of the equation (7.1) can be reduced to the determination of zeros of the function $T^m(x_0)$, or (in the case when $T^m(x_0)$ has no zeros) to finding the solutions of the inequality (7.36).

In some cases, the initial values of the periodic solutions of eqn.(7.1) can be found exactly. This can be done, for instance, when the right-hand side of (7.1) is such that $T^m(x) = 0$ for all $m = 0, 1, \dots$ and some fixed $x = x_0$.

The following statement establishes a class of equation for which $T^m(x) = 0$ ($m = 0$, 1,...).

Theorem 1.11. (Martinyuk, 1967). *Suppose that the right-hand side of equation*

$$\frac{d^2x(t)}{dt^2} = f\left(t, x(t), x\left(t - \frac{\omega}{2}\right), \dot{x}(t), \dot{x}\left(t - \frac{\omega}{2}\right)\right) \tag{7.37}$$

satisfies conditions of Theorem1.7 and condition

$$f(-t, x, x_\Delta, y, y_\Delta) = -f(t, x, x_\Delta, y, y_\Delta) \tag{7.38}$$

in the region (7.2). Then the equation (7.37) possesses the ω-periodic solution defined by the relation $x(t, 0) = x_\infty(t, 0)$.

The proof of Theorem 1.11 is quite similar to the proof of Theorem 1.5.

§8. Periodic Solutions for Countable Systems of Differential Equations with Lag

We note that the dimensionality of the system is of no importance for the investigation of periodic solutions for nonlinear systems with lag by the numerical-analytic method. Therefore, one can easily extend this method for infinite systems of differential equations with lag.

Let $x = (x_1, x_2, ..., x_n, ...)$ be a point of the space m of bounded number sequences with norm $|x| = \sup |x_n|$. Consider a countable system of differential equations with lag

$$\frac{dx(t)}{dt} = f\big(t, x(t), x(t - \Delta)\big), \tag{8.1}$$

where $f(t, x, y)$ $(f_1(t, x, y), f_2(t, x, y), ..., f_n(t, x, y))$ is a continuous function of the variables t, x, and y belonging to the region

$$-\infty < t < \infty, \quad x \in D, \quad y \in D \tag{8.2}$$

(D is a bounded closed region of the space m). Let $f(t, x, y)$ be a function periodic in t with period ω which satisfies the inequalities

$$|f(t, x, y)| \le M; \tag{8.3}$$

$$|f(t, x', y') - f(t, x'', y'')| \le K_1 |x' - x''| + K_2 |y' - y''|, \tag{8.4}$$

where $M = (M_1, M_2, ..., M_n, ...)$ and $K_1 = \{k'_{ij}\}$, $K_2 = \{k''_{ij}\}$ $(i, j = 1, 2, ..., n, ...)$ are an infinite dimensional vector and matrices with nonnegative elements, respectively.

We call the function $x(t) = (x_1(t), x_2(t), ..., x_n(t), ...)$, defined for all t from the interval (a, b) and taking values from the space m, a solution of the system of equations (8.1), if it is continuously differentiable and satisfies the equation (8.1).

As is known (Kolmogorov and Fomin, 1976), a matrix $K = \{k_{ij}\}$ $(i, j = 1, 2, ..., n, ...)$ generates the operator K acting in the space m, if

$$\sup_i \sum_{i=1}^{\infty} k_{ij} < \infty,$$

moreover, the norm of the operator K is given by

$$|K| = \sup_i \sum_{i=1}^{\infty} k_{ij}.$$

The operator K is called completely regular if $|K| \leq q < 1$. Assume that the right-hand side of the system (8.1) satisfies the following conditions:

(i) the region $D - M\omega/2$ is nonempty;

(ii) the operator $Q = (K_1 + K_2)(\omega/3 + 3/2\Delta(1 - \Delta/\omega)^2)$ is completely regular.

For the system (8.1), one can prove the statements similar to Theorems 1.1 and 1.5. Before solving the problem whether ω-periodic solutions of the system (8.1) exist, we prove the following auxiliary statement.

Lemma 1.4. (Samoilenko and Ronto, 1976). *Suppose that D is a closed boun-ded region in the space m, and that A_0 and A are continuous mappings of D onto m such that $|Ax - A_0x| \leq \varepsilon$. Assume also that A_0 is a topological mapping. Then AD contains the set $A_0D \setminus \Pi(\varepsilon)$ which consists of the points belonging to A_0D together with their ε-neighborhoods.*

Proof. Since A_0 is a topological mapping, the continuous mapping AA_0^{-1}, which maps A_0D onto AD, is defined on A_0D. By setting $AD = D_1$ and $AA_0^{-1} = A_1$, we re-duce the proof of Lemma 1.4 to the case when $A_0 = I$, where I is the identical mapping. We now prove that $D \setminus \Pi(\varepsilon) \subset AD$. Let us choose $x_0 \in D \setminus \Pi(\varepsilon)$, then the sphere $T_\varepsilon(x_0): |x - x_0| \leq \varepsilon$ belongs to $D: T_\varepsilon(x_0) \subset D$. On $T_\varepsilon(x_0)$, we define the continuous operator

$$A^* x = x_0 + (x - Ax). \tag{8.5}$$

We shall consider this operator as a mapping on a certain space C^∞, which is defined (see (Alexandrov, 1947)) as a space of number sequences $x = (x_1, x_2, ..., x_n, ...)$ uni-formly bounded by ε (i.e., $|x_i| \leq \varepsilon$) with norm

$$|x| = \sum_{k=1}^{\infty} \frac{1}{2^k} |x_k|. \tag{8.6}$$

For $x_0 = 0$, the sphere $T_\varepsilon(x_0)$ coincides with the space C^∞, and the operator A^* maps C^∞ onto itself, i.e.,

$$|A^* x - x_0| = |x - Ax| < \varepsilon,$$

moreover, the family of kth coordinates $\{A^* x\}_k$ of the mapping A^* is compact for all $k = 1, 2, \ldots$ By virtue of the theorem on compactness of a set in the space C^∞ (Alexandrov, 1947), this implies that the set of points $\{A^* x\}_k$ is compact as a set of points in C^∞.

By employing Schauder's theorem (Alexandrov, 1947), we find that the operator A^* has a fixed point x^* in C^∞, i.e.,

$$A^* x^* = x^*. \tag{8.7}$$

It follows from the formulas (8.5) and (8.7) that

$$A^* x = x_0. \tag{8.8}$$

This equality means that x_0 is the image of x^* under the mapping A. Taking into account the fact that the equality (8.8) holds for every point $x_0 \in D \setminus \Pi(\varepsilon)$ and that $x^* \in T_\varepsilon(x_0)$ (and hence, $x^* \subset D$), we conclude that the whole of the set $D \setminus \Pi(\varepsilon)$ is the image of the set D under the mapping A. Consequently, $D \setminus \Pi(\varepsilon) \subset AD$. The case when $x_0 \neq 0$ is reduced to the already considered case by the translation of coordinates. Lemma 1.4 is thus proved.

The problem concerning the existence of periodic solutions for the countable system of differential equations (8.1) is solved by the following theorem.

Theorem 1.12. (Martinyuk, 1968a). *Suppose that the system (8.1), given in the region D of the space m, satisfies the inequalities (8.3) and (8.4), conditions (i), and (ii) in this section, and the following conditions:*

(a) for some integer m the mapping

$$T^m: T^m = T^m(x_0) = \overline{f(t, x_m(t, x_0), x_m(t - \Delta, x_0))} \tag{8.9}$$

of the region $D - M\omega/2$ onto the region $T^m(D - M\omega/2)$ has a weak point $x_0 = x^0$, i.e.,

$$T^m(x^0) = 0; \tag{8.10}$$

(b) there exists a closed bounded region D_1 belonging to $D - M\omega/2$, such that $x^0 \in D_1$, and the operator T^m topologically maps D_1 onto $T^m D_1$;

c) the following inequality

$$\inf_{x \in \Gamma_{D_1}} \left\| T^m(x) \right\| \geq \frac{q^m}{1-q} \left(\|K_1\| + \|K_2\| \right) \frac{\|M\|\omega}{3} \tag{8.11}$$

holds on the boundary Γ_{D_1} of the region D_1.

Then the system (8.1) has an ω--periodic solution $x = x(t)$ such that $x(o) \in D_1$.

Proof. By virtue of the estimate (3.9), which in this case has the form

$$\left\| T^m(x) - T(x_0) \right\| \geq \frac{q^m}{1-q} \left(\|K_1\| + \|K_2\| \right) \frac{\|M\|\omega}{3} \tag{8.12}$$

and Lemma 1.4, we find that the set $T^m D_1 \setminus \Pi\left(\frac{q^m}{1-q} \left(\|K_1\| + \|K_2\| \right) \frac{\|M\|\omega}{3} \right)$ is a subset of $T D_1$. This means that if the set under consideration contains the origin of the coordinate system $T = (T_1, \ldots, T_n, \ldots)$, then the set $T D_1$ contains this origin too. The last statement is sufficient in order that the system (8.1) possesses a periodic solution with period ω.

To complete the proof of Theorem 1.12 it remains to show that

$$0 \in T^m D_1 \setminus \Pi\left(\frac{q^m}{1-q} \left(\|K_1\| + \|K_2\| \right) \frac{\|M\|\omega}{3} \right)$$

The mapping $T^m(x)$ is topological, therefore $T^m D_1$ is the image of the boundary Γ_{D_1} of the region, i.e.

$$\Gamma_{T^m D_1} = T^m \Gamma_{D_1} \tag{8.13}$$

By virtue of condition a), the set $T^m D_1$ contains the zero point of the system $T = (T_1, T_2, \ldots, T_n, \ldots)$. This point belongs to the set $T D_1 \setminus \Pi(r)$ if the distance between it and the boundary $\Gamma_{T^m D_1}$ of $T^m D_1$ is not less than r. Consequently, this zero point belongs to the set

$$T^m D_1 \setminus \Pi\left(\frac{q^m}{1-q}(\|K_1\|+\|K_2\|)\frac{\|M\|\omega}{3}\right)$$

if

$$\inf_{z\in\Gamma_{T^m D_1}} \|z\| \geq \left(\frac{q^m}{1-q}(\|K_1\|+\|K_2\|)\frac{\|M\|\omega}{3}\right) \tag{8.14}$$

By virtue of the inequality (8.13), we have

$$z = T^m(x)\Big|_{x\in\Gamma_{D_1}} \tag{8.15}$$

Inserting (8.15) into (8.14) we obtain

$$\inf_{x\in\Gamma_{D_1}} \|T^m(x)\| \geq \left(\frac{q^m}{1-q}(\|K_1\|+\|K_2\|)\frac{\|M\|\omega}{3}\right) \tag{8.16}$$

The above argument implies that the system (8.1) has an ω-periodic solution provided that the inequality (8.16) holds. Theorem 1.12 is thus proved.

§9. Periodic Solutions for Nonlinear Systems of Differential Equations of the Neutral Type

Consider a system of differential equations of the form

$$\frac{dx(t)}{dt} = f\left(t, x(t), x(t-\Delta), \frac{dx(t-\Delta)}{dt}\right), \tag{9.1}$$

where $f(t, x, y, z)$ is a vector function periodic in t with period ω and defined for all

$$-\infty < t < \infty, \quad x \in D, \quad y = x(t-\Delta) \in D, \, z = dx(t-\Delta)/dt \in D_1 \tag{9.2}$$

(D is a closed bounded region of the space E_n, and $D_1 = \{z: |z| \leq 2M\}$).

Assume that the vector function $f(t, x, y, z)$ is continuous in the region (9.2) with respect to all its variables $t, x, y,$ and z, and satisfies the inequalities

$$|f(t, x, y, z)| \leq M; \tag{9.3}$$

$$|f(t, x_1, y_1, z_1) - f(t, x_2, y_2, z_1)| \leq K_1 |x_1 - x_2| + K_2 |y_1 - y_2| + K_3 |z_1 - z_2|, \tag{9.4}$$

where K_1, K_2, K_3 and M are $(n \times n)$-dimensional matrices with nonnegative elements and an n-dimensional vector, respectively. The region D, matrices K_1, K_2, K_3, lag Δ, and period ω are such that the following conditions hold:

(i) the set $D - M\omega/2$ is nonempty;
(ii) the greatest eigenvalue λ_{max} of the matrix

$$Q = (K_1 + K_2) \frac{\omega}{2} + 2K_3$$

does not exceed 1.

Assume that the system (9.1) has an ω-periodic solution, and that the point $x_0 \in D - M\omega/2$, through which this solution passes at $t = t_0 = 0$, is known. Then the algorithm for finding this solution is established by the following theorem.

Theorem 1.13 (Tkach, 1969). *Let $x = \varphi(t)$ be a solution of the system (9.1) periodic in t with period ω and satisfying the conditions (9.3), (9.4), and (i), (ii). Then*

$$\varphi(t) = \lim_{m \to \infty} x_m(t, x_0) \tag{9.5}$$

uniformly with respect to $-\infty < t < \infty$, $x_0 \in D - M\omega/2$, *and*

$$|\varphi(t) - x_m(t, x_0)| \leq Q^m (E - Q)^{-1} (M\omega/2) \tag{9.6}$$

for all $m = 0, 1, 2, ...,$ where $x_m(t, x_0)$ $(x_0(t, x_0) = x_0)$ are ω-periodic functions defined by

$$x_{m+1}(t, x_0) = x_0 + \int_0^t [[f(t, x_m(t, x_0), x_m(t - \Delta, x_0), \dot{x}_m(t - \Delta, x_0))$$

$$\overline{-f(t, x_m(t, x_0), x_m(t - \Delta, x_0), \dot{x}_m(t - \Delta, x_0))}] dt.. \tag{9.7}$$

Proof. It is clear that all the functions (9.7) are ω-periodic. Moreover, according to Lemma 1.1,

$$|x_m(t, x_0) - x_0| \le M \, \alpha_1(t) \le M \, \omega/2. \tag{9.8}$$

By differentiating (9.7), we obtain

$$|\dot{x}_m(t, x_0)| \le |f(t, x_{m-1}(t, x_0), x_{m-1}(t-\Delta, x_0), \dot{x}_{m-1}(t-\Delta, x_0))|$$

$$+ \overline{|f(t, x_{m-1}(t, x_0), x_{m-1}(t-\Delta, x_0), \dot{x}_{m-1}(t-\Delta, x_0))|} \le 2M. \tag{9.9}$$

The periodicity of the functions $x_m(t, x_0)$ and $\dot{x}_m(t, x_0)$ implies that $x_m(t, x_0)$ belongs to the region (9.2) for all m.

We now prove the convergence of the sequences $\{x_m(t, x_0)\}$ and $\{\dot{x}_m(t, x_0)\}$. For this purpose, we estimate the differences $x_{m+1}(t, x_0) - x_m(t, x_0)$ and $\dot{x}_{m+1}(t, x_0) - \dot{x}_m(t, x_0)$. Let us write the difference $x_2(t, x_0) - x_1(t, x_0)$ as follows

$$|x_2(t, x_0) - x_1(t, x_0)|$$

$$\le \left(1 - \frac{t}{\omega}\right) \int_0^t [K_1 |x_1(t, x_0) - x_0| + K_2 |x_1(t-\Delta, x_0) - x_0| + K_3 |\dot{x}_1(t-\Delta, x_0)|] dt$$

$$+ \frac{t}{\omega} \int_t^\omega [K_1 |x_1(t, x_0) - x_0| + K_2 |x_1(t-\Delta, x_0) - x_0| + K_3 |\dot{x}_1(t-\Delta, x_0)|] dt. \tag{9.10}$$

Taking (9.8) and (9.9) into account, we can rewrite the inequality (9.10) as follows

$$|x_2(t, x_0) - x_1(t, x_0)| \le \alpha_1(t) \, QM \le Q(M \, \omega/2). \tag{9.11}$$

By differentiating (9.7), one can easily get

$$|\dot{x}_2(t, x_0) - \dot{x}_1(t, x_0)| \le 2 \, |f(t, x_1(t, x_0), x_1(t-\Delta, x_0), \dot{x}_1(t-\Delta, x_0))$$

$$-f(t, x_0, x_0)| < 2\left[(K_1 + K_2)\frac{\omega}{2} + 2K_3\right]M \le 2QM.$$

It is easy to prove by induction that

$$|x_{m+1}(t, x_0) - x_m(t, x_0)| \le Q^m(M \, \omega/2),$$

$$|\dot{x}_{m+1}(t, x_0) - \dot{x}_m(t, x_0)| \le 2Q^m M \tag{9.12}$$

for all $m \geq 1$. It follows from these inequalities and conditions imposed on the matrix Q that the sequence of functions (9.7) converges uniformly with respect to t and x_0. Passing to the limit in (9.7) as $m \to \infty$, we find that the limiting function $x_\infty(t, x_0)$ is a periodic solution of the equation

$$
x_\infty(t, x_0) = x_0 + \int\limits_0^t [f(t, x_\infty(t, x_0), x_\infty(t - \Delta, x_0), \dot{x}_\infty(t - \Delta, x_0))
$$

$$
- \overline{f(t, x_\infty(t, x_0), x_\infty(t - \Delta, x_0), \dot{x}_\infty(t - \Delta, x_0))}] dt. \tag{9.13}
$$

The uniqueness of this solution, which can be easily proved by contradiction, yields the relation (9.5).

The theorem on the existence of a periodic solution to (9.1) can be formulated and proved by analogy with Theorem 1.2.

§10. Investigation of Periodic Solutions for Some Classes of Systems of Integro-Differential Equations

Consider a system of integro-differential equations

$$
\frac{dx(t)}{dt} = f\left(t, x, \int\limits_t^{t+T} \varphi(t, s, x(s)) \, ds\right) \tag{10.1}
$$

where $f(t, x, y)$ and $\varphi(t, x, y)$ are continuous functions periodic in t with period ω defined in the region

$$
-\infty < t < \infty, \quad -\infty < s < \infty, \quad x = (x_1, ..., x_n) \in D, \quad y = (y_1, ..., y_n) \in D_1
$$

(D and D_1 are bounded regions in the Euclidean space E_n, and T is a fixed number).

Assume that the functions $f(t, x, y)$ and $\varphi(t, x, y)$ satisfy the inequalities

$$
|f(t, x, y)| \leq M, \quad |\varphi(t, x, y)| \leq N,
$$

$$
|f(t, x', y') - f(t, x'', y'')| \leq K_1 |x' - x''| + K_2 |y' - y''|,
$$

$$
|\varphi(t, s, x') - \varphi(t, s, x'')| \leq K_3 |x' - x''|, \tag{10.2}
$$

where M and N are n-dimensional vectors with nonnegative coordinates and K_1, K_2, and K_3 are $(n \times n)$-dimensional matrices with nonnegative elements. In what follows, we consider only those systems of the form (10.1) for which the following conditions are satisfied:

(i) the set $D - M\,\omega/2$ is nonempty;

(ii) the greatest eigenvalue of the matrix $Q = \dfrac{\omega}{3}\left[K_1 + \dfrac{3T}{2}K_2 K_3\right]$ is less than 1.

In this case, the statement similar to Theorem 1.1 has the following form.

Theorem 1.14. (Nurzhanov, 1977a). *Suppose that the system of integro-differential equations satisfies the inequalities* (10.2) *and conditions* (i) *and* (ii). *Suppose also that the system* (10.1) *has the* ω-*periodic solution* $x = x(t)$ *which takes a value* $x_0 \in D - M\,\omega/2$ *at* $t = 0.$ *Then*

$$x(t) = x_\infty(t, x_0), \qquad (10.3)$$

where $x_\infty(t, x_0)$ *is the limiting function of the sequence* $\{x_m(t, x_0)\}$ *of* ω-*periodic functions defined by*

$$x_m(t, x_0) = x_0 + \int_0^t \left[f\left(t, x_{m-1}(t, x_0), \int_t^{t+T} \varphi(t, s, x_{m-1}(s, x_0))\,ds\right)\right.$$

$$\left. - \frac{1}{\omega}\int_0^\omega f\left(t, x_{m-1}(t, x_0), \int_t^{t+T} \varphi(t, s, x_{m-1}(s, x_0))\,ds\right)dt\right]dt, \quad m = 1, 2, \ldots, \qquad (10.4)$$

Proof. It is quite obvious that each function (10.4) is periodic in t with period ω. Moreover, by virtue of Lemma 1.1, we have

$$|x_1(t, x_0) - x_0| \le \left(1 - \frac{t}{\omega}\right)\int_0^t \left|f\left(t, x_0, \int_t^{t+T}\varphi(t, s, x_0)\,ds\right)\right|$$

$$+ \frac{t}{\omega}\int_t^\omega \left|f\left(t, x_0, \int_t^{t+T}\varphi(t, s, x_0)\,ds\right)\right|dt \le 2t\left(1 - \frac{t}{\omega}\right)M \qquad (10.5)$$

for all $x_0 \in D$ and $0 \le t \le \omega$.
   Denote

$$\alpha_1(t) = \begin{cases} 2t\left(1-\dfrac{t}{\omega}\right), & \text{for } t \in [0, \omega], \\ \alpha_1(t - k\omega), & \text{for } t \in [k\omega, (k+1)\omega], \end{cases} \tag{10.6}$$

$$(k = \pm 1, \pm 2, \pm 3, \ldots).$$

Taking (10.6) into account, we can rewrite the inequality (10.5) as follows

$$|x_1(t, x_0) - x_0| \le \alpha_1(t) M \le M \omega/2. \tag{10.7}$$

This implies that $x_1(t, x_0) \in D$ if $x_0 \in D - M \omega/2$. If we now assume that $x_{m-1}(t, x_0) \in D$, then by use of (10.4), we easily obtain the following inequality (by analogy with (10.7))

$$|x_m(t, x_0) - x_0| \le M \omega/2.$$

It follows from this inequality that $x_m(t, x_0) \in D$ if $x_0 \in D - M \omega/2$.

We can conclude by induction that for all $m \ge 0$, $t \in (-\infty, \infty)$ and $x_0 \in D - M \omega/2$. the functions $x_m(t, x_0)$ given by (10.4) exist, are periodic in t with period ω, and belong to the region D.

To prove the convergence of the sequence of functions (10.4), we estimate the difference

$$x_2(t, x_0) - x_1(t, x_0) = \left(1 - \frac{t}{\omega}\right) \int_0^t \left[f\left(t, x_1(t, x_0), \int_t^{t+T}\varphi(t, s, x_1(s, x_0))\,ds\right) \right.$$

$$\left. - f\left(t, x_0, \int_t^{t+T}\varphi(t, s, x_0)\,ds\right) \right] dt \ - \frac{1}{\omega} \int_t^\omega \left[f\left(t, x_1(t, x_0), \int_t^{t+T}\varphi(t, s, x_1(t, x_0))\,ds\right) \right.$$

$$\left. - f\left(t, x_0, \int_t^{t+T}\varphi(t, s, x_0)\,ds\right) \right] dt \ . \tag{10.8}$$

Hence,

$$|x_2(t, x_0) - x_2(t, x_0)|$$

$$\le \left(1 - \frac{t}{\omega}\right) \int_0^t \left[K_1|x_1(t, x_0) - x_0| + K_2 K_3 \int_t^{t+T}|x_1(s, x_0) - x_0|\,ds \right] dt$$

$$+ \frac{t}{\omega} \int_t^{\omega} \left[K_1 |x_1(t, x_0) - x_0| + K_2 K_3 \int_t^{t+T} |x_1(s, x_0) - x_0| ds \right] dt. \tag{10.9}$$

The inequalities (10.7) and (10.9) yield

$$|x_2(t, x_0) - x_1(t, x_0)| \le \left[\left(1 - \frac{t}{\omega}\right) \int_0^t \alpha_1(t) dt + \frac{t}{\omega} \int_t^{\omega} \alpha_1(t) dt \right] K_1 M$$

$$+ \left[\left(1 - \frac{t}{\omega}\right) \int_0^t dt + \frac{t}{\omega} \int_t^{\omega} dt \right] \frac{T\omega}{2} K_2 K_3 M$$

$$= \alpha_2(t) K_1 M + \alpha_1(t) \frac{T\omega}{2} K_2 K_3 M, \tag{10.10}$$

where

$$\alpha_2(t) = \left(1 - \frac{t}{\omega}\right) \int_0^t \alpha_1(t) dt + \frac{t}{\omega} \int_t^{\omega} \alpha_1(t) dt = \alpha_1(t) \left[\frac{\omega}{6} + \frac{\alpha_1(t)}{3} \right] \le \frac{\omega}{3} \alpha_1(t)$$

The inequality (10.10) can be rewritten in the form

$$|x_2(t, x_0) - x_1(t, x_0)| \le \alpha_1(t) R M, \tag{10.11}$$

where

$$R = \frac{\omega}{3} \left(K_1 + \frac{3T}{2} K_2 K_3 \right), \quad \Delta \le t \le \omega.$$

Assuming that the difference $x_m(t, x_0) - x_{m-1}(t, x_0)$ satisfies the inequality

$$|x_m(t, x_0) - x_{m-1}(t, x_0)| \le \alpha_1(t) R^{m-1} M, \tag{10.12}$$

we show that $x_{m+1}(t, x_0) - x_m(t, x_0)$ satisfies the inequality

$$|x_{m+1}(t, x_0) - x_m(t, x_0)| \le \alpha_1(t) R^m M. \tag{10.13}$$

Let us represent the difference $x_{m+1}(t, x_0) - x_m(t, x_0)$ in the form

$$x_{m+1}(t, x_0) - x_m(t, x_0) = \left(1 - \frac{t}{\omega}\right) \int_0^t \left[f\left(t, x_m(t, x_0), \int_t^{t+T} \varphi(t, s, x_m(s, x_0))\,ds\right) \right.$$

$$\left. - f\left(t, x_{m-1}(t, x_0), \int_t^{t+T} \varphi(t, s, x_{m-1}(s, x_0))\,ds\right) \right] dt$$

$$- \frac{t}{\omega} \int_t^\omega \left[f\left(t, x_m(t, x_0), \int_t^{t+T} \varphi(t, s, x_m(s, x_0))\,ds\right) \right.$$

$$\left. - f\left(t, x_{m-1}(t, x_0), \int_t^{t+T} \varphi(t, s, x_{m-1}(s, x_0))\,ds\right) \right] dt . \qquad (10.14)$$

Taking into account (10.12), we find

$$| x_{m+1}(t, x_0) - x_m(t, x_0) |$$

$$\leq \left(1 - \frac{t}{\omega}\right) \int_0^t \left[K_1 |x_m(t, x_0) - x_{m-1}(t, x_0)| + K_2 K_3 \int_t^{t+T} |x_m(s, x_0) - x_{m-1}(s, x_0)|\,ds \right] dt$$

$$+ \frac{t}{\omega} \int_t^\omega \left[K_1 |x_m(t, x_0) - x_{m-1}(t, x_0)| + K_2 K_3 \int_t^{t+T} |x_m(s, x_0) - x_{m-1}(s, x_0)|\,ds \right] dt$$

$$\leq \left(1 - \frac{t}{\omega}\right) \int_0^t \left[K_1 \alpha_1(t) R^{m-1} M + K_2 K_3 \int_t^{t+T} \frac{\omega R^{m-1}}{2} M\,ds \right] dt$$

$$+ \frac{t}{\omega} \int_t^\omega \left[K_1 \alpha_1(t) R^{m-1} M + K_2 K_3 \int_t^{t+T} \frac{\omega R^{m-1}}{2} M\,ds \right] dt = \alpha_2(t) K_1 R^{m-1} M$$

$$+ \alpha_1(t) \frac{T\omega}{2} K_2 K_3 R^{m-1} M \leq \alpha_1(t) \frac{\omega}{3} \left(K_1 + \frac{3T}{2} K_2 K_3 \right) R^{m-1} M,$$

i.e., we have proved the inequality (10.13).

Since $x_m(t, x_0)$ is a function periodic in t with period ω and $\alpha_1(t) \leq \omega/2$, we conclude by induction that

$$| x_{m+1}(t, x_0) - x_m(t, x_0) | \leq R^m M \omega/2 \qquad (10.15)$$

for all $-\infty < t < \infty$ and $m = 0, 1, 2, \ldots$. The inequality (10.15) yields

$$| x_{m+1}(t, x_0) - x_m(t, x_0) | \leq \frac{\omega}{2} \sum_{i=0}^{k-1} R^{m+i} M.$$ (10.16)

As we have assumed, the eigenvalues of the matrix $R = \frac{\omega}{3}\left(K_1 + \frac{3T}{2}K_2 K_3\right)$ belong to the disk with radius 1. This implies that

$$R^m \to 0 \quad \text{as} \quad m \to \infty;$$ (10.17)

$$\sum_{i=0}^{k-1} R^{m+i} \leq R^m (E - R)^{-1} \quad \ldots \ldots \ldots \ldots$$ (10.18)

The relations (10.17) and (10.18) involve the convergence of the sequence (10.4) (uniform with respect to $x(t, x_0) \in (-\infty, \infty) \times D - M\,\omega/2$).

Denoting the limiting function of the sequence (10.4) by $x_\infty(t, x_0)$ and passing to the limit in (10.4) as $m \to \infty$, we find that the limiting function $x_\infty(t, x_0)$ is a solution of the equation

$$x(t, x_0) = x_0 + \int_0^t \left[f\left(t, x(t, x_0), \int_t^{t+T} \varphi(t, s, x(s, x_0))\,ds \right) \right.$$

$$\left. - \frac{1}{\omega} \int_0^\omega f\left(t, x(t, x_0), \int_t^{t+T} \varphi(t, s, x(s, x_0))\,ds \right) dt \right] dt.$$ (10.19)

Then the following estimate holds for the difference

$$| x_\infty(t, x_0) - x_\infty(t, x_0) | \leq R^m (E - R)^{-1} M\,\omega/2.$$ (10.20)

Since $x(t)$ is a solution of eqn.(10.1), we have

$$x(t) = x_0 + \int_0^t f\left(t, x(t), \int_t^{t+T} \varphi(t, s, x(s))\,ds \right) dt,$$ (10.21)

By virtue of the periodicity of $x(t)$, its integral average with respect to time vanishes, i.e.,

$$\overline{f\left(t, x(t), \int_t^{t+T} \varphi(t, s, x(s))\,ds\right)} = 0. \tag{10.22}$$

It follows from the equations (10.21) and (10.22) that $x(t)$ and $x_\infty(t, x_0)$ are solutions of the same equation (10.19). Thus, in order to prove that $x(t) = x_\infty(t, x_0)$, it suffices to show that eqn.(10.19) cannot possess two different periodic solutions. As before, one can easily prove this by contradiction.

Denote

$$S(x_0) = \frac{1}{\omega} \int_0^\omega f\left(t, x_\infty(t, x_0), \int_t^{t+T} \varphi(t, s, x_\infty(s, x_0))\,ds\right) dt ; \tag{10.23}$$

and

$$S_m(x_0) = \frac{1}{\omega} \int_0^\omega f\left(t, x_m(t, x_0), \int_t^{t+T} \varphi(t, s, x_m(s, x_0))\,ds\right) dt . \tag{10.24}$$

The theorem on existence of a periodic solution of the system (10.1) can be now easily proved by analogy with Theorem 1.2. Thus, in the case of systems of integro-differential equations of the standard form

$$\frac{dx(t)}{dt} = \varepsilon X\left(t, x, \int_t^{t+T} \varphi(t, s, x(s))\,ds\right), \tag{10.25}$$

where ε is a small positive parameter, the corresponding statement has the following form.

Theorem 1.15. (Nurzhanov, 1977a). *Suppose that the right-hand side of the system of integro-differential equations* (10.25) *is defined in the region*

$$-\infty < t < \infty, \quad (x, y) \in D \times D_1. \tag{10.26}$$

Suppose also that it is periodic in t and s with period ω, *continuous with respect to all its variables t, s, x, and y, and satisfies the inequalities* (10.2). *Assume that the averaged system*

$$\frac{d\xi(t)}{dt} = \varepsilon X_0(\xi) \tag{10.27}$$

where

$$X_0(\xi) = \frac{1}{\omega} \int_0^\omega X(t, \xi, y(t, \xi)) \, dt,$$

$$y(t, \xi) = \int_t^{t+T} \varphi(t, s, \xi) \, ds,$$

has the isolated equilibrium point $\xi = \xi_0$, *i.e.,*

$$X_0(\xi_0) = 0 \tag{10.28}$$

and that the index of this point is not zero. Then for sufficiently small ε *the system* (10.25) *has a periodic solution with period* ω.

In the general case, the point x_0, through which periodic solutions pass at the initial time $t = t_0 = 0$, can be found by employing the numerical method described in §3. However, in some cases this problem can be solved easily. One of these cases is established by the following theorem.

Theorem 1.16. (Nurzhanov, 1977a). *Let the right-hand side of the system of integro-differential equations*

$$\frac{dx(t)}{dt} = f\left(t, x, \int_t^{t+T} \varphi(t, s, x(s)) \, ds\right) \tag{10.29}$$

satisfy the inequalities (10.2) *and conditions* (i), (ii). *Suppose also that the following identities*

$$f(t, x, y) \equiv -f(-t, x, y); \tag{10.30}$$

$$\varphi(t, t, x) \equiv -\varphi(-t, -t, x); \tag{10.31}$$

$$\varphi(t, t+T, x) \equiv -\varphi(-t, -t+T, x) \tag{10.32}$$

hold for all $(x, y) \in D \times D$ *and* $-\infty < t < \infty$. *Then, for an arbitrarily chosen point* x_0 *$\in D - M$ $\omega/2$, there exists the ω-periodic solution $x = x(t)$ of the system* (10.29) *which passes through this point x_0 at $t = 0$.*

Proof. The function

$$f\left(t, x_0, \int_t^{t+T} \varphi(t, s, x_0)\, ds\right)$$

is odd in t, and therefore, the function $x_1(t, x_0)$, given by (10.4) with $m = 1$, is even as an integral of an odd function. Let us show that the function

$$f\left(t, x_1(-t, x_0), \int_t^{t+T} \varphi(t, s, x_1(s, x_0))\, ds\right)$$

is odd. Taking (10.30)-(10.32) into account, we find

$$f\left(-t, x_1(-t, x_0), \int_{-t}^{-t+T} \varphi(-t, s, x(s, x_0))\, ds\right) = -f\left(t, x_1(t, x_0), \int_t^{t+T} \varphi(t, s, x(s, x_0))\, ds\right)$$

This is why the second function $x_2(t, x_0)$ from the sequence of function (10.4) is even, and

$$f\left(t, x_2(t, x_0), \int_t^{t+T} \varphi(t, s, x_2(s, x_0))\, ds\right)$$

is odd. It is easy to prove by induction that all the approximations (10.4) to the solutions of the system of integro-differential equations (10.29) are even functions in t, whereas

$$f\left(t, x_m(t, x_0), \int_t^{t+T} \varphi(t, s, x_m(s, x_0))\, ds\right)$$

are odd functions, and hence, $S_m(x_0) = 0$ for all $x_0 \in D - M \,\omega/2$, $m = 0, 1, 2, \ldots$. This means that the limiting function $x_\infty(t, x_0)$ is a periodic solution of the system (10.29) such that $x(0) = x_0$.

The system (10.1) can be considered as a nonlinear system of Volterra type integro-differential equations with finite aftereffect.

Consider now a system of Volterra type integro-differential equations with infinite aftereffect

$$\frac{dx(t)}{dt} = f\left(t, x, \int_{-\infty}^t R(t-s)\varphi(t, s, x(t), x(s))\, ds\right), \qquad (10.33)$$

where $x = (x_1, x_2, ..., x_n)$, $K = (K_1, K_2, ..., K_n)$, $f = (f_1, f_2, ..., f_n)$, and $\varphi = (\varphi_1, \varphi_2, ..., \varphi_n)$. Assume that the function $f(t, x, z)$ is defined and continuous on the set

$$D_f: \{R \times D_1 \times D_2; \quad D_1: \|x\| \leq d_1, \quad D_2: \|z\| \leq d_2, \quad R: (\infty, \infty)\}; \qquad (10.34)$$

and that it is ω-periodic in t. The function φ is assumed to be defined and continuous on the set $D_\varphi: R \times R \times D_1 \times D_1$; and ω-periodic in t and s. Furthermore, we assume that both functions f and φ are bounded

$$\|f(t, x, z)\| \leq M_1, \quad \|\varphi(t, s, x, z)\| \leq M_2 \qquad (10.35)$$

and satisfy the Lipschitz conditions

$$\|f(t, x_1, z_1) - f(t, x_2, z_2)\| \leq N_1 \|x_1 - x_2\| + N_2 \|z_1 - z_2\|,$$

$$\|\varphi(t, s, x_1, z_1) - \varphi(t, s, x_2, z_2)\| \leq N_3 \|x_1 - x_2\| + N_4 \|y_1 - y_2\|. \qquad (10.36)$$

Suppose also that the function $\int_{-\infty}^{t} R(t-s)\varphi(t, s, x(t), x(s))\, ds$ is defined, continuous, and ω-periodic for any ω-periodic function $x(t)$ from the region D_1, and that the kernel $R(t-s)$ satisfy the conditions

$$\int_{-\infty}^{t} \|R(t-s)\|\, ds \leq K, \quad KM_2 \leq d_2 \qquad (10.37)$$

for all $t \in R$.

In addition, we suppose that the following conditions are satisfied

(a) the set $D_1': \|x\| \leq d_1 - M_1 \omega/2$ is nonempty;

(b) $$q = \frac{\omega}{2}(N_1 + N_2 K(N_3 + N_4)) < 1.$$

For systems of the form (10.33), one can establish the following statement, the proof of which is analogous to the proof of Theorem 1.14.

Theorem 1.17. (Vuitovich, 1982). *Let the functions $f(t, x, z)$ and $\varphi(t, s, x, z)$ be defined and continuous in the region D_f and D_φ, respectively. Suppose that they are ω-periodic in t and s, respectively, and satisfy the conditions (10.36) and (10.37), (a) and (b). Suppose also that eqn.(10.33) has an ω-periodic solution $x(t) =$*

$\psi(t)$ *which passes through the point* $x_0 \in D_1'$ *at* $t = 0$. *Then this solution satisfies*
the relation

$$\psi(t) = \lim_{m \to \infty} x_m(t, x_0), \tag{10.38}$$

where $x_m(t, x_0)$ *are functions periodic in* t *with period* ω, *defined by the recursion*
relations

$$x_{m+1}(t, x_0) = x_0 + \int_0^t \left[f\left(l, x_m(l, x_0), \int_{-\infty}^l R(l - s)\varphi(l, s, x_m(l, x_0), x_m(s, x_0))\, ds \right) \right. $$
$$\left. - \frac{1}{\omega} \int_0^\omega f\left(t, x_m(l, x_0), \int_{-\infty}^t R(t - s)\varphi(t, s, x_m(t, x_0), x_m(s, x_0))\, ds \right) dt \right] dl,$$

$$(m = 0, 1, 2, ..., x_0(t, x_0) = x_0).$$

§11. Periodic Solutions of Nonlinear Systems of Difference Equations

Consider a system of difference equations

$$x_{n+1} = x_n + f_n(x_n), \quad n = 0, \pm 1, \pm 2, ..., \tag{11.1}$$

where

$$x_n = (x_n^{(1)}, ..., x_n^{(m)}), \qquad f_n = (f_n^{(1)}, ..., f_n^{(m)});$$

$f_n(x)$ is an m-dimensional vector function periodic with period N. It is defined in a
closedbounded region D of the Euclidean space E_m and satisfies the inequalities

$$|f_n(x)| \leq M; \tag{11.2}$$

$$|f_n(x') - f_n(x'')| \leq K |x' - x''|. \tag{11.3}$$

Here, M is a vector with positive coordinates, and K is a matrix with nonnegative
elements.

Below we consider only the systems for which the following conditions are satisfied:

(i) the set $D - MN/2$ is nonempty;

(ii) the eigenvalues of the matrix $Q = NK/2$ are contained in the unit disk.

Let us denote the average value of a function f_n periodic with period N by $\bar{f}_n$, i.e.,

$$\bar{f}_n = \frac{1}{N} \sum_{j=0}^{N-1} f_j.$$

The next statement is similar to Lemma 1.1.

Lemma 1.5 (Martinyuk, Mironov, and Kharabovskaya, 1971b; Mitropolsky, Mikhailovskaya, 1972). *Suppose that $n \in [0, N-1]$. Then*

$$\left| \sum_{i=0}^{n} (f_i - \bar{f}_n) \right| \le 2(n+1)\left(1 - \frac{n+1}{N}\right) |f_n|_0. \tag{11.4}$$

Proof. We have

$$\sum_{i=0}^{n} \left(f_i - \frac{1}{N} \sum_{j=0}^{N-1} f_j \right) = \left(1 - \frac{n+1}{N}\right) \sum_{i=0}^{n} f_i - \frac{n+1}{N} \sum_{i=n+1}^{N-1} f_i.$$

This yields

$$\left| \sum_{i=0}^{n} (f_i - \bar{f}_n) \right| \le \left| \left(1 - \frac{n+1}{N}\right) \sum_{i=1}^{n} f_i \right| + \frac{n+1}{N} \left| \sum_{i=n+1}^{N-1} f_i \right|$$

$$\le (n+1)\left(1 - \frac{n+1}{N}\right) |f_n|_0 + \frac{n+1}{N}(N-1-n)|f_n|_0$$

$$= 2(n+1)\left(1 - \frac{n+1}{N}\right) |f_n|_0 = \alpha_n |f_n|_0,$$

where $\alpha_n = 2(n+1)\left(1 - \frac{n+1}{N}\right)$. Assume that the system (11.1) has a periodic solution with period N and that the point x_0 which this solution passes through at $n = n_0 = 0$ is known. Then the following theorem is valid.

Theorem 1.18. (Martinyuk, 1972; Martinyuk, Mironov, and Kharabovskaya, 1971b). *Let φ_n be the N-periodic solution of the system (11.1) which passes through the point $x_0 \in D - MN/2$. Then*

$$\varphi_n = \lim_{k \to \infty} x_n^{(k)}(x_0) \tag{11.5}$$

uniformly in n and x_0, and

$$|\varphi_n - x_n^{(k)}(x_0)| \le Q^k(E - Q)^{-1} MN/2, \tag{11.6}$$

where $x_n^{(k)}(x_0)$ are functions periodic in n. They are defined by

$$x_n^{(k)}(x_0) = x_0 + \sum_{i=0}^{n-1} \left[f_i(x_i^{(k-1)}(x_0)) - \overline{f_n(x_n^{(k-1)}(x_0))} \right],$$

$$x_i^0(x_0) = x_0, \quad k = 1, 2, \ldots. \tag{11.7}$$

Proof. All the functions (11.7) are periodic in n with period N. By virtue of Lemma 1.5, we have

$$|x_n^{(k)}(x_0) - x_0| \le 2Mn \left(1 - \frac{n}{N} \right) \le \frac{N}{2} M, \tag{11.8}$$

i.e., $x_n^{(k)}(x_0) \in D$ for all n and k provided that $x_0 \in D - MN/2$.

Let us prove the convergence of the sequence of functions (11.7). Estimating the difference $x_n^{(2)}(x_0) - x_n^{(1)}(x_0)$, we obtain

$$|x_n^{(2)}(x_0) - x_n^{(1)}(x_0)| \le \left(1 - \frac{n}{N} \right) \sum_{i=0}^{n-1} K|x_i^{(1)}(x_0) - x_0| + \frac{n}{N} \sum_{i=n}^{N-1} K|x_i^{(1)}(x_0) - x_0|$$

$$= \left(1 - \frac{n}{N} \right) \sum_{i=0}^{N-1} K M \alpha_i + \frac{n}{N} \sum_{i=n}^{N-1} K M \alpha_i = \beta_n K M, \tag{11.9}$$

where

$$\beta_n = \frac{\alpha_n \alpha_{n-1}}{n^2} + \frac{(n+1)(n+2)}{6N} \alpha_{n-1} + n \left(\frac{N}{3} - \frac{1}{3N} \right)$$

$$- \frac{n^2}{6N} \alpha_n - \frac{n^2(n+1)(n+2)}{3N^2}. \tag{11.10}$$

It is easy to show that $\beta_n \le N^2/4$ and, hence,

$$|x_n^{(2)}(x_0) - x_n^{(1)}(x_0)| \le (NK/2)(MN/2). \tag{11.11}$$

One can prove by induction that

$$|x_n^{(k+1)}(x_0) - x_n^{(k)}(x_0)| \le (NK/2)^k (MN/2)$$ (11.12)

for all n and $k = 0, 1, 2, \dots$. It follows from the last inequality and condition (ii) that the sequence of periodic functions (11.5) converges uniformly with respect to x_0 and n. By passing to the limit in (11.7) as $k \to \infty$, we find that the limiting function $x_n(x_0)$ satisfies the equation

$$x_n(x_0) = x_0 + \sum_{i=0}^{n-1} \left[f_i(x_i(x_0)) - \overline{f_n(x_n(x_0))} \right].$$ (11.13)

Taking the inequality (11.12) into account, we obtain

$$|x_n^{(k+s)}(x_0) - x_n^{(k)}(x_0)| \le \sum_{i=0}^{s-1} (NK/2)^{k+i} (MN/2),$$

furthermore, by employing condition (ii), we find

$$|x_n(x_0) - x_n^{(k)}(x_0)| \le Q^k (E-Q)^{-1} MN/2.$$ (11.14)

According to condition of Theorem 1.18, the function φ_n is a periodic solution of the equation (11.1) (and consequently, the relations $\varphi_n = x_0 + \sum_{i=0}^{n-1} f_i(\varphi_i)$ hold); moreover, it possesses the property $\overline{f_n(\varphi_n)} = 0$. Hence, φ_n (together with $x_n(x_0)$) is a periodic solution of eqn.(11.13). It is easy to prove by contradiction that eqn.(11.13) cannot have two different periodic solutions, and this means that the relation (11.5) is valid.

Let us now investigate the problem of existence of periodic solutions for the system (11.1). If it is known that the periodic solution of the system (11.1) exists and passes through the point x_0 which is also known, then, according to Theorem 1.18, the problem of determination of this solution is reduced to the calculation of the function $x_n^{(k)}(x_0)$. Let us denote

$$S(x_0) = \overline{f_n(x_n(x_0))} = \frac{1}{N} \sum_{i=0}^{N-1} f_i(x_i(x_0)),$$ (11.15)

where $x_n(x_0)$ satisfies eqn.(11.13). If we take a solution of the equation $S(x_0) = 0$ as the initial point x_0, then eqn.(11.13) turns into

$$x_n(x_0) = x_0 + \sum_{i=0}^{n-1} f_i(x_i(x_0)).$$ (11.16)

This means that the function $x_n(x_0)$ is a periodic solution of the equation (11.1) provided that the initial point x_0 is a solution of the equation $S(x_0) = 0$.

Thus, the problem of existence of periodic solutions to the system of difference equations (11.1) is reduced to the problem of existence of zeros of the function $S(x_0)$. In the general case, it is impossible to find the limiting function $x_n(x_0)$ of the sequence (11.7), and consequently, it is impossible to determine $S(x_0)$. Therefore, by using the approximate solutions $x^{(k)}(x_0)$, we calculate the function

$$S^{(k)}(x_0) = \frac{1}{N} \sum_{i=0}^{N-1} f_i(x_i^{(k)}(x_0)) \qquad (11.17)$$

and raise the following question: how to solve the problem of existence of zeros of the mapping (11.15) and, hence, of existence of periodic solutions to the system (11.1), by use of the mapping (11.17). The answer to this question is given by the theorem similar to Theorem 1.2.

§12. Bilateral Approximations to Periodic Solutions of Systems with Lag

Consider a system of differential equations of the form

$$\frac{dx(t)}{dt} = g(t, x(t), x(t-\Delta)), \qquad (12.1)$$

where $x = (x_1, x_2, ..., x_n)$ and $g(t, x, y) = (g_1(t, x, y), g_2(t, x, y), ..., g_n(t, x, y))$ are elements of the Euclidean space E_n; $0 \le \Delta \le \omega$. The function $g(t, x, y)$ is periodic in t with period ω. We set

$$g(t, x(t), x(t-\Delta)) = f(t, x(t), x(t-\Delta), x(t), x(t-\Delta))$$

and consider the system of equations

$$\frac{dx(t)}{dt} = f(t, x(t), x(t-\Delta), x(t), x(t-\Delta)). \qquad (12.2)$$

Assume that the right-hand side of this system $f(t, x, y, u, v)$ is defined in the region

$$t \in (-\infty, \infty), \quad x, y, u, v \in [a, b],$$

$$a = (a_1, a_2, ..., a_n) \in E_n, \quad b = (b_1, b_2, ..., b_n) \in E_n,$$

(12.3)

and that it is continuous with respect to all its variables $t, x, y, u,$ and v, and ω-periodic in t. Assume also that

$$m \le f(t, x, y, u, v) \le M,$$

(12.4)

where

$$m = (m_1, m_2, ..., m_n) \in E_n, \quad M = (M_1, M_2, ..., M_n) \in E_n;$$

and that

$$f(t, x, y, u, v) \le f(t, \bar{x}, \bar{y}, \bar{u}, \bar{v})$$

(12.5)

for $x \le \bar{x}, y \le \bar{y}, u \ge \bar{u}$, and $v \ge \bar{v}$. Furthermore, we suppose that the constants $\omega, M, m, b,$ and a satisfy the inequality

$$\frac{\omega}{2}(M - m) \le b - a.$$

(12.6)

Let us define two sequences of functions ($\{u_n(t, x_0)\}$ and $\{v_n(t, x_0)\}$ ($n = 0, 1, ...$)) by the recursion relations

$$u_{n+1}(t, x_0) = x_0 + \left(1 - \frac{t}{\omega}\right) \int_0^t f\left(s, u_n(s, x_0), u_n(s - \Delta, x_0), v_n(s, x_0), v_n(s - \Delta, x_0)\right) ds$$

$$- \frac{t}{\omega} \int_t^\omega f(s, v_n(s, x_0), v_n(s - \Delta, x_0), u_n(s, x_0), u_n(s - \Delta, x_0)) ds,$$

(12.7)

$$v_{n+1}(t, x_0) = x_0 + \left(1 - \frac{t}{\omega}\right) \int_0^t f\left(s, v_n(s, x_0), v_n(s - \Delta, x_0), u_n(s, x_0), u_n(s - \Delta, x_0)\right) ds$$

$$- \frac{t}{\omega} \int_t^\omega f(s, u_n(s, x_0), u_n(s - \Delta, x_0), v_n(s, x_0), v_n(s - \Delta, x_0)) ds,$$

with the zeroth approximation given by

$$u_0(t, x_0) = x_0 - \frac{1}{2}\alpha_1(t)(M - m), \quad v_0(t, x_0) = x_0 + \frac{1}{2}\alpha_1(t)(M - m),$$

$$u_0(t - \Delta, x_0) = x_0 - \frac{M - m}{2}\begin{cases} \alpha_1(t - \Delta + \omega) & \text{for } 0 \le t \le \Delta, \\ \alpha_1(t - \Delta) & \text{for } 0 \le t \le \omega, \end{cases} \tag{12.8}$$

$$v_0(t - \Delta, x_0) = x_0 + \frac{M - m}{2}\begin{cases} \alpha_1(t - \Delta + \omega) & \text{for } 0 \le t \le \Delta, \\ \alpha_1(t - \Delta) & \text{for } 0 \le t \le \omega, \end{cases}$$

where

$$\alpha_1(t) = 2t\left(1 - \frac{t}{\omega}\right), \quad a + \frac{\omega}{4}(M - m) \le x_0 \le b - \frac{\omega}{4}(M - m).$$

It is clear that for all $n = 0, 1, 2, \ldots$, the functions $u_n(t, x_0)$ and $v_n(t, x_0)$ satisfy the conditions

$$u_n(0) = u_n(\omega) = x_0, \quad a \le u_n(t, x_0) \le b,$$

$$v_n(0) = v_n(\omega) = x_0, \quad a \le v_n(t, x_0) \le b. \tag{12.9}$$

Let us show that

$$u_0(t, x_0) \le u_1(t, x_0) \le \ldots u_n(t, x_0) \le \ldots v_n(t, x_0) \le \ldots \le v_1(t, x_0) \le v_0(t, x_0). \tag{12.10}$$

In fact, taking (12.4), (12.5), and (12.8) into account, we obtain the following bounds for the functions given by (12.7) with $n = 0$

$$u_1(t, x_0) \ge x_0 + \left(1 - \frac{t}{\omega}\right)\int_0^t m\,ds - \frac{t}{\omega}\int_t^\omega M\,ds = u_0(t, x_0),$$

$$v_1(t, x_0) \le x_0 + \left(1 - \frac{t}{\omega}\right)\int_0^t M\,ds - \frac{t}{\omega}\int_t^\omega m\,ds = v_0(t, x_0),$$

$$u_2(t, x_0) \ge x_0 + \left(1 - \frac{t}{\omega}\right)\int_0^t f\left(s, u_0(s, x_0), u_0(s - \Delta, x_0), v_0(s, x_0), v_0(s - \Delta, x_0)\right)ds$$

$$-\frac{t}{\omega}\int_t^\omega f\left(s, v_0(s, x_0), v_0(s - \Delta, x_0), u_0(s, x_0), u_0(s - \Delta, x_0)\right)ds = u_1(t, x_0),$$

$$v_2(t, x_0) \leq x_0 + \left(1 - \frac{t}{\omega}\right) \int_0^t f\left(s, v_0(s, x_0), v_0(s - \Delta, x_0), u_0(s, x_0), u_0(s - \Delta, x_0)\right) ds$$

$$- \frac{t}{\omega} \int_t^\omega f(s, u_0(s, x_0), u_0(s - \Delta, x_0), v_0(s, x_0), v_0(s - \Delta, x_0)) ds = v_1(t, x_0).$$

It is easy to prove by induction that the inequalities

$$u_n(t, x_0) \leq u_{n+1}(t, x_0), \quad v_n(t, x_0) \geq v_{n+1}(t, x_0) \tag{12.11}$$

hold for all $n = 0, 1, 2, \ldots$.

By using the obvious inequality,

$$u_n(t, x_0) \leq v_n(t, x_0),$$

we get the inequality (12.10) from (12.11).

Assume that $x^*(t, x_0) \in [a, b]$ is a solution of the equation

$$x(t, x_0) = x_0 + \left(1 - \frac{t}{\omega}\right) \int_0^t f\left(s, x(s, x_0), x(s - \Delta, x_0), x(s, x_0), x(s - \Delta, x_0)\right) ds$$

$$- \frac{t}{\omega} \int_t^\omega f(s, x(s, x_0), x(s - \Delta, x_0), x(s, x_0), x(s - \Delta, x_0)) ds. \tag{12.12}$$

Taking (12.5), (12.8), and (12.12) into account, we obtain

$$x^*(t, x_0) \geq x_0 + \left(1 - \frac{t}{\omega}\right) \int_0^t f(s, a, a, b, b) ds - \frac{t}{\omega} \int_t^\omega f(s, b, b, a, a) ds$$

$$\geq x_0 + t\left(1 - \frac{t}{\omega}\right) m - \frac{t}{\omega}(\omega - t) M = x_0 - \frac{1}{2}\alpha_1(t)(M - m) = u_0(t, x_0),$$

$$x^*(t, x_0) \geq x_0 + \left(1 - \frac{t}{\omega}\right) \int_0^t f(s, b, b, a, a) ds - \frac{t}{\omega} \int_t^\omega f(s, a, a, b, b) ds$$

$$\leq x_0 + t\left(1 - \frac{t}{\omega}\right) M - \frac{t}{\omega}(\omega - t) m = x_0 + \frac{1}{2}\alpha_1(t)(M - m) = v_0(t, x_0).$$

Thus, we have

$$u_n(t, x_0) \le x^*(t, x_0) \le v_n(t, x_0),. \tag{12.13}$$

By virtue of continuity of the function $f(t, x, y, u, v)$, the sequences $\{u_n(t, x_0)\}$ and $\{v_n(t, x_0)\}$ are equicontinuous and uniformly bounded. Moreover, they are monotonic. The monotonicity and boundedness (both from above and from below) imply that these sequences are convergent. We denote the corresponding limiting functions by $u_\infty(t, x_0)$ and $v_\infty(t, x_0)$. It follows from (12.13) that

$$u_\infty(t, x_0) \le x^*(t, x_0) \le v_\infty(t, x_0).$$

In addition, we assume that in the region (12.3) the function $f(t, x, y, u, v)$ satisfies the condition

$$|f(t, x', y', u', v') - f(t, x'', y'', u'', v'')|$$

$$\le K_1 (x' - x'') + K_2(y' - y'') + K_3(u'' - u') + K_4(v'' - v') \tag{12.14}$$

for $x' \ge x''$, $y' \ge y''$, $u'' \ge u'$, and $v'' \ge v'$, where K_1, K_2, K_3, and K_4 are $(n \times n)$-dimensional matrices with nonnegative elements such that the eigenvalues of the matrix

$$Q = (K_1 + K_2 + K_3 + K_4) \left[\frac{\omega}{3} + \frac{3\Delta^2}{2\omega} \left(1 - \frac{\Delta}{\omega} \right)^2 \right]$$

are contained in the unit circle. By using the inequality (12.14), we now estimate the difference $v_n(t, x_0) - u_n(t, x_0)$. The relations (12.7) and (12.8) yield

$$v_0(t, x_0) - u_0(t, x_0) = \alpha_1(t) (M - m),$$

$$v_1(t, x_0) - u_1(t, x_0) = \left(1 - \frac{t}{\omega} \right) \int_0^t \{ f(s, v_0(s, x_0), v_0(s - \Delta, x_0), u_0(s, x_0), u_0(s - \Delta, x_0))$$

$$- f(s, u_0(s, x_0), u_0(s - \Delta, x_0), v_0(s, x_0), v_0(s - \Delta, x_0)) \} ds$$

$$+ \frac{t}{\omega} \int_t^\omega \{ f(s, v_0(s, x_0), v_0(s - \Delta, x_0), u_0(s, x_0), u_0(s - \Delta, x_0))$$

$$- f(s, u_0(s, x_0), u_0(s - \Delta, x_0), v_0(s, x_0), v_0(s - \Delta, x_0)) \} ds$$

$$\le \left(1 - \frac{t}{\omega} \right) \int_0^t \{ (K_1 + K_3)(v_0(s, x_0) - u_0(s, x_0)) + (K_2 + K_4)(v_0(s - \Delta, x_0)$$

$$-u_0(s-\Delta,x_0))\} \, ds + \frac{t}{\omega}\int_t^\omega \{(K_1+K_3)(v_0(s,x_0)-u_0(s,x_0)) + (K_2$$

$$+K_4)(v_0(s-\Delta,x_0)-u_0(s-\Delta,x_0))\} \, ds \; \le \; (K_1+K_3)\left\{\left(1-\frac{t}{\omega}\right)\int_0^t \alpha_1(t)\,dt\right.$$

$$+\frac{t}{\omega}\int_t^\omega \alpha_1(t)\,dt\right\}(M-m) + (K_2+K_4)\left\{\left(1-\frac{t}{\omega}\right)\int_{-\Delta}^{t-\Delta}[v_0(s_1,x_0)\right.$$

$$-u_0(s_1,x_0)]\,ds_1 + \frac{t}{\omega}\int_{t-\Delta}^{\omega-\Delta}[v_0(s_1,x_0)-u_0(s_1,x_0)]\,ds_1\right\}(M-m). \qquad (12.15)$$

Taking the inequalities (12.8) into account, we can rewrite (12.15) as follows

$$v_1(t,x_0)-u_1(t,x_0) \; \le \; \alpha_1(t)\,(K_1+K_3)\,(M-m) + \beta_1(t)\,(K_2+K_4)\,(M-m).$$

By employing the estimates (2.9), (2.10), (2.14), and (2.18), we finally obtain

$$v_1(t,x_0)-u_1(t,x_0) \; \le \; \alpha_1(t)\,Q(M-m).$$

One can easily prove by induction that

$$v_n(t,x_0)-u_n(t,x_0) \; \le \; \alpha_1(t)\,Q^n(M-m) \qquad (12.16)$$

for all n 0, 1, 2, ... Since the eigenvalues of the matrix Q belong to the unit disk, the inequality (12.16) implies that

$$u_\infty(t,x_0) = v_\infty(t,x_0) = x_\infty(t,x_0). \qquad (12.17)$$

The relation (12.17) means that the sequences $\{u_n(t,x_0)\}$ and $\{v_n(t,x_0)\}$ have the same limit as $n \to \infty$; moreover, the limiting function $x_\infty(t,x_0)$ is a periodic solution of the equation (12.12).

Let $x = \varphi(t)$ be an ω-periodic solution of the system (12.2), which passes through the point

$$x_0 \in \left[a+\frac{\omega}{4}(M-m), \; b-\frac{\omega}{4}(M-m)\right]$$

at $t = 0$ and satisfies the equation (12.12). One can easily prove by contradiction that this solution is unique and, consequently, that the following equality holds

$$\varphi(t) = x^*(t, x_0).\tag{12.18}$$

Thus, we have proved the following statement.

Theorem 1.19. (Tsydilo, 1973). *Let the right-hand side of the system of equations* (12.2) *be defined in the region* (12.3). *Suppose that it is continuous with respect to all its variables and periodic in t with period* ω. *Suppose also that it satisfies the inequalities* (12.4), (12.5), *and* (12.6). *Then the sequences* $\{u_n(t, x_0)\}$ *and* $\{v_n(t, x_0)\}$, *defined by* (12.7) *and* (12.8), *satisfy the conditions*

$$u_n(0, x_0) = u_n(\omega, x_0) = x_0, \quad v_n(0, x_0) = v_n(\omega, x_0) = x_0$$

and inequalities

$$a \le u_n(t, x_0) \le b, \quad a \le v_n(t, x_0) \le b,$$

$$u_0(t, x_0) \le u_1(t, x_0) \le \dots u_n(t, x_0) \le \dots v_n(t, x_0) \le \dots \le v_1(t, x_0) \le v_0(t, x_0).$$

Moreover, these sequences converge uniformly in t and x_0, *and the limiting functions* $u_\infty(t, x_0)$ *and* $v_\infty(t, x_0)$ *satisfy the equation* (12.12).

If $x^*(t, x_0) \in [a, b]$ is a periodic solution of eqn.(12.12), then

$$u_n(t, x_0) \le u_\infty(t, x_0) \le x^*(t, x_0) \le v_\infty(t, x_0) \le v_n(t, x_0).$$

Under the assumption, that the right-hand side of the equation (12.2) satisfies the conditions (12.14), we have

$$v_n(t, x_0) - u_n(t, x_0) \le \alpha_1(t) Q^n (M - m).$$

This means that the limiting functions $u_\infty(t, x_0)$ and $v_\infty(t, x_0)$ coincide and that the following relation

$$\varphi(t) = u_\infty(t, x_0) = v_\infty(t, x_0) = x_\infty(t, x_0)$$

holds for the ω–periodic solution $x = \varphi(t)$ of (12.2) which passes through the point

$$x_0 \in \left[a + \frac{\omega}{4}(M - m), \ b - \frac{\omega}{4}(M - m) \right]$$

at $t = 0$.

Let us now investigate the problem of existence of ω-periodic solutions of the system (12.2). For this purpose, we denote

$$S(x_0) = \frac{1}{\omega} \int_0^t f(t, x_\infty(t, x_0), x_\infty(t - \Delta, x_0), x_\infty(t, x_0), x_\infty(t - \Delta, x_0)) dt,$$

$$\underline{S}_m(x_0) = \frac{1}{\omega} \int_0^t f(t, u_m(t, x_0), u_m(t - \Delta, x_0), v_m(t, x_0), v_m(t - \Delta, x_0)) dt, \quad (12.19)$$

$$\overline{S}_m(x_0) = \frac{1}{\omega} \int_0^t f(t, v_m(t, x_0), v_m(t - \Delta, x_0), u_m(t, x_0), u_m(t - \Delta, x_0)) dt.$$

Let $x_\infty(t, x_0)$ be a solution of eqn.(12.12) periodic in t with period ω. Then, clearly, the periodic solution of eqn.(12.12) is, at the same time, the periodic solution of the system (12.2), provided that $S(x_0) = 0$. Therefore, the problem of existence of periodic solutions of eqn.(12.2) is reduced to the problem of existence of zeros of the function $S(x_0)$. The latter problem can be solved by using the functions $\underline{S}_m(x_0)$ or $\overline{S}_m(x_0)$.

Taking (12.5) and (12.10) into account, we find

$$\underline{S}_m(x_0) \le S(x_0) \le \overline{S}_m(x_0). \tag{12.20}$$

By virtue of the continuity of $S(x_0)$, the inequality (12.20) implies the following statement.

Theorem 1.20. (Tsydilo, 1973). *If for some integer m we have $\underline{S}_m(x_0) > 0$ or $\overline{S}_m(x_0) > 0$, then the system of equations (12.2) has no ω-periodic solutions (which pass through the point $(0, x_0)$) in the region $[a, b]$.*

In other cases, the problem of existence of periodic solutions of the system (12.2) is solved by the theorem similar to Theorem 1.2.

We now give some examples of transition from the equation (12.1) to the equation (12.2) whose right-hand side satisfies the inequality (12.5).

1. Assume that a function $g(t, x, y)$ satisfies the conditions

$$-M \le \frac{\partial g(t, x, y)}{\partial x} \le M, \quad -N \le \frac{\partial g(t, x, y)}{\partial y} \le N$$

in some region. Then the function

$$f(t, x, y, u, v) = \frac{1}{2}[g(t, x, y) + Mx + Ny] + \frac{1}{2}[g(t, u, v) - Mu - Nv]$$

satisfies conditions (12.5) and

$$f(t, x(t), x(t - \Delta), x(t), x(t - \Delta)) = g(t, x(t), x(t - \Delta)) \tag{12.21}$$

in the same region.

2. Assume that the inequalities

$$-h(t, x, y) + h(t, u, v) \leq g(t, x, y) - g(t, u, v) \leq h(t, x, y) - h(t, u, v)$$

hold in some region such that $x \geq u$ and $y \geq v$. Then the function

$$f(t, x, y, u, v) = \frac{1}{2}[g(t, x, y) + h(t, x, y)] + \frac{1}{2}[g(t, u, v) - h(t, u, v)]$$

possesses the properties (12.5) and (12.21).

We now illustrate the general aspects of the method given above with the help of the same example as in §6, i.e.,

$$\frac{dx(t)}{dt} = \varepsilon[x^2(t) - \sin t - x(t - \Delta)\cos t + 0.001], \quad 0 < \varepsilon \leq 0.06, \ 0 \leq \Delta \leq 2\pi. \tag{12.22}$$

Assume that the right-hand side of this equation is defined in the region

$$-\infty < t < \infty, \quad |x| \leq 0.25, \quad |y| \leq 0.25. \tag{12.23}$$

It has been shown in §6 that for $\varepsilon = 0.01$ and for all Δ from the intervals

$$0 \leq \Delta \leq 5\pi/18, \quad 31\pi/18 \leq \Delta \leq 2\pi,$$

there exists a 2π-periodic solution of eqn.(12.22) which passes through the point x_0 close to the point

$$x_0^{(1)} = \frac{(4 + \sin \Delta)\varepsilon + \sqrt{(4 + \sin \Delta)^2 \varepsilon^2 - 4(\varepsilon^2 + 2)(0.002 + 3\varepsilon^2 - \varepsilon \cos \Delta)}}{2 + \varepsilon^2}.$$

at $t = 0$. We now want to construct the bilateral approximations to this periodic solution. Let us denote the right-hand side of eqn.(12.22) by $g(t, x, y)$,

$$g(t, x, y) = \varepsilon(x^2 - \sin t - y \cos t + 0.001).$$

Since $g(t, x, y)$ satisfies the inequalities

$$\left|\frac{\partial g(t, x, y)}{\partial x}\right| \leq 0.5\varepsilon, \quad \left|\frac{\partial g(t, x, y)}{\partial y}\right| \leq \varepsilon$$

in the region (12.23), the function

$$f(t, x, y, u, v) = \frac{\varepsilon}{2}(x^2 - \sin t - y \cos t + 0.001 + 0.5x + y)$$

$$+ \frac{\varepsilon}{2}(u^2 - \sin t - v \cos t + 0.001 - 0.5u - v)$$

satisfies the condition (12.5); moreover, the equality (12.21) holds in the region (12.23), and thus,

$$|f(t, x, y, u, v)| \leq 1.696\varepsilon.$$

The recursion relation (12.7) implies that the zeroth and the first bilateral approximations have the following analytic form for $0 \leq t \leq 2\pi$:

$$u_0(t, x_0) = x_0 - 3.392\varepsilon(1 - t/2\pi), \quad v_0(t, x_0) = x_0 + 3.392\varepsilon(1 - t/2\pi);$$

$$u_1(t, x_0) = \begin{cases} u_1'(t, x_0) & \text{for } 0 \leq t \leq \Delta, \\ u_1'''(t, x_0) & \text{for } \Delta \leq t \leq 2\pi; \end{cases}$$

$$v_1(t, x_0) = \begin{cases} v_1'(t, x_0) & \text{for } 0 \leq t \leq \Delta, \\ v_1'''(t, x_0) & \text{for } \Delta \leq t \leq 2\pi, \end{cases}$$

where

$$u_1'(t, x_0) = x_0 + \varepsilon\left(1 - \frac{t}{2\pi}\right)\left\{(0.001 + x_0^2)t + \cos t - 1 + \frac{3.392^2}{3}\varepsilon^3 t^3\right.$$

$$+ \frac{1.696^2}{5\pi^2}\varepsilon^2 t^5 - \frac{1.696}{\pi}\varepsilon^2 t^4 - x_0 \sin t - 0.848\varepsilon t^2$$

$$+ \frac{0.283}{\pi}\varepsilon t^3 + 1.696\varepsilon[(2\pi - \Delta)^2 - (t - \Delta + 2\pi)^2]$$

$$- \frac{0.566}{\pi}\varepsilon[(2\pi - \Delta)^2 - (t - \Delta + 2\pi)^2] + \frac{0.566}{\pi}\varepsilon[(2\pi - \Delta)^3$$

$$-(t-\Delta+2\pi)^3]\}-\varepsilon\frac{t}{2\pi}\{(0.001+x_0^2)(2\pi-\Delta)+1-\cos t$$

$$+\left(\frac{3.392}{3}-\frac{0.283}{\pi\varepsilon}\right)\varepsilon^2(8\pi^3-t^3)+\frac{1.696^2\varepsilon^2}{5\pi^2}(32\pi^5-t^5)$$

$$-\frac{1.696}{\pi}\varepsilon^2(16\pi^4-t^4)+x_0\sin t+0.848\varepsilon(4\pi^2-t^2)$$

$$+1.696\varepsilon[4\pi^2+(2\pi-\Delta)^2-(t-\Delta+2\pi)^2]$$

$$-\frac{0.566}{\pi}\varepsilon[8\pi^3+(2\pi-\Delta)^3-(t-\Delta+2\pi)^3]\};$$

$$u_1''(t,x_0)=x_0+\varepsilon\left(1-\frac{t}{2\pi}\right)\{(0.001+x_0^2)t+\cos t-1+\frac{3.392^2}{3}\varepsilon^3t^3$$

$$+\frac{1.696^2}{5\pi^2}\varepsilon^2t^5-\frac{1.696}{\pi}\varepsilon^2t^4-x_0\sin t-0.848\varepsilon t^2$$

$$+\frac{0.283}{\pi}\varepsilon t^3-1.696\varepsilon[(t-\Delta)^2+4\pi^2-(\pi-\Delta)^2]$$

$$+\frac{0.566}{\pi}\varepsilon[(t-\Delta)^3+8\pi^3-(2\pi-\Delta)^3]\}-\varepsilon\frac{t}{2\pi}\{(0.001+x_0^2)(2\pi-t)+1-\cos t$$

$$+\left(\frac{3.392^2}{3}-\frac{0.283}{\pi\varepsilon}\right)\varepsilon^2(8\pi^3-t^3)+\frac{1.696^2\varepsilon^2}{5\pi^2}(32\pi^5-t^5)$$

$$-\frac{1.696^2}{\pi}\varepsilon^2(16\pi^4-t^4)+x_0\sin t+0.848\varepsilon(4\pi^2-t^2)$$

$$+1.696\varepsilon[(2\pi-\Delta)^2-(t-\Delta)^2]-\frac{0.566}{\pi}\varepsilon[(2\pi-\Delta)^2-(t-\Delta)^2]\};$$

$$v_1'(t,x_0)=x_0+\varepsilon\left(1-\frac{t}{2\pi}\right)\{(0.001+x_0^2)t+\cos t-1+\left(\frac{3.392}{3}-\frac{0.283}{\pi\varepsilon}\right)\varepsilon^2t^3$$

$$+\frac{1.696^2}{5\pi^2}\varepsilon^2t^5-\frac{1.696}{\pi}\varepsilon^2t^4-x_0\sin t+0.848\varepsilon t^2$$

$$+ 1.696\varepsilon[(t-\Delta+2\pi)^2 - (2\pi-\Delta)^2] - \frac{0.566}{\pi}\varepsilon\left[(t-\Delta+2\pi)^3 - (2\pi-\Delta)^3\right]\Big\}$$

$$- \varepsilon\frac{t}{2\pi}\Big\{(0.001+x_0^2)(2\pi-t)+1-\cos t$$

$$+ \left(\frac{3.392^2}{3}+\frac{0.283}{\pi\varepsilon}\right)\varepsilon^3(8\pi^3-t^3)+\frac{1.696\varepsilon^2}{5\pi^2}(32\pi^5-t^5)$$

$$- \frac{1.696^2}{\pi}\varepsilon^2(16\pi^4-t^4)+x_0\sin t-0.848\varepsilon(4\pi^2-t^2)$$

$$- 1.696\varepsilon[4\pi^2+(2\pi-\Delta)^2-(t-\Delta+2\pi)^2]$$

$$+ \frac{0.566}{\pi}\varepsilon[8\pi^3+(2\pi-\Delta)^3-(t-\Delta+2\pi)^3]\Big\};$$

$$v_1''(t,x_0) = x_0+\varepsilon\left(1-\frac{t}{2\pi}\right)\Big\{(0.001+x_0^2)t+\cos t-1+\left(\frac{3.392^2}{3}-\frac{0.283}{\pi\varepsilon}\right)\varepsilon^2 t^3$$

$$+ \frac{1.696^2}{5\pi^2}\varepsilon^2 t^5-\frac{1.696^2}{\pi}\varepsilon^2 t^4-x_0\sin t+0.848\varepsilon t^2$$

$$+ 1.696\varepsilon[(t-\Delta)^2+4\pi^2-(2\pi-\Delta)^2]$$

$$- \frac{0.566}{\pi}\varepsilon\left[(t-\Delta)^3+8\pi^3-(2\pi-\Delta)^3\right]\Big\}$$

$$- \varepsilon\frac{t}{2\pi}\Big\{(0.001+x_0^2)(2\pi-t)+1-\cos t$$

$$+ \left(\frac{3.392}{3}+\frac{0.283}{\pi\varepsilon}\right)\varepsilon^3(8\pi^3-t^3)+\frac{1.696\varepsilon^2}{5\pi^2}(32\pi^5-t^5)$$

$$- \frac{1.696^2}{\pi}\varepsilon^2(16\pi^4-t^4)+x_0\sin t-0.848\varepsilon(4\pi^2-t^2)$$

$$- 1.696\varepsilon[(2\pi-\Delta)^2-(t-\Delta)^2]+\frac{0.566}{\pi}\varepsilon[(2\pi-\Delta)^3-(t-\Delta)^3]\Big\}.$$

§13. Periodic Solutions of Operator-Differential Equations

Consider an operator-differential equation of the form

$$\frac{dx}{dt} = f(t, x, Ax), \tag{13.1}$$

where A is an operator defined on the space of continuous functions. Assume that $x = (x_1, x_2, ..., x_n)$ is an n-dimensional vector; $y = (y_1, y_2, ..., y_s)$ is either a finite dimensional $(s < \infty)$ vector, or an infinite dimensional $(s = \infty)$ vector, and $f(t, x, y) = (f_1(t, x, y), f_2(t, x, y), ..., f_n(t, x, y))$ is an n-dimensional vector function defined for $(t, x, y) \in R \times D \times D_1$, where $R = (-\infty, \infty)$, D is a region in the n-dimensional Euclidean space E_n, and D_1 is a region either in a finite dimensional space, or in an infinite dimensional space. Furthermore, let the function $f(t, x, y)$ be periodic in t with period ω, i.e.,

$$f(t, x, y) = f(t + \omega, x, y), \tag{13.2}$$

continuous with respect to all its variables $t, x,$ and y in the region $R \times D \times D_1$, and bounded

$$|f(t, x, y)| \leq M, \; M = (M_1, M_2, ..., M_n), \; M_j \geq 0, \; j = 1, ..., n. \tag{13.3}$$

Also let $f(t, x, y)$ satisfy the Lipschitz condition with respect to x and y, i.e.,

$$|f(t, x_1, y_1) - f(t, x_2, y_2)| \leq K |x_1 - x_2| + Q |y_1 - y_2| \tag{13.4}$$

for all $(t, x, y) \in R \times D \times D_1$, where $|f| = (|f_1|, |f_2|, ..., |f_n|)$, $|x| = (|x_1|, |x_2|, ..., |x_n|)$, $|y| = (|y_1|, |y_2|, ..., |y_s|)$, $K = \{k_{ij}\}$ is an $(n \times n)$-dimensional matrix with nonnegative elements $k_{ij} \geq 0$, $i, j = 1, 2, ..., n$, and $Q = \{q_{ij}\}$ is an $(n \times n)$-dimensional matrix with nonnegative elements $q_{ij} \geq 0$, $i, j = 1, 2, ..., s$.

Assume that each coordinate A_j $(j = 1, ..., s)$ of the operator $A = (A_1, A_2, ..., A_j, ..., A_s)$ is defined on the class of functions continuous on (a_j, b_j) and maps this class onto the class of functions continuous on (c_j, d_j). Further, we assume that the operator A transforms every continuous ω-periodic function $x(t) = (x_1(t), ..., x_n(t))$ into a continuous ω-periodic function $y(t) = (y_1(t), ..., y_s(t))$ with a finite or infinite number of coordinates. In addition, let $Ax(t) \in D_1$ for every continuous ω-periodic function $x(t)$ with values in the region D_1, and let

$$|Ax_1(t) - Ax_2(t)|_0 \leq R|x_1(t) - x_2(t)|_0 \tag{13.5}$$

for any pair of functions $x_1(t)$ and $x_2(t)$. Here $R = \{r_{ij}\}$ is an $(s \times n)$-dimensional matrix with nonnegative elements $r_{ij} \geq 0$, $i = 1, ..., s$, $j = 1, 2, ..., n$.

Suppose also that the vector M and matrices K, Q and R satisfy the following conditions:

(i) the set $D - M\,\omega/2$ is nonempty;

(ii) the greatest eigenvalue λ of the matrix $P = (K + QR)\omega/2$ is less that 1.

Theorem 1.21. (Zavalykut, 1983). *Suppose that eqn.(13.1) satisfies conditions (i) and (ii), and has an ω-periodic solution $x = x(t)$ which takes the values $x(0) = x_0$* $\in D - M\,\omega/2$ *at $t = 0$. Then*

$$x(t) = \lim_{m \to \infty} x_m(t, x_0) \tag{13.6}$$

and

$$|x(t) - x_m(t, x_0)| \leq P^m(E - P)^{-1}(M\omega/2) \tag{13.7}$$

where $x_m(t, x_0)$ are the functions periodic in t with period ω defined by

$$x_m(t, x_0) = x_0 + \int_0^t [f(t, x_{m-1}(t, x_0), Ax_{m-1}(t, x_0))$$

$$- \overline{f(t, x_{m-1}(t, x_0), Ax_{m-1}(t, x_0))}]dt, \tag{13.8}$$

$$\left(\overline{f(t, x_j(t), Ax_j(t))} = \frac{1}{\omega} \int_0^\omega f(t, x_j(t), Ax_j(t))dt \right).$$

Proof. First, we prove that the recursion relation (13.8) defines the functions $x_m(t,x_0)$ for all $m \geq 1$ and that each of these functions takes values in the region D. In fact, the operator A is defined for $x_0 \in D - M\,\omega/2$, and hence, the function $f(t, x_0, Ax_0)$ is also defined; moreover, $|f(t, x_0, Ax_0)| \leq M$. According to Lemma 1.1, the difference $x_1(t, x_0) - x_0$ can be estimated as follows

$$|x_1(t, x_0) - x_0| \leq |f(t, x_0, Ax_0)|_0 \, \alpha_1(t) \leq M\,\omega/2.$$

This implies that $|x_1(t, x_0) - x_0| \leq M\,\omega/2$, and therefore, the function $x_1(t, x_0)$ takes values in the region D and is periodic in t with period ω. One can easily prove by induction that for all $m \geq 1$ the recursion relation (13.8) defines continuous functions $x_m(t, x_0)$ periodic in t with period ω, which take values in the region D.

Let us prove that the sequence (13.8) is convergent. We have

$$| x_{m+1}(t, x_0) - x_m(t, x_0) |$$

$$= \left| \int_0^t [f(t, x_m(t, x_0), Ax_m(t, x_0)) - \overline{f(t, x_m(t, x_0), Ax_m(t, x_0))}] dt \right.$$

$$- \left. \int_0^t [f(t, x_{m-1}(t, x_0), Ax_{m-1}(t, x_0)) - \overline{f(t, x_{m-1}(t, x_0), Ax_{m-1}(t, x_0))}] dt \right|$$

$$\leq \left(1 - \frac{t}{\omega}\right) \int_0^t |f(t, x_m(t, x_0), Ax_m(t, x_0)) - f(t, x_{m-1}(t, x_0), Ax_{m-1}(t, x_0))| dt$$

$$+ \frac{t}{\omega} \int_0^\omega |f(t, x_m(t, x_0), Ax_m(t, x_0)) - f(t, x_{m-1}(t, x_0), Ax_{m-1}(t, x_0))| dt$$

$$\leq \left(1 - \frac{t}{\omega}\right) \int_0^t (K + QR)|x_m(t, x_0) - x_{m-1}(t, x_0)|_0 \, dt$$

$$+ \frac{t}{\omega} \int_t^\omega (K + QR)|x_m(t, x_0) - x_{m-1}(t, x_0)|_0 \, dt$$

$$\leq (K + QR)|x_m(t, x_0) - x_{m-1}(t, x_0)|_0 \alpha_1(t)$$

$$\leq \frac{(K + QR)\omega}{2} |x_m(t, x_0) - x_{m-1}(t, x_0)|_0 \leq P |x_m(t, x_0) - x_{m-1}(t, x_0)|. \qquad (13.9)$$

This yields

$$| x_{m+1}(t, x_0) - x_m(t, x_0) |_0 \leq P^m |x_1(t, x_0) - x_0)| \leq P^m M \, \omega/2. \qquad (13.10)$$

By virtue of condition (ii), the inequality (13.8) involves the uniform convergence of the sequence of functions (13.8) to a certain continuous function $x_1(t, x_0)$ periodic in t with period ω. By passing to the limit in (13.8) as $m \to \infty$, we find that the limiting function $x_\infty(t, x_0)$ is an ω-periodic solution of the integral equation

$$x_{\infty}(t) = x_0 + \int_0^t [f(t, x_{\infty}(t, x_0), Ax_{\infty}(t, x_0)) - \overline{f(t, x_{\infty}(t, x_0), Ax_{\infty}(t, x_0))}] dt. \quad (13.11)$$

Since $x(t)$ is a periodic solution of the equation .(13.1), we have

$$\overline{f(t, x(t), Ax(t))} = 0; \quad (13.12)$$

$$x(t) = x_0 + \int_0^t [f(t, x(t), Ax(t)) dt. \quad (13.13)$$

These relations imply that $x(t)$ is also a solution of the equation (13.11) (together with $x_{\infty}(t, x_0)$). This is why, to prove (13.6), it suffices to show that eqn.(13.11) cannot have two different periodic solutions. Assume the opposite, i.e., that eqn.(13.11) has two different periodic solutions $x = x(t)$ and $x_1 = x_1(t)$ which take values in D and satisfy the same initial conditions. Then for the difference $x(t) - x_1(t)$, we have

$$|x(t) - x_1(t)|_0 = |\int_0^t \{[f(t, x(t), Ax(t)) dt - \overline{f(t, x(t), Ax(t))}$$

$$- [f(t, x_1(t), Ax_1(t)) - \overline{f(t, x_1(t), Ax_1(t))}]\} dt|_0 \le P |x(t) - x_1(t)|_0. \quad (13.14)$$

By virtue of conditions (ii), this estimate implies that $|x(t) - x_1(t)|_0 \le 0$, i.e., $x(t) = x_1(t)$, and these relations contradict the assumption made above.

The existence of the periodic solution of the system (13.1) is established by the theorem similar to Theorem 1.2.

Remark. Consider now the equation (13.1) in which x is a scalar. Let A be a translation operator

$$Ax(t) = (x(t - \Delta_1), x(t - \Delta_2), ...), \quad (13.15)$$

where $\Delta_1, \Delta_2, ..., \Delta_j, ...$ is a number sequence. We consider a scalar differential equation with infinite number of lags

$$\frac{dx(t)}{dt} = f(t, x(t), x(t - \Delta_1), x(t - \Delta_2), ...), \quad (13.16)$$

where the function $f(t, x, y_1, y_2, ..., y_s, ...)$ is periodic in t with period ω and continuously differentiable with respect to $x, y_1, y_2, ..., y_s, ...$ in the region

$$|x| \le d, \quad |y_j| \le d, \quad j = 1, 2, \dots \tag{13.17}$$

Assume also that

$$|f(t, x, y_1, y_2, \dots, y_s, \dots)| \le M, \quad \left|\frac{\partial f}{\partial x}\right| \le k, \quad \left|\frac{\partial f}{\partial y_j}\right| \le q_j, \quad j = 1, 2, \dots \tag{13.18}$$

In this case, the constants K, Q, and R are defined by

$$K = k, \quad Q = f(q_1, q_2, \dots, q_j, \dots), \quad R = (1, 1, \dots, 1, \dots),$$

and conditions (i) and (ii) take the form

$$M\,\omega/2 < d, \quad P = \omega/2\left(k + \sum_{j=1}^{\infty} q_j\right) < 1.$$

Under these assumptions the statements similar to Theorem 1.1 and 1.2 are valid for the differential equation with infinite number of lags.

2. INVESTIGATION OF PERIODIC SOLUTIONS OF SYSTEMS WITH AFTEREFFECT BY BUBNOV-GALERKIN'S METHOD

§1. Preliminary Remarks. Auxiliary Statements

Consider a system of differential equations

$$\frac{dx(t)}{dt} = f(t, x(t), x(t - \Delta)), \tag{1.1}$$

where $x = (x_1, x_2, \ldots, x_n)$ is an n-dimensional vector, $f(t, x, y) = (f_1(t, x, y), \ldots, f_n(t, x, y))$ is an n-dimensional vector function periodic in t with period 2π, and Δ is a constant. We denote by $C^r(\mho_1 \times D \times D)$ the space of functions $f(t, x, y)$ periodic in t with period 2π and r times continuously differentiable with respect to $(t, x, y) \in \mho_1 \times D \times D$, where $\mho_1 = [0, 2\pi]$, and D is some bounded convex region in the Euclidean space E_n. In the space $C^r(\mho_1 \times D \times D)$, we introduce the differential norm $|\cdot|_r$ by the relation

$$|f|_r = \max_{0 \le \nu \le r} |D^\nu f|_0, \quad |f|_0 = \max_{t_1 \times D \times D} \|f(t, x, y)\|$$

and the integral norm $\|f\|_0$ by

$$\|f\|_0 = \left| \frac{1}{2\pi} \int_0^{2\pi} \|f\|^2 \, dt \right|^{\frac{1}{2}},$$

where $D^\nu f$ is an arbitrary partial derivative of $f(t, x, y)$ of the νth order with respect to the arguments t, x, and y; $\|\cdot\|$ is the Euclidean norm of a vector.

77

Let $g(t)$ be a continuous 2π-periodic vector-function, and let

$$g(t) = \frac{c_0}{2} + \sum_{n=1}^{\infty}(c_n \cos nt + d_n \sin nt) \tag{1.2}$$

be its expansion into the Fourier series. We denote

$$S_m g(t) = \frac{c_0}{2} + \sum_{n=1}^{m}(c_n \cos nt + d_n \sin nt),$$

$$R_m g(t) = \sum_{n=m+1}^{\infty}(c_n \cos nt + d_n \sin nt).$$

We want to find an approximate periodic solution of the system (1.1) in the form of a trigonometric polynomial

$$x_m(t) = \frac{a_0}{2} + \sum_{n=1}^{m}(a_n \cos nt + b_n \sin nt). \tag{1.3}$$

The coefficients of this polynomial are determined from the system of algebraic equations

$$\frac{dx_m(t)}{dt} = S_m f(t, x_m(t), x_m(t-\Delta)). \tag{1.4}$$

The trigonometric polynomial $x_m(t)$ satisfying (1.4) is called Bubnov-Galerkin's approximation of the mth order. The system (1.4) can be rewritten as follows

$$F^{(m)}(\alpha) = 0, \tag{1.5}$$

where α is an $(2m + 1)$-dimensional vector whose components are the coefficients of the polynomial $x_m(t)$, i.e. $\alpha = (a_0/2, a_1, a_2, ..., a_m, a_m)$. The equations (1.5) are called the determining equations for Bubnov-Galerkin's approximations. Clearly, we have $\|x_m\|_0 = \frac{1}{\sqrt{2}}\|\alpha\|$, since

$$\|x_m\|_0^2 = \frac{1}{2\pi}\int_0^{2\pi}\|x_m(t)\|^2\,dt = \frac{1}{2\pi}\int_0^{2\pi}\sum_k\left[\frac{a_{0k}}{2}\right.$$

$$+ \sum_{n=1}^{m} (a_{nk} \cos nt + b_{nk} \sin nt) \Big]^2 \, dt = \sum_{k} \left[\frac{a_{0k}^2}{4} + \frac{1}{2} \sum_{n=1}^{m} (a_{nk}^2 + b_{nk}^2) \right]$$

$$= \frac{1}{2} \left[\frac{\|a_0\|^2}{2} + \sum_{n=1}^{m} (\|a\|^2 + \|b\|^2) \right] = \frac{1}{2} \|\alpha\|^2.$$

We now prove the theorem which establishes the criterion on the solubility of the system of equations (1.5)

Theorem 2.1. (Urabe, 1966). *Consider a real system of equations*

$$F(\alpha) = 0, \tag{1.6}$$

where $F(\alpha)$ and α are vectors of the same dimensionality; $F(\alpha)$ is a continuously differentiable function of α defined in a certain region Ω. Suppose that (1.6) has an approximate solution $\alpha = \hat{\alpha}$ such that the determinant of Jacobian matrix $J(\alpha)$ of $F(\alpha)$ with respect to α is nonzero for $\alpha = \hat{\alpha}$. Suppose also that there exists a positive constant δ and a nonnegative constant $\kappa < 1$ such that

(a) $\Omega_\delta = \{\alpha \mid \|\alpha - \hat{\alpha}\| \le \delta\} \subset D$;
(b) $\| J(\alpha) - J(\hat{\alpha}) \| \le \kappa/M'$ for all $\alpha \in \Omega_\delta$;
(c) $M'l/(1 - \kappa) \le \delta$,
where l and M' are positive numbers such that

$$\| F(\hat{\alpha}) \| \le l, \quad \| J^{-1}(\hat{\alpha}) \| \le M'. \tag{1.8}$$

Then the system (1.6) possesses the unique solution $\alpha = \overline{\alpha}$ in the region Ω_δ and, furthermore,

$$\|\overline{\alpha} - \hat{\alpha}\| = M'l/(1 - \kappa). \tag{1.9}$$

Proof. Let $J_m(\hat{\alpha}) = \dfrac{\partial F^{(m)}}{\partial a} \Big|_{\alpha = \hat{\alpha}}$. We set $A = J^{-1}(\hat{\alpha})$ and consider an iterative process

$$\alpha_{k+1} = \alpha_k - A\, F(\alpha_k), \quad k = 0, 1, 2, \dots, \tag{1.10}$$

where $\alpha_0 = \hat{\alpha}$. First, we show that the approximations α_k exist for all k and that $\alpha_k \in$

Ω_δ.

By using the equation (1.10), for $k = 0$, we find

$$\|\alpha_1 - \alpha_0\| = \|AF(\hat{\alpha})\| \le M'l \le (1-\kappa)\delta < \delta, \tag{1.11}$$

and hence, $\alpha_1 \in \Omega_\delta$. For the difference $\alpha_2 - \alpha_1$, we have

$$\alpha_2 - \alpha_1 = (\alpha_1 - \alpha_0) - A[F(\alpha_1) - F(\alpha_0)]$$

$$= A \int_0^1 \{J(\alpha_0) - J[\alpha_0 + \theta(\alpha_1 - \alpha_0)]\}(\alpha_1 - \alpha_0)\,d\theta, \quad 0 \le \theta \le 1, \tag{1.12}$$

since $\alpha_0 + \theta(\alpha_1 - \alpha_0) \in \Omega_\delta$. By using the relation (1.12) and conditions (1.7), we obtain the following estimate for this difference

$$\|\alpha_2 - \alpha_1\| \le M'\frac{\kappa}{M'}\|\alpha_1 - \alpha_0\| \le \kappa\|\alpha_1 - \alpha_0\|. \tag{1.13}$$

It is easy to prove by induction that the inequality

$$\| \alpha_{k+1} - \alpha_k \| \le \kappa^k \| \alpha_1 - \alpha_0 \| \tag{1.14}$$

holds for all k. Moreover, by virtue of the inequality $\| \alpha_{k+1} - \alpha_0 \| \le \| \alpha_{k+1} - \alpha_k \| + \ldots + \| \alpha_1 - \alpha_0 \|$, the inequalities (1.11) and (1.14) yield

$$\| \alpha_{k+1} - \alpha_0 \| \le M'l/(1-\kappa) \le \delta, \tag{1.15}$$

and this implies that $\alpha_{k+1} \in \Omega_\delta$ for all k.

It follows from (1.14) and (1.15) that the iterative process can be infinitely continued. As a result, we obtain the infinite sequence $\{\alpha_k\}$ which is convergent by virtue of the inequalities (1.14) and $\kappa < 1$. It is now obvious that $\overline{\alpha}$ is a solution of the equation (1.6), and the inequality (1.9) follows from (1.15). The uniqueness of $\overline{\alpha}$ can be easily proved by contradiction if we assume that there exists another solution $\overline{\alpha}' \in \Omega_\delta$.

Indeed, taking into account that $\overline{\alpha}$ and $\overline{\alpha}'$ satisfy the relations

$$\overline{\alpha} = \overline{\alpha} - AF(\overline{\alpha}),$$

$$\overline{\alpha}' = \overline{\alpha}' - AF(\overline{\alpha}'),$$

we obtain (by analogy with (1.12) and (1.13))

$$\| \bar{\alpha} - \bar{\alpha}' \| = \kappa \| \bar{\alpha} - \bar{\alpha}' \|.$$

Since $0 \leq \kappa < 1$, this yields the identity $\bar{\alpha} = \bar{\alpha}'$.

Consider a system of equations

$$\frac{dx}{dt} = f(t), \tag{1.16}$$

where $f(t)$ is a 2π-periodic continuous function which can be expanded in the Fourier series

$$f(t) = \frac{c_0}{2} + \sum_{n=1}^{\infty} (c_n \cos nt + d_n \sin nt).$$

Clearly, the equation (1.16) has a periodic solution if $c_0 = 0$. Let us estimate the difference $x(t) - S_m x(t)$ between the exact solution of the equation (1.16) and its approximation.

Lemma 2.1. (Urabe, 1966). *The following estimates hold for a periodic solution of the equation (1.16)*

$$|x(t) - x_m(t)|_0 \leq \sigma(m) \|f\|_0 \leq \sigma(m) |f|_0, \tag{1.17}$$

$$\| x(t) - x_m(t) \|_0 \leq \sigma_1(m) \| f \|_0, \tag{1.18}$$

where

$$\sigma(m) = \sqrt{2} \left[\frac{1}{(m+1)^2} + \frac{1}{(m+2)^2} + \ldots \right]^{\frac{1}{2}}; \tag{1.19}$$

$$\sigma_1(m) = \frac{1}{m+1} \tag{1.20}$$

and

$$\frac{\sqrt{2}}{m+1} < \sigma(m) < \frac{\sqrt{2}}{m}. \tag{1.21}$$

Proof. By using the expansion of $f(t)$ in the Fourier series, we obtain

$$x(t) - x_m(t) = \sum_{n=m+1}^{\infty} \frac{1}{n}(-d_n \cos nt + c_n \sin nt) \qquad (1.22)$$

Applying here the Schwartz inequality, we find

$$\|x(t) - x_m(t)\|^2 = \left\| \sum_{n=m+1}^{\infty} \frac{1}{n}(-d_n \cos nt + c_n \sin nt) \right\|^2$$

$$\leq \sum_{n=m+1}^{\infty} \frac{1}{n^2} \sum_{n=m+1}^{\infty} \left(\|-d_n \cos nt + c_n \sin nt\| \right)^2$$

Inserting the Bessel inequality

$$\sum_{n=m+1}^{\infty} \left(\|c_n\|^2 + \|d_n\|^2 \right) \leq \|f\|_0^2 \qquad (1.24)$$

in (1.23), we get

$$|x(t) - x_m(t)|_0 \leq \sigma(m)\|f\|_0 \leq \sigma(m)|f|_0 .$$

The estimate (1.17) is thus proved. By applying the Parseval formula to (1.22), we obtain

$$\|x(t) - x_m(t)\|^2 = \sum_{n=m+1}^{\infty} \frac{1}{n^2} \left(\|c_n\|^2 + \|d_n\|^2 \right)$$

$$\leq \sum_{n=m+1}^{\infty} \frac{1}{(m+1)^2} \left(\|c_n\|^2 + \|d_n\|^2 \right) .$$

This implies the inequality (1.18)

$$|x(t) - x_m(t)|_0 \leq \frac{1}{m+1} \|f\|_0 .$$

§2. Green's Function for the Problem of Periodic Solutions of Linear Systems with Lag. Properties of This Function

Consider a system of linear differential equations of the form

$$\frac{dz(t)}{dt} = A(t)z(t) + B(t)z(t - \Delta) + c(t) \tag{2.1}$$

where z is an n-dimensional vector, $c(t)$ is an n-dimensional vector function, $A(t)$ and $B(t)$ are $(n \times n)$-dimensional square matrices; $A(t)$, $B(t)$, and $c(t)$ belong to $C^0(\mathcal{T}_1)$.

We call a matrix function $G_0(t, \tau)$ Green's function if it is defined for $t \neq \tau$, continuous in t and τ, and satisfies the system of equations

$$\frac{dz(t)}{dt} = A(t)z(t) + B(t)z(t - \Delta) ,$$

the condition

$$G_0(t,\tau)\big|_{t=\tau+0} = - G_0(t,\tau)\big|_{t=\tau-0} = E \tag{2.2}$$

and the inequality

$$\int_{-\infty}^{\infty} \|G_0(t, \tau)\| \, dt \leq M^0 < +\infty, \quad t, \tau \in R , \ t \neq \tau.$$

Here E is the unit matrix, $R = (-\infty, \infty)$ and M^0 is a positive constant.

Assume that the system of equations (2.1) has Green's function satisfying the condition

$$\|G_0(t, \tau)\| \leq M_0 e^{-\lambda_0|t-\tau|}, \quad t, \tau \in R , \ t \neq \tau, \tag{2.3}$$

where λ_0 and M_0 are positive constants. Then the system (2.1) has a unique 2π-periodic solution $z = z(t)$ which can be written in the form

$$z(t) = \int_{-\infty}^{\infty} G_0(t, \tau)c(\tau) \, d\tau . \tag{2.4}$$

The properties of Green's function $G_0(t, \tau)$ enable us to establish the following statements.

Lemma 2.2. (Samoilenko and Nurzhanov, 1979). *Suppose that the system of differential equation* (2.1) *has Green's function $G_0(t, \tau)$ satisfying the condition* (2.3). *Then the periodic solution* (2.4) *of this system satisfies the inequality*

$$\|z\|_0 \leq \frac{2M_0}{\lambda_0}\|c\|_0. \tag{2.5}$$

Proof. According to the definition of the norm $\|\cdot\|$, we have

$$\|z\|_0^2 = \frac{1}{2\pi}\int\limits_0^{2\pi}\left\|\int\limits_{-\infty}^{\infty}G_0(t, \tau)c(\tau)\,d\tau\right\|^2 dt. \tag{2.6}$$

Using here the Bunyakovsky-Schwartz inequality, we find

$$\|z\|_0^2 \leq \frac{1}{2\pi}\int\limits_0^{2\pi}\left[\int\limits_{-\infty}^{\infty}\|G_0(t, \tau)\|d\tau\int\limits_{-\infty}^{\infty}\|G_0(t, \tau)\|\,\|c(\tau)\|^2\,d\tau\right]dt$$

$$\leq \frac{M_0^2}{2\pi}\int\limits_0^{2\pi}\left[\int\limits_{-\infty}^{\infty}e^{-\lambda_0|t-\tau|}\,d\tau\int\limits_{-\infty}^{\infty}e^{-\lambda_0|t-\tau|}\|c(\tau)\|^2\,d\tau\right]dt$$

$$\leq \frac{M_0^2}{\pi\lambda_0}\int\limits_0^{2\pi}\left[\int\limits_{-\infty}^{\infty}e^{-\lambda_0|t-\tau|}\|c(\tau)\|^2\,d\tau\right]dt.$$

Changing the variables in the last inequality $(\tau - t = \theta)$, we obtain

$$\|z\|_0^2 \leq \frac{M_0^2}{\pi\lambda_0}\int\limits_0^{2\pi}dt\int\limits_{-\infty}^{\infty}e^{-\lambda_0|\theta|}\|c(t+\theta)\|^2\,d\theta.$$

According to the Fubini theorem (see (Kolmogorov and Fomin, 1976)), we can change here the order of integration. Thus, we have

$$\|z\|_0^2 \leq \frac{M_0^2}{\pi\lambda_0}\int\limits_{-\infty}^{\infty}e^{-\lambda_0|\theta|}\int\limits_0^{2\pi}\|c(t+\theta)\|^2\,dt\,d\theta.$$

Taking into account that

$$\frac{1}{2\pi} \int_0^{2\pi} \| c(\tau + \theta)\|^2 \, dt = \frac{1}{2\pi} \int_0^{2\pi+\theta} \| c(s)\|^2 \, ds = \|c\|_0^2$$

for $c(t) \in C^0(\mathcal{T}_1)$, we get

$$\|z\|_0^2 \le \frac{2M_0^2}{\lambda_0} \|c\|_0^2 \int_{-\infty}^{\infty} e^{-\lambda_0 |\theta|} d\theta = \frac{2M_0^2}{\lambda_0^2} \|c\|_0^2 \, .$$

This inequality immediately yields the inequality (2.5). Lemma 2.2 is thus proved.

Lemma 2.3. (Samoilenko and Nurzhanov, 1979). *Suppose that the system of differential equation* (2.1) *has Green's function* $G_0(t, \tau)$ *which satisfies the condition* (2.3). *Then the perturbed linear system*

$$\frac{dz(t)}{dt} = [A(t) + A_0(t)]z(t) + [B(t) + B_0(t)] + c(t) \tag{2.7}$$

has Green's function $G_0(t, \tau)$ *satisfying the inequality*

$$\|G(t, \tau)\| \le M e^{-\lambda_0 |t - \tau|}, \quad t, \tau \in R, \ t \ne \tau, \tag{2.8}$$

where

$$M = M_0 \left\{ 1 + \frac{M_0}{(1-\rho)\lambda_0} \left[|A_0|_0 + |B_0|_0 \, e^{\lambda \Delta} \right] \sup_{t \ge 0} \left[(1 + \lambda_0 t) e^{-(\lambda_0 - \lambda)t} \right] \right\}$$

provided that the matrices $A_0(t) \in C^0(\mathcal{T}_1)$ *and* $A_0(t) \in C^0(\mathcal{T}_1)$ *are so small that*

$$\frac{2M_0}{\lambda_0 - \lambda} \left[|A_0|_0 + |B_0|_0 \, e^{\lambda \Delta} \right] \le \rho < 1 \tag{2.9}$$

for $0 < \lambda < \lambda_0$.

Proof. Let us represent Green's function $G(t, \tau)$ of the perturbed system (2.7) in the form

$$G(t, \tau) = G_0(t, \tau) + Z(t, \tau). \tag{2.10}$$

Inserting (2.10) in the system (2.7), we find that the matrix $Z(t, \tau)$ should be a solution of the matrix equation

$$\frac{dZ(t)}{dt} = A(t)\,Z(t,\tau) + B(t)Z(t-\Delta,\tau) + A_0(t)\,G_0(t,\tau)$$

$$+ B_0(t)G_0(t-\Delta,\tau) + A_0(t)\,Z(t,\tau) + B_0(t)Z(t-\Delta,\tau). \qquad (2.11)$$

Moreover, $Z(t,\tau)$ should be continuous in t and τ and such that

$$\|Z(t,\tau)\| \leq M_0' e^{-\lambda(t-\tau)}, \qquad t,\tau \in \mathbb{R},\ t \neq \tau. \qquad (2.12)$$

Taking into account the properties of the matrix $G_0(t,\tau)$, we can take

$$Z(t,\tau) = \int_{-\infty}^{\infty} G_0(t,s)\Big[A_0(s)\,Z(s,\tau) + B_0(s)Z(s-\Delta,\tau)\Big]\,ds + \Phi(t,\tau), \quad (2.13)$$

where

$$\Phi(t,\tau) = \int_{-\infty}^{\infty} G_0(t,s)\Big[A_0(s)\,G_0(s,\tau) + B_0(s)G_0(s-\Delta,\tau)\Big]\,ds. \qquad (2.14)$$

The matrices $A_0(t)$ and $B_0(t)$ are bounded. Therefore, by virtue of the inequality (2.3), the integral on the right-hand side of (2.14) converges uniformly, and moreover, the following estimate is valid for the matrix $\Phi(t,\tau)$

$$\|\Phi(t,\tau)\| \leq M_0'' e^{-\lambda(t-\tau)},$$

where

$$M_0'' = \frac{M_0}{\lambda_0}\Big[|A_0|_0 + |B_0|_0\, e^{\lambda\Delta}\Big]\sup_{t\geq 0}\Big[(1+\lambda_0\,t)e^{-(\lambda_0-\lambda)t}\Big], \quad 0<\lambda<\lambda_0.$$

If we set $W(t,\tau) = e^{\lambda(t-\tau)}\,Z(t,\tau)$, then we obtain the following integral equation

$$W(t,\tau) = \int_{-\infty}^{\infty} G_0(t,s)\Big[e^{\lambda(|t-\tau|-|s-\tau|)}\,A_0(s)\,W(s,\tau)$$

$$+ e^{\lambda(|t-\tau|-|s-\tau-\Delta|)}\,B_0(s)\,W(s-\Delta,\tau)\Big]\,ds + \Phi(t,\tau)e^{\lambda|t-\tau|}. \qquad (2.15)$$

for the matrix $W(t,\tau)$.

By applying the principle of contracting mappings, one can easily prove that the equation (2.15) has a solution continuous in t and τ provided that

$$\frac{2M_0}{\lambda_0 - \lambda} \left[|A_0|_0 + |B_0|_0 e^{\lambda \Delta} \right] \le \rho < 1. \tag{2.16}$$

Thus, if we set $M_0' = \dfrac{1}{1-\rho} M_0''$, then the matrix $Z(t, \tau) = e^{-\lambda(t-\tau)} W(t, \tau)$ is a continuous solution of the equation (2.13) satisfying the inequality (2.12).

§3. The Main Properties of the Jacobian Matrix of Determining Equations for Galerkin's Approximations

Let $\hat{x}(t)$ be the exact periodic solution of the system (1.1), and let

$$\hat{x}(t) = \frac{\hat{a}_0}{2} + \sum_{n=1}^{\infty} (\hat{a}_n \cos nt + \hat{b}_n \sin nt)$$

be its expansion into the Fourier series. The Bubnov-Galerkin's approximation to this solution of the mth order is given by

$$\hat{x}_m(t) = \frac{\hat{a}_0}{2} + \sum_{n=1}^{m} (\hat{a}_n \cos nt + \hat{b}_n \sin nt).$$

In order to determine the coefficients $\left(\dfrac{\hat{a}_0}{2}, \hat{a}_1, \hat{b}_1, ..., \hat{a}_n, \hat{b}_n \right) = \hat{\alpha}$, one must solve the system of determining equations

$$F^{(m)}(\hat{\alpha}) = 0. \tag{3.1}$$

Let $J_m(\hat{\alpha}) = \left. \dfrac{\partial F^{(m)}(\alpha)}{\partial \alpha} \right|_{\alpha = \hat{\alpha}}$ be the Jacobian matrix of the left-hand side of the system (3.1). We now establish some properties of this matrix which are necessary to prove the existence and convergence of Bubnov-Galerkin's approximations. For this purpose, we consider the auxiliary linear system

$$J_m(\hat{\alpha})\xi + \gamma = 0, \tag{3.2}$$

where

$$\hat{\alpha} = \left(\frac{\hat{a}_0}{2}, \hat{a}_1, \hat{b}_1, ..., \hat{a}_m, \hat{b}_m \right),$$

$$\gamma = \left(\frac{c_0}{2}, c_1, d_1, ..., c_m, d_m \right),$$

$$\xi = \left(\frac{v_0}{2}, v_1, w_1, ..., v_m, w_m \right).$$

We set

$$z_m(t) = \frac{v_0}{2} + \sum_{n=1}^{m}(v_n \cos nt + w_n \sin nt); \tag{3.3}$$

$$\varphi_m(t) = \frac{c_0}{2} + \sum_{n=1}^{m}(c_n \cos nt + d_n \sin nt). \tag{3.4}$$

By using the definition of $J_m(\hat{\alpha})$ and the relations (3.2)–(3.4), we find that $z_m(t)$ satisfies the system of differential equations

$$\frac{dz_m(t)}{dt} = S_m\left[\frac{\partial \hat{f}_m}{\partial x} z_m(t) \right] + S_m\left[\frac{\partial \hat{f}_m}{\partial y} z_m(t-\Delta) \right] + \varphi_m(t) \tag{3.5}$$

where

$$\hat{f}_m = f(t, \hat{x}_m(t), \hat{x}_m(t-\Delta)).$$

Lemma 2.4. (Martinyuk, 1982a). *Suppose that the right-hand side of the system of equations* (1.1) *satisfies the following conditions:*

(i) $f(t, x, y) \in C(\mathcal{T}_1 \times D \times D)$;

(ii)   there exists a 2π-periodic solution $x = \hat{x}(t)$ of this system which belongs to the region D together with some its vicinity δ_0;

(iii)   the system of variational equations which corresponds to this solution, namely,

$$\frac{dz(t)}{dt} = A(t)z(t) + B(t)z(t - \Delta),$$ (3.6)

where

$$A(t) = \frac{\partial f(t, \hat{x}(t), \hat{x}(t - \Delta))}{\partial x}, \quad B(t) = \frac{\partial f(t, \hat{x}(t), \hat{x}(t - \Delta))}{\partial y},$$

possesses Green's function $G_0(t, \tau)$ *satisfying the condition* (2.3).

Then, for any fixed λ $(0 < \lambda < \lambda_0)$, *one can find a sufficiently large number* m_0 *such that for all* $m \geq m_0$ *there exist the reciprocal matrices* $J_m^{-1}(\alpha)$ *which satisfy the inequality*

$$\|J_m^{-1}(\alpha)\| \leq M'$$ (3.7)

where M' *is a positive constant independent of* m.

Proof. Let us represent the system (3.5) as follows

$$\frac{dz_m(t)}{dt} = \frac{\partial \hat{f}}{\partial x} z_m(t) - \left(\frac{\partial \hat{f}}{\partial x} - \frac{\partial \hat{f}_m}{\partial x} \right) z_m(t) + \frac{\partial \hat{f}}{\partial y} z_m(t - \Delta)$$

$$- \left(\frac{\partial \hat{f}}{\partial y} - \frac{\partial \hat{f}_m}{\partial y} \right) z_m(t) - R_m \left[\frac{\partial \hat{f}_m}{\partial x} z_m(t) \right] - R_m \left[\frac{\partial \hat{f}_m}{\partial y} z_m(t - \Delta) \right] + \varphi_m(t). \quad (3.8)$$

Taking the inequality (1.17) into account, we obtain

$$\left| \frac{\partial \hat{f}}{\partial x} - \frac{\partial \hat{f}_m}{\partial x} \right|_0 \leq |f|_2 |\hat{x} - \hat{x}_m|_0 \leq |f|_2 |f|_0 \sigma(m),$$

$$\left| \frac{\partial \hat{f}}{\partial y} - \frac{\partial \hat{f}_m}{\partial y} \right|_0 \leq |f|_2 |f|_0 \sigma(m).$$ (3.9)

It follows from (3.8) that $z_m(t)$ is a periodic solution of a system of the type (2.7), one should only set

$$A_0(t) = -\left(\frac{\partial \hat{f}}{\partial x} - \frac{\partial \hat{f}_m}{\partial x} \right), \quad B_0(t) = -\left(\frac{\partial \hat{f}}{\partial y} - \frac{\partial \hat{f}_m}{\partial y} \right)$$

and regard the expressions with R_m as known functions. This means that if we choose m_0 sufficiently large such that the inequality

$$\frac{2M_0}{\lambda_0 - \lambda}\left(1 + e^{\lambda\Delta}\right) |f|_2 |f|_0 \, \sigma(m) < 1 \tag{3.10}$$

holds for all $m > m_0$, then Lemma 2.3 is valid for the system (3.8). Consequently, this system possesses Green's function $G_m(t, \tau)$ satisfying the inequality

$$\|G_m(t, \tau)\| \leq M e^{-\lambda|t-\tau|}, \quad t, \tau \in R, \ t \neq \tau. \tag{3.11}$$

The periodic solution $z = z_m(t)$ of (3.8) can thus be written in the form

$$z_m(t) = \int_{-\infty}^{\infty} G_m(t, \tau)\left\{\varphi_m(s) - R_m\left[\frac{\partial \hat{f}_m}{\partial x} z_m(s)\right] - R_m\left[\frac{\partial \hat{f}_m}{\partial y} z_m(s-\Delta)\right]\right\} ds. \tag{3.12}$$

We now show that the operator T_m defined by the right-hand side of (3.12) on the class of functions of the form (3.3) maps this class of functions onto itself. Differentiating (3.12) and taking into account the definition of $G_m(t, \tau)$, we get

$$\frac{dz_m(t)}{dt} \equiv \varphi_m(t) + S_m\left[\frac{\partial \hat{f}_m}{\partial x} z_m(t)\right] + S_m\left[\frac{\partial \hat{f}_m}{\partial y} z_m(t-\Delta)\right]. \tag{3.13}$$

This implies that $\dfrac{dz_m(t)}{dt}$ and, hence, $z_m(t)$ are trigonometric polynomials of the mth order.

Estimating (3.13) with the help of (1.17) and (1.18), we obtain

$$\left\|R_m\left[\frac{\partial \hat{f}_m}{\partial x} z_m(t)\right]\right\|_0 \leq L_3 \sigma_1(m)\|z_m(t)\|_0,$$

$$\left\|R_m\left[\frac{\partial \hat{f}_m}{\partial y} z_m(t-\Delta)\right]\right\|_0 \leq L_4 \sigma_1(m)\|z_m(t)\|_0. \tag{3.14}$$

Inserting these inequalities in the right-hand side of (3.12), we find

$$\|z_m(t)\|_0 \leq \frac{2M}{\lambda}\|\varphi_m\|_0 + \frac{2M}{\lambda}(L_3 + L_4)\sigma_1(m)\|z_m(t)\|_0. \tag{3.15}$$

It follows from this estimate that if we choose m_0 so large that

$$\rho_m = \frac{2M}{\lambda}(L_3 + L_4)\sigma_1(m) < 1 \tag{3.16}$$

for all $m \geq m_0$, then the operator T_m^0 defined by the right-hand side of the system (3.12) satisfies the inequality

$$\|T_m^0 z_m\|_0 \leq \rho_m \|z_m(t)\|_0, \qquad \rho_m < 1 \tag{3.17}$$

provided that $\varphi_m(t) \equiv 0$.

By virtue of the principle of contracting mappings the equation (3.12) has a unique periodic solution $z = z_m(t)$ satisfying the inequality

$$\|z_m\|_0 \leq M' \|\varphi_m\|_0 \tag{3.18}$$

where

$$M' = \frac{2M[1+|f|_1 \sigma_1(m_0)]}{\lambda - 2M(L_3 + L_4)\sigma_1(m_0)}, \qquad m \geq m_0. \tag{3.19}$$

Since the system (3.8) is equivalent to (3.2), we have $\xi = - J_m^{-1}(\hat{\alpha})\gamma$, moreover, taking into account that $\frac{1}{\sqrt{2}}\|\xi\| = \|z_m\|_0$ and $\|\varphi_m\|_0 = \frac{1}{\sqrt{2}}\|\xi\|$, we find that the inequality (3.18) is equivalent to the inequality

$$\|\xi\| \leq M' \|\gamma\|, \qquad m \geq m_0.$$

These statements imply the relation (3.7). The proof of Lemma 2.4 is thus completed.

The following statement is valid for the difference $J_m(\alpha) - J_m(\hat{\alpha})$.

Lemma 2.5. *Suppose that the right-hand side of the system of equations* (1.1) *satisfies conditions (i) and (ii) of Lemma 2.4. Assume also that, for an arbitrary vector* $\alpha = \left(\frac{a_0}{2}, a_1, b_1, ..., a_m, b_m\right)$, *the polynomial*

$$x_m(t) = \frac{a_0}{2} + \sum_{n=1}^{m}(a_n \cos nt + b_n \sin nt)$$

belongs to the region D for all $t \in \mathfrak{T}_1$. Then

$$\|J_m(\alpha) - J_m(\hat{\alpha})\| \le 2|f|_2 \, |x_m - \hat{x}_m|_0 \le 2|f|_2 \sqrt{m + \frac{1}{2}} \|\alpha - \hat{\alpha}\|.$$

Proof. Let $\xi = \left(\dfrac{v_0}{2}, v_1, w_1, ..., v_m, w_m \right)$ be an arbitrary vector, and let

$$\gamma' = -J_m(\alpha)\,\xi, \quad \gamma = -J_m(\hat{\alpha})\,\xi, \tag{3.20}$$

$$\gamma' = \left(\frac{c_0'}{2}, c_1', d_1', ..., c_m', d_m' \right), \quad \gamma = \left(\frac{c_0}{2}, c_1, d_1, ..., c_m, d_m \right).$$

We construct the corresponding trigonometric polynomials

$$z_m(t) = \frac{v_0}{2} + \sum_{n=1}^{m} (v_n \cos nt + w_n \sin nt); \tag{3.21}$$

$$\varphi_m'(t) = \frac{c_0'}{2} + \sum_{n=1}^{m} (c_n' \cos nt + d_n' \sin nt)$$

$$\varphi_m(t) = \frac{c_0}{2} + \sum_{n=1}^{m} (c_n \cos nt + d_n \sin nt). \tag{3.22}$$

It follows from the definition of the matrices $J_m(\alpha)$ and $J_m(\hat{\alpha})$, and from relations (3.21) and (3.22) that

$$\frac{dz_m(t)}{dt} = S_m \left[\frac{\partial f_m}{\partial x} z_m(t) \right] + S_m \left[\frac{\partial f_m}{\partial y} z_m(t - \Delta) \right] + \varphi_m'(t),$$

$$\frac{dz_m(t)}{dt} = S_m \left[\frac{\partial \hat{f}_m}{\partial x} z_m(t) \right] + S_m \left[\frac{\partial \hat{f}_m}{\partial y} z_m(t - \Delta) \right] + \varphi_m(t), \tag{3.23}$$

where

$$f_m = f(t, x_m(t), x_m(t - \Delta)), \quad \hat{f}_m = f(t, \hat{x}_m(t), \hat{x}_m(t - \Delta)).$$

This yields

$$\|\varphi'(t) - \varphi(t)\|_0 \le \left\| \left(\frac{\partial f}{\partial x} - \frac{\partial \hat{f}}{\partial x} \right) z_m(t) \right\|_0 + \left\| \left(\frac{\partial f}{\partial y} - \frac{\partial \hat{f}}{\partial y} \right) z_m(t - \Delta) \right\|_0$$

$$\leq \; 2|f|_2 \; |x_m - \hat{x}_m|_0 \, \|z_m(t)\|_0$$

and

$$\|\gamma' - \gamma\| \leq \; 2|f|_2 \; |x_m - \hat{x}_m|_0 \, \|z_m(t)\|_0. \tag{3.24}$$

However,

$$\|\gamma' - \gamma\| = \left\| \left[J_m(\alpha) - J_m(\hat{\alpha}) \right] \xi \right\| = \sqrt{2} \|\varphi'_m - \varphi_m\|_0,$$

$$\|\xi\| = \sqrt{2} \|z\|_0,$$

and therefore, the inequality (3.24) is equivalent to the following one

$$\left\| \left[J_m(\alpha) - J_m(\hat{\alpha}) \right] \xi \right\| \leq 2|f|_2 \|x_m - \hat{x}_m\|_0 \; \|\xi\|.$$

The last inequality implies

$$\left\| J_m(\alpha) - J_m(\hat{\alpha}) \right\| \leq 2|f|_2 \|x_m - \hat{x}_m\|_0$$

$$\leq 2|f|_2 \sqrt{2m+1} \|x_m - \hat{x}_m\|_0 = 2|f|_2 \sqrt{m + \frac{1}{2}} \|\alpha - \hat{\alpha}\|.$$

This completes the proof.

§4. Existence and Convergence of Bubnov-Galerkin's Approximations

In this section, we prove the following statement:

Assume that the variational system which corresponds to the exact periodic solution $\hat{x}(t)$ of the system (1.1) possesses Green's function satisfying the condition (2.3). Then, for all sufficiently large m, Bubnov-Galerkin's approximations exist and converge uniformly as $m \to \infty$ to the exact periodic solution $\hat{x}(t)$.

Theorem 2.2. (Martinyuk, 1982a; Samoilenko and Nurzhanov, 1979). *Suppose that the right-hand side of the system of differential equations* (1.1) *satisfies conditions* (i) – (iii) *of Lemma 2.4. Then one can always find sufficiently large* m_0 *such* *that Bubnov-Galerkin's approximations* $x = \bar{x}_m(t)$ *exist for all* $m \geq m_0$ *and converge uniformly, as* $m \to \infty$, *to the exact periodic solution* $x = \hat{x}(t)$. *Furthermore*,

$$\|\bar{x}_m(t) - \hat{x}(t)\|_0 \leq M_1 \frac{\sqrt{2m+1}}{m+1} + |f|_0 \sigma(m), \tag{4.1}$$

where M_1 *is a positive constant independent of m.*

Proof. By setting $S_m \hat{x}(t) = \hat{x}_m(t)$, we obtain

$$\frac{d\hat{x}_m(t)}{\partial t} = S_m f(t, \hat{x}(t), \hat{x}(t-\Delta)) = S_m \hat{f}_m + S_m(\hat{f} - \hat{f}_m) = S_m \hat{f}_m + R_m(t). \tag{4.2}$$

Assume that

$$\hat{x}(t) = \frac{\hat{a}_0}{2} + \sum_{n=1}^{\infty}(\hat{a}_n \cos nt + \hat{b}_n \sin nt) \tag{4.3}$$

and

$$R_m(t) = \frac{r_0^{(m)}}{2} + \sum_{n=1}^{\infty}(r_n^{(m)} \cos nt + s_n^{(m)} \sin nt). \tag{4.4}$$

By virtue of (4.3) and (4.4), the equality (4.2) is equivalent to the following system of equations

$$F_0^{(m)}(\hat{\alpha}) = \frac{1}{2\pi} \int_0^{2\pi} f(t, \hat{x}_m(t), \hat{x}_m(t-\Delta))dt = -\frac{r_0^{(m)}}{2},$$

$$F_n^{(m)}(\hat{\alpha}) = \frac{1}{2\pi} \int_0^{2\pi} f(t, \hat{x}_m(t), \hat{x}_m(t-\Delta))\cos nt \, dt - n\hat{b}_m = -r_n^{(m)},$$

$$G_n^{(m)}(\hat{\alpha}) = \frac{1}{2\pi} \int_0^{2\pi} f(t, \hat{x}_m(t), \hat{x}_m(t-\Delta))\sin nt \, dt + n\hat{a}_m = -s_n^{(m)}, \tag{4.5}$$

$$n = 1, 2, ..., m.$$

For brevity, we rewrite this system in the form

$$F^{(m)}(\hat{\alpha}) = -q^{(m)}, \tag{4.6}$$

where

$$\hat{\alpha} = \left(\frac{\hat{a}_0}{2}, \hat{a}_1, \hat{b}_1, ..., \hat{a}_m, \hat{b}_m \right), \quad q^{(m)} = \left(\frac{r_0^{(m)}}{2}, r_1^{(m)}, s_1^{(m)}, ..., r_m^{(m)}, s_m^{(m)} \right).$$

Since

$$\left\| S_m\left(\hat{f} - \hat{f}_m \right) \right\|_0 \leq 2 |f|_1 \|f\|_0 \, \sigma_1(m),$$

the relation (4.6) implies that

$$\left\| F^{(m)}(\hat{\alpha}) \right\| = \| q^{(m)} \| \leq 2\sqrt{2} \, |f|_1 \|f\|_0 \, \sigma_1(m) = l_m.$$

Let us take $\hat{\alpha}$ as an approximate solution of the equation

$$F^{(m)}(\alpha) = 0. \tag{4.7}$$

Then according to Lemma 2.4, the Jacobian matrix $J_m(\alpha)$ has the reciprocal matrix $J_m^{-1}(\hat{\alpha})$ for $\alpha = \hat{\alpha}$; moreover, the latter satisfies the inequality (3.7). Hence, for all sufficiently large $m \geq m_0$, where $m_0 = m_0 \, (l_m, M')$, we have

$$\left\| F^{(m)}(\hat{\alpha}) \right\| \leq l_m, \quad \left\| J_m^{-1}(\hat{\alpha}) \right\| \leq M^1. \tag{4.8}$$

Suppose that

$$u = \{ x / \|x - \hat{x}(t)\| \leq \delta_0, \ t \in t_1 \} \subset D$$

for a sufficiently small positive number δ_0. This is possible, since we have assumed that $x = \hat{x}(t)$ belongs to the region D. For $m \geq m_0$, where m_0 is sufficiently large, we define the region

$$V_m = \{ x / \|x - \hat{x}(t)\| \leq \delta_0 - |f|_0 \, \sigma(m), \ t \in t_1 \}.$$

Then, by virtue of (1.17), we have

$$V_m \subset u \subset D$$

for all $m \geq m_0$. Further, we define the region Ω_m as the set of points

$$\alpha = \left(\frac{a_0}{2}, a_1, b_1, \dots, a_m, b_m \right)$$

which satisfy the inequality

$$\| \alpha - \hat{\alpha} \| \leq \frac{\delta_0 - |f|_0 \, \sigma(m)}{\sqrt{m + 1/2}}.$$

Let us construct the polynomial

$$x = x(t) = \frac{a_0}{2} + \sum_{n=1}^{m} (a_n \cos nt + b_n \sin nt)$$

which corresponds to some vector $\alpha \in \Omega_{m_0}$. Then the inequality

$$|x - \hat{x}_0|_0 \leq |x - \hat{x}_m|_0 + |\hat{x} - \hat{x}_m|_0 \leq \sqrt{m + 1/2} \|\alpha - \hat{\alpha}\| + |f|_0 \, \sigma(m) \leq \delta_0$$

implies that $x = x(t) \in D$ for all $t \in \mathcal{T}_1$. Moreover, by virtue of Lemma 2.5, we have

$$\left\| J_m(\alpha) - J_m(\hat{\alpha}) \right\| \leq 2\sqrt{m + 1/2} \, |f|_2 \, \|\alpha - \hat{\alpha}\| \tag{4.9}$$

for all $\alpha \in \Omega_{m_0}$ and $m \geq m_0$. Having fixed an arbitrary number $\kappa \ (0 < \kappa < 1)$, we put

$$\min\left(\frac{\kappa}{2|f|_2 \, M'}, \ \delta_0 - |f|_0 \, \sigma(m) \right) = \delta_1. \tag{4.10}$$

Let $m_1 \geq m_0$ be such that for all $m \geq m_1$ the inequality

$$\frac{2\sqrt{2} \, M' |f|_1 \|f\|_0}{(1 - \kappa)(m + 1)} < \frac{\delta_1}{\sqrt{2m + 1}} \tag{4.11}$$

holds. This is possible since $\dfrac{\sqrt{2}(m + 1)}{\sqrt{2m + 1}} \to \infty$ as $m \to \infty$. It follows from the relation (4.11) that one can always choose a positive number δ_m such that

$$\frac{2\sqrt{2} \, M' |f|_1 \|f\|_0 \, \sigma_1(m)}{1 - \kappa} \leq \delta_m \leq \frac{\delta_1}{\sqrt{m + 1/2}}. \tag{4.12}$$

Consider a set Ω_{δ_m} which consists of the points α such that $\|\alpha - \hat{\alpha}\| \leq \delta_m$. For all $\alpha \in \Omega_{\delta_m}$, we have

$$\| \alpha - \hat{\alpha} \| \le \frac{\delta_1}{\sqrt{m+1/2}} \le \frac{\delta_0 - |f|_0 \, \sigma(m_0)}{\sqrt{m+1/2}}$$

$$\le \frac{\delta_0 - |f|_0 \, \sigma(m)}{\sqrt{m+1/2}} \quad (m \ge m_1 \ge m_0) \ .$$

This yields

$$\Omega_{\delta_m} \subset \Omega_m. \tag{4.13}$$

Then, for any $\alpha \in \Omega_{\delta_m}$, the inequality (4.9) takes the form

$$\left\| J_m(\alpha) - J_m(\hat{\alpha}) \right\| \le 2 |f|_2 \, \sqrt{m+1/2} \, \delta_m. \tag{4.14}$$

We now choose κ so small that

$$\frac{\kappa}{2|f|_2 M'} \le \delta_0 - |f|_0 \, \sigma(m)$$

and set

$$\delta_m = \frac{\kappa}{2|f|_2 M' \sqrt{m+1/2}} \ .$$

Then by virtue of (4.12)-(4.14), all conditions of Theorem 2.1 are satisfied for every $\alpha \in \Omega_{\delta_m}$. This means that the equation (4.7) has the unique solution $\alpha = \bar{\alpha}$ in the region Ω_{δ_m} such that

$$\| \bar{\alpha} - \hat{\alpha} \| \le \frac{2\sqrt{2} \, M' |f|_1 \|f\|_0 \, \sigma_1(m)}{1 - \kappa} \tag{4.15}$$

and Bubnov-Galerkin's approximations to this solution

$$x = \bar{x}_m(t) = \frac{\bar{a}_0}{2} + \sum_{n=1}^{m} (\bar{a}_n \cos nt + \bar{b}_n \sin nt)$$

exist.

Since the inequality (4.15) is equivalent to the inequality

$$\|\bar{x}_m - \hat{x}_m\|_0 \leq \frac{2\,|f|_1\|f\|_0\,\sigma_1(m)M'}{1-\kappa}\,,$$

we have

$$|\bar{x}_m - \hat{x}_m|_0 \leq \sqrt{2m+1}\,\|\bar{x}_m - \hat{x}_m\|_0 \leq \frac{2\,|f|_1\|f\|_0\,M'}{1-\kappa}\,\frac{\sqrt{2m+1}}{m+1}.$$

This yields

$$|\bar{x}_m(t) - \hat{x}(t)|_0 \leq |\bar{x}_m(t) - \hat{x}_m(t)|_0 + |\hat{x}(t) - \hat{x}_m(t)|_0$$

$$\leq \frac{2\,|f|_1\|f\|_0\,M'}{1-\kappa}\,\frac{\sqrt{2m+1}}{m+1} + |f|_0\,\sigma(m).$$

The proof of Theorem 2.2 is thus completed.

§5. Existence of Periodic Solutions for Systems of Differential Equations with Lag

Let us clarify under what conditions the existence of Bubnov-Galerkin's approximations $\bar{x}_m(t)$ of an arbitrary order $m \geq m_0$ involves the existence of periodic solutions to the system of differential equations (1.1).

Theorem 2.3. (Martinyuk, 1982a; Samoilenko and Nurzhanov, 1979). *Suppose that for the system of differential equations* (1.1):

(a) *condition (i) of Lemma 2.4 is satisfied;*

(b) *Bubnov-Galerkin's approximations* $x = \bar{x}_m(t)$ *of the m-th order exist for all* $m \geq m_0$ *and belong to the region* D *together with their* ε*-neighborhood;*

(c) *each linear system of differential equations*

$$\frac{dz(t)}{dt} = \frac{\partial \bar{f}_m}{\partial x}\,z(t) + \frac{\partial \bar{f}_m}{\partial y}\,z(t-\Delta)\,, \quad m \geq m_0, \tag{5.1}$$

where $\bar{f}_m = f(t, \bar{x}_m(t), \bar{x}_m(t-\Delta))$, *has Green's function* $\bar{G}_m(t, \tau)$ *satisfying the in-*

equality

$$\|\bar{G}_m(t, \tau)\| \le \bar{M} e^{-\bar{\lambda}|t-\tau|}, \quad t, \tau \in R, \ t \ne \tau \tag{5.2}$$

with positive constants $\bar{M}$ and $\bar{\lambda}$ which do not depend on m.

Then the system (1.1) possesses the periodic solution $x = \hat{x}_m(t)$ which is unique in a certain vicinity of $\bar{x}_m(t)$. This solution belongs to the region D and satisfies the inequality

$$|\hat{x}(t) - \bar{x}_m(t)|_0 \le \frac{2\bar{M} |f|_1}{\bar{\lambda}(1-\kappa)} \sigma(m), \ 0 < \kappa < 1. \tag{5.3}$$

Proof. According to the definition of Bubnov-Galerkin's approximations, the function $\bar{x}_m(t)$ satisfies the equation

$$\frac{d\bar{x}_m(t)}{dt} = S_m f(t, \bar{x}_m(t), \bar{x}_m(t - \Delta))$$

which can be rewritten as follows

$$\frac{d\bar{x}_m(t)}{dt} = \frac{\partial \bar{f}_m}{\partial x} \bar{x}_m(t) + \frac{\partial \bar{f}_m}{\partial y} \bar{x}_m(t - \Delta)$$

$$+ \left[\bar{f}_m - \frac{\partial \bar{f}_m}{\partial x} \bar{x}_m(t) - \frac{\partial \bar{f}_m}{\partial y} \bar{x}_m(t - \Delta) - R_m \bar{f}_m \right]. \tag{5.4}$$

Taking condition (c) and (5.4) into account, we can write the mth Bubnov-Galerkin's approximation in the form

$$\bar{x}_m(t) = \int_{-\infty}^{\infty} \bar{G}_m(t, \tau) \left[\bar{f}_m - \frac{\partial \bar{f}_m}{\partial x} \bar{x}_m(\tau) - \frac{\partial \bar{f}_m}{\partial y} \bar{x}_m(\tau - \Delta) - R_m \bar{f}_m \right] d\tau. \tag{5.5}$$

Let us show that the equation

$$x(t) = \int_{-\infty}^{\infty} \bar{G}_m(t, \tau) \left[f(\tau, x(\tau), x(\tau - \Delta)) - \frac{\partial \bar{f}_m}{\partial x} x(\tau) - \frac{\partial \bar{f}_m}{\partial y} x(\tau - \Delta) \right] d\tau \tag{5.6}$$

has a periodic solution. For this purpose, we employ the method of successive approximations. Let us take the mth Bubnov-Galerkin's approximation (for sufficiently large m) as the zeroth approximation, i.e., $x_0 = \bar{x}_0(t)$. Then all the subsequent approximations are determined by

$$x_{k+1}(t) = \int_{-\infty}^{\infty} \overline{G}_m(t, \tau) \left[f(\tau, x_k(\tau), x_k(\tau - \Delta)) - \frac{\partial \overline{f}_m}{\partial x} x_k(\tau) - \frac{\partial \overline{f}_m}{\partial y} x_k(\tau - \Delta) \right] d\tau. \quad (5.7)$$

Denote by D_δ the region, which consists of the points x such that $\| x(t) - \overline{x}_m(t) \| \le \delta$ for all $t \in \mathcal{T}_1$. Let us choose δ and κ sufficiently small such that

$$0 < \delta \le \frac{\kappa \overline{\lambda}}{4 \overline{M} |f|_1} \le \varepsilon < 1. \quad (5.8)$$

We now estimate the difference $x_1(t) - x_m(t)$. It follows from the relations (5.5) and (5.7) that

$$x_1(t) - x_0(t) = - \int_{-\infty}^{\infty} \overline{G}(t, \tau) R_m \overline{f}_m \, d\tau.$$

This yields

$$|x_1(t) - x_0(t)|_0 \le \frac{2\overline{M}}{\overline{\lambda}} |R_m \overline{f}_m|_0 \le \frac{2\overline{M}}{\overline{\lambda}} |f|_1 \sigma(m). \quad (5.9)$$

Let us now choose the number m_1 large enough such that

$$\frac{2\overline{M} |f|_1}{\overline{\lambda}(1 - \kappa)} \frac{\sqrt{2}}{\sqrt{m_1}} \le \delta. \quad (5.10)$$

Then, by virtue of (5.10), the inequality (5.9) implies that

$$|x_1(t) - x_0(t)|_0 \le \delta$$

for all $m \ge m_1$. Assume that the inequalities

$$|x_k(t) - x_{k-1}(t)|_0 \le \kappa^{k-1} |x_1(t) - x_0(t)|_0; \quad (5.11)$$

$$|x_k(t) - x_0(t)|_0 \le \delta \quad (5.12)$$

are valid for all k. Then, by using (5.7), we obtain the following estimate for the difference $x_{k+1}(t) - x_k(t)$:

$$\|x_{k+1}(t) - x_k(t)\| \le \int_{-\infty}^{\infty} \|\overline{G}_m(t, \tau)\| \left\| f(\tau, x_k(\tau), x_k(t - \Delta)) \right.$$

$$- f(\tau, x_{k-1}(\tau), x_{k-1}(t-\Delta)) - \frac{\partial \bar{f}_m}{\partial x}(x_k(\tau) - x_{k-1}(\tau))$$

$$- \frac{\partial \bar{f}_m}{\partial y}(x_k(\tau-\Delta) - x_{k-1}(\tau-\Delta)) \Big\| \, d\tau$$

$$\leq \frac{4\overline{M}}{\lambda} |f|_1 \, \delta |x_k(t) - x_{k-1}(t)|_0 \leq \kappa^k |x_1(t) - x_0(t)|_0. \tag{5.13}$$

Moreover, by virtue of (5.11), we get

$$|x_{k+1}(t) - x_0(t)|_0 \leq \frac{2\overline{M} |f|_1 \, \sigma(m)}{\lambda (1-\kappa)} < \delta \tag{5.14}$$

for all $m \geq m_1 \geq m_0$. Thus, we have proved by induction that the inequalities (5.11) and (5.12) hold for all $k = 1, 2, \ldots$ and $m \geq m_1$, and consequently, that the sequence of functions $\{x_k(t)\}$ defined by (5.7) uniformly converges in the region D_δ for all all $k = 1, 2, \ldots$ and $m \geq m_1$. By proceeding to the limit as $k \to \infty$ in (5.7), we find that the limiting function $\hat{x}(t)$ of the sequence $\{x_k(t)\}$ is a periodic solution of the equation (5.6) (and hence, of the system (5.1)) satisfying the inequality (5.3). The last statement follows from (5.14).

Finally, the uniqueness of this periodic solution in a δ-neighborhood of $\bar{x}_m(t)$ follows from the existence of Green's function $\overline{G}_m(t, \tau)$.

§6. Application of Bubnov-Galerkin's Method to the Investigation of Periodic Solutions for Some Classes of Systems of Integro-Differential Equations

Consider a system of integro-differential equations of the form

$$\frac{dx(t)}{dt} = f\left(t, x(t), \int_t^{t+T} \varphi(t, \theta, x(\theta)) \, d\theta \right), \tag{6.1}$$

where $f(t, x, u)$ and $\varphi(t, \theta, x)$ are n-dimensional vector functions periodic in t and θ

with period 2π, and T is a positive constant.

Let us examine the linear system of integro-differential equations

$$\frac{dz(t)}{dt} = A(t)z(t) + \int_t^{t+T} B(t, \theta)\, z(\theta)\, d\theta + c(t), \qquad (6.2)$$

where $A(t), c(t) \in C^0(\mathcal{T}_1)$ and $B(t, \theta) \in C^0(\mathcal{T}_1 \times \mathcal{T}_1)$. Assume that the system (6.2) possesses Green's function $G_0(t, \tau)$ satisfying the inequality

$$\|G_0(t, \tau)\| \le M_0\, e^{-\lambda_0|t-\tau|}, \quad t \ne \tau, \quad t, \tau \in R. \qquad (6.3)$$

For Green's function $G_0(t, \tau)$, we can formulate the statements analogous to Lemmas 2.2 and 2.3, and prove them similarly. One of these statements is given below.

Lemma 2.6. (Samoilenko and Nurzhanov, 1979). *Suppose that the system of linear integro-differential equations (6.2) possesses Green's function $G_0(t, \tau)$ satisfying the inequality (6.3). Assume also that the matrices $A_0(t) \in C^0(\mathcal{T}_1)$ and $B_0(t, \theta) \in C^0(\mathcal{T}_1 \times \mathcal{T}_1)$ are sufficiently small such that*

$$\frac{M_0}{\lambda_0 - \lambda}\left[2|A_0|_0 + \frac{|B_0|_0}{\lambda e^{\lambda T}}(e^{2\lambda T} - 1)\right] \le \rho < 1, \qquad (6.4)$$

for λ, $0 < \lambda < \lambda_0$. Then the perturbed system of integro-differential equations

$$\frac{dz(t)}{dt} = [A(t) + A_0(t)]z(t) + \int_t^{t+T}[B(t, \theta) + B_0(t, \theta)]z(\theta)\, d\theta + c(t)$$

possesses Green's function $G(t, \tau)$ satisfying the inequality

$$\|G(t, \tau)\| \le M e^{-\lambda|t-\tau|}, \quad t, \tau \in R, \quad t \ne \tau,$$

where

$$M = M_0\left\{1 + \frac{M_0}{(1-\rho)\lambda_0}\left[|A_0|_0 + \frac{|B_0|_0}{\lambda_0}(e^{\lambda_0 T} - 1)\right]\Delta(\lambda_0, \lambda)\right\};$$

$$\Delta(\lambda_0, \lambda) = \sup_{t \ge 0}\left[(1 + \lambda_0 t)e^{-(\lambda_0 - \lambda)t}\right].$$

By using the above-mentioned lemmas, one can prove the statements similar to Theorems 2.2 and 2.3.

Theorem 2.4. (Samoilenko and Nurzhanov, 1979). *Suppose that the system of integro-differential equations* (6.1) *satisfies the following conditions:*

(a) $\varphi(t, \theta, x) \in C^2(\mathfrak{T}_1 \times \mathfrak{T}_1 \times D)$; $f(t, x, u) \in C^2(\mathfrak{T}_1 \times D \times D_1)$,

where D is a certain bounded convex region on the Euclidean space E_n, *and* D_1: ‖ *u* ‖ ≤ *d* *is a sphere on* E_n;

b) $d \geq T \max\limits_{t_1 \times t_1 \times D} \|\varphi(t, \theta, x)\|$;

c) there exists a 2π-periodic solution $x = \hat{x}(t)$ *which belongs to the region D together with a certain its* δ_0*-neighborhood;*

d) the system of variational equations for the solution $x = \hat{x}(t)$

$$\frac{dz(t)}{dt} = \frac{\partial \hat{f}}{\partial x} z(t) + \frac{\partial \hat{f}}{\partial u} \int\limits_t^{t+T} \frac{\partial \hat{f}}{\partial x} z(\theta_1) d\theta_1,$$

where

$$\hat{f} = f(t, \hat{x}(t), \hat{u}(t)), \quad \hat{u}(t) = \int\limits_t^{t+T} \varphi(t, \theta, \hat{x}(\theta)) d\theta, \quad \hat{\varphi} = \varphi(t, \theta_1, \hat{x}(\theta_1))$$

possesses Green's function $G_0(t, \tau)$ *satisfying the condition* (6.3).

Then one can always find sufficiently large m_0 *such that Bubnov-Galerkin's approximations* $x = \bar{x}_m(t)$ *exist for al* $m \geq m_0$ *and converge to the exact periodic solution* $x = \hat{x}(t)$ *as* $m \to \infty$. *Moreover, these approximations satisfy the inequality*

$$|\bar{x}_m(t) - \hat{x}(t)|_0 \leq M_1 \frac{\sqrt{2m+1}}{m+1} + |f|_0 \sigma(m)$$

where M_1 *is a positive constant independent of m.*

Consider now a system of Volterra type integro-differential equations with infinite aftereffect

$$\frac{dx(t)}{dt} = f\left(t, x(t), \int_{-\infty}^{t} R(t-\theta)\, \varphi(t, \theta, x(\theta))\, d\theta\right) \le \delta, \tag{6.5}$$

where the vector functions $f(t, x, z) = (f_1(t, x, z), f_2(t, x, z), ..., f_n(t, x, z))$ and $\varphi(t, \theta, x) = (\varphi_1(t, \theta, x), \varphi_2(t, \theta, x), ..., \varphi_n(t, \theta, x))$ are periodic in t and θ with period 2π; the kernel $R(t-\theta)$ is a piecewise continuous function such that

$$\|R(t-\theta)\| \le K_0\, e^{-\gamma_0|t-\theta|} \tag{6.6}$$

for all real t and θ; K_0 and γ_0 are some positive constants.

We also assume that $f(t, x, y) \in C^2(\mathcal{T}_1 \times D \times D_1)$ and $\varphi(t, \theta, x) \in C^2(\mathcal{T}_1 \times \mathcal{T}_1 \times D)$, where D is a certain bounded convex region of the Euclidean space E_n, $D_1: \|z\| \le d$ is a sphere in this space, and $\mathcal{T}_1 = [0, 2\pi]$.

Consider a linear system of integro-differential equations of the form

$$\frac{dy(t)}{dt} = A(t)y(t) + \int_{-\infty}^{t} R(t-\theta)\, B(t, \theta)\, y(\theta)\, d\theta + c(t), \tag{6.7}$$

where $A(t) \in C^0(\mathcal{T}_1)$ and $B(t, \theta) \in C^0(\mathcal{T}_1 \times \mathcal{T}_1)$ are matrices, $c(t) \in C^0(\mathcal{T}_1)$ is a vector function, the matrix $R(t-\theta)$ is piecewise continuous and satisfies (6.6). As before, the matrix function $G_0(t, \tau)$ defined and continuous in t and τ for $t \ne \tau$ is called Green's function for the system of linear integro-differential equations (6.7) if

(i) it is a solution of the homogeneous $(c(t) = 0)$ system, which corresponds to (6.7);

(ii) it satisfies the inequality

$$\int_{-\infty}^{\infty} \|G_0(t, \tau)\|\, d\tau \le M^0 < \infty, \qquad t, \tau \in R, \qquad t \ne \tau,$$

where E is the unit matrix and $R = (-\infty, \infty)$;

(iii) it has a jump on E at $t = \tau$.

Let the system of linear integro-differential equations (6.7) have Green's function $G_0(t, \tau)$ satisfying the condition

$$\|G_0(t, \tau)\| \leq M_0\, e^{-\lambda_0|t-\tau|}, \quad t, \tau \in R, \quad t \neq \tau, \tag{6.8}$$

where m_0 and λ_0 are some positive constants, moreover, $\lambda_0 \leq \gamma_0$. Under this assumption, one can easily prove that the system (6.7) always possesses the unique ω-periodic solution $y = y(t)$ which can be written in the form

$$y(t) = \int_{-\infty}^{\infty} G_0(t, \tau)\, c(\tau) d\tau\ . \tag{6.9}$$

For Green's function $G_0(t, \tau)$ of the linear system of integro-differential equations (6.7) we can also establish the statements similar to Lemmas 2.2 and 2.3. The proofs of these statements are completely analogous with the proofs of the corresponding lemmas.

As shown by Vuitovich and Nurzhanov (1982), for the systems of integro-differential equations of the Volterra type with infinite aftereffect, one can also formulate and prove the statements similar to the theorem given above. Here we present one of these statements.

Theorem 2.5. *Suppose that the right-hand side of the system of integro-differential equations* (6.5) *satisfies the following conditions:*

(i) $f(t, x, z) \in C^2(\mathcal{T}_1 \times D \times D_1)$, $\varphi(t, \theta, x) \in C^2(\mathcal{T}_1 \times \mathcal{T}_1 \times D\,)$;

(ii) $R(t - \theta)$ *is a piecewise continuous matrix satisfying the inequality*

$$\|R(t - \theta)\| \leq K_0\, e^{-\gamma|t-\theta|}$$

for all real t and θ; K_0 and γ_0 are positive constants;

(iii) *for given* $\varepsilon > 0$ *and all* $m \geq m_0$, *where* m_0 *is a positive integer, there exist Bubnov-Galerkin's approximations* $x = \bar{x}_m(t)$ *which belong to the region D together with their ε-neighborhoods;*

(iv) *each linear system of integro-differential equations*

$$\frac{dy(t)}{dt} = \frac{\partial \bar{f}}{\partial x} y(t) + \int_{-\infty}^{\infty} R(t - \theta_1) \frac{\partial \bar{f}_m}{\partial z} \frac{\partial \varphi_m}{\partial x} y(\theta_1) d\theta_1, \quad m \geq m_0,$$

where

$$\bar{z}_m(t) = \int\limits_{-\infty}^{t} R(t-\theta)\varphi(t, \theta, \bar{x}_m(\theta))d\theta, \quad \bar{\varphi}_m = \varphi(t, \theta_1, \bar{x}_m(\theta_1)),$$

possesses Green's function $G_m(t, \tau)$ *satisfying the inequality*

$$\|G_m(t, \tau)\| \le \bar{M}e^{-\bar{\lambda}|t-\tau|}, \quad t, \tau \in R, \ t \ne \tau,$$

with positive constants $\bar{M}$ *and* $\bar{\lambda}$ *both independent of m.*

    Then the system of equation (6.5) has the 2π-periodic solution $\hat{x} = \hat{x}(t)$ *which belongs to the region* D. *Furthermore, this solution is unique in a certain neighborhood of* $\bar{x}_m(t)$ *and satisfies the inequality*

$$|\hat{x} - \bar{x}_m|_0 \le \frac{2Mc_6|f|_1}{\bar{\lambda}(1-\kappa)},$$

where κ *is some fixed number* $0 < \kappa < 1$.

3. QUASIPERIODIC SOLUTIONS OF SYSTEMS WITH LAG. BUBNOV-GALERKIN'S METHOD

§1. Definitions and Auxiliary Statements

Let $\mathcal{T}_m: 0 \leq \varphi_j \leq 2\pi, j = 1, ..., m$ be an m-dimensional torus with $\varphi = (\varphi_1, \varphi_2, ..., \varphi_n)$ being the coordinates on it. Consider a function $f(\varphi) = (f_1(\varphi), f_2(\varphi), ..., f_s(\varphi))$ periodic in $\varphi_j, j = 1, ..., m$, with period 2π, i.e., $f(\varphi + 2\pi k) = f(\varphi)$ for all integer k_j except zero, $k = (k_1, k_2, ..., k_m)$. A norm for these functions is defined by

$$\| f \|_0 = \max_{\varphi \in \mathcal{T}_m} \| f(\varphi) \|, \tag{1.1}$$

where $\| \cdot \|$ denotes, as before, the standard Euclidean norm. The set of all continuous 2π-periodic functions with norm (1.1) forms a complete normed space $C(\mathcal{T}_m)$. By $C^r(\mathcal{T}_m)$, where r is a natural number, we denote a complete space of functions $f(\varphi)$ with the norm

$$\| f \|_r = \max_{0 \leq |\rho| \leq r} \max_{\varphi \in \mathcal{T}_m} \| D^\rho f(\varphi) \|,$$

where $\rho = (\rho_1, \rho_2, ..., \rho_n)$ is a vector with nonnegative integer coordinates, $|\rho| = \sum_{i=1}^{m} \rho_i$, and

$$D^\rho = \frac{\partial^{|\rho|}}{\partial \varphi_1^{\rho_1} \partial \varphi_2^{\rho_2} ... \partial \varphi_m^{\rho_m}}.$$

We say that a collection of numbers $\omega = (\omega_1, \omega_2, ..., \omega_m)$ forms a basis or that it is incommensurable if $(k, \omega) = k_1 \omega_1 + k_2 \omega_2 + ... + k_m \omega_m \neq 0$ for all integer $k_j, j = 1, 2,$

107

..., m, except zero.

Let $f \in C(\mathcal{T}_m)$ and let ω be some basis. The function $F(t) = f(\omega t)$ is called a quasi-periodic function with a frequency basis ω. The definition of a quasiperiodic function implies that it is defined for all $-\infty < t < \infty$. The set of all these functions forms a linear space which is denoted by $C(\omega)$. We introduce a norm in $C(\omega)$ by

$$|F|_0 = \sup_{-\infty < t < \infty} |F(t)|;$$

moreover,

$$|F|_0 = \sup_{-\infty < t < \infty} |f(\omega t)| = |f|_0.$$

We denote by $T(\mathcal{T}_m)$ a set of real trigonometric polynomials of the form

$$P(\varphi) = \sum_{\|n\| \le N} P^{(n)} e^{i(n, \varphi)},$$

where N is an arbitrary nonnegative number with complex-valued coefficients satisfying the condition $P^{(n)} = \overline{P}^{(-n)}$, $P = (P_1, P_2, ..., P_s)$. We introduce a scalar product of two arbitrarily chosen polynomials $P \in T(\mathcal{T}_m)$ and $Q \in T(\mathcal{T}_m)$ by (see (Bers, John, and Schechter, 1966))

$$(P, Q)_r = \sum_n (1 + \|n\|^2)^r \langle P^{(n)}, Q^{(-n)} \rangle, \tag{1.2}$$

where $r > 0$ is a natural number and $\langle P^{(n)}, Q^{(-n)} \rangle = \sum_{i=1}^s P_i^{(n)} Q_i^{(-n)}$.

The product (1.2) satisfies all the axioms of scalar product and, consequently, induces the norm

$$\|P\|_r^2 = (P, P)_r = \sum_{\|n\| \le N} (1 + \|n\|^2)^r |P^{(n)}|^2 .$$

Denote by K the operator $1 - \Delta_1$, where $\Delta_1 = \sum_{i=1}^m \dfrac{\partial^2}{\partial \varphi_i^2}$. Then

$$KP(\varphi) = \sum_{\|n\| \le N} (1 + \|n\|^2) P^{(n)} e^{i(n, \varphi)},$$

$$K^r P(\varphi) = \sum_{\|n\| \le N} (1 + \|n\|^2)^r P^{(n)} e^{i(n, \varphi)},$$

and

$$(P, Q)_r = \frac{1}{(2\pi)^m} \int_0^{2\pi} \cdots \int_0^{2\pi} \langle KP, Q \rangle \, d\varphi, \tag{1.3}$$

because $P^{(n)}$ is defined by

$$P^{(n)} = \frac{1}{(2\pi)^m} \int_0^{2\pi} \cdots \int_0^{2\pi} P e^{-i(n, \varphi)} \, d\varphi.$$

Thus, by completing the space $T(\mathfrak{T}_m)$ in norm $\|\cdot\|_r$, we obtain the (separable) Hilbert space $H^r(\mathfrak{T}_m)$. In this space, the theorem on compactness is valid which states that from any infinite sequence bounded in norm of $H^r(\mathfrak{T}_m)$ one can always extract a subsequence convergent in norm of $H^r(\mathfrak{T}_m)$, where $s < r$.

For arbitrary functions $f(\varphi) \in H^r(\mathfrak{T}_m)$ and $g(\varphi) \in H^r(\mathfrak{T}_m)$, the Schwartz inequality

$$|(f, g)_r| \leq \|f\|_r \|g\|_r$$

and its generalization

$$|(f, g)_r| \leq \|f\|_{r-s} \|g\|_{r+s}, \tag{1.4}$$

are valid, provided that $0 \leq s \leq r$ and $g(\varphi) \in H^{r+s}(\mathfrak{T}_m)$. If $f(\varphi) \in H^{r+s}(\mathfrak{T}_m)$, then the inequality (1.4) yields

$$\|f\|_r^2 \leq \|f\|_{r-s} \|f\|_{r+s}. \tag{1.5}$$

For $s < r$, this implies

$$\|f\|_s \leq \|f\|_r. \tag{1.6}$$

It follows from the inequality

$$(1 + \|n\|)^s \leq \varepsilon (1 + \|n\|^2)^{s_1} + \varepsilon^{(s_0 - s)/(s_1 - s)} (1 + \|n\|^2)^{s_0}$$

that

$$\|f\|_s \leq \varepsilon \|f\|_{s_1} + \varepsilon^{(s_0 - s)/(s_1 - s)} \|f\|_{s_0} \tag{1.7}$$

for all $s_1 \geq s \geq s_0, \varepsilon > 0$, and $f(\varphi) \in H^{s_1}(\mathfrak{T}_m)$. If we set $s_1 = r$ and $s_0 = 0$, then for appropriate choice of ε, the inequality (1.7) yields

$$\|f\|_s \leq 2\|f\|_0^{1-s/r}\,\|f\|_r^{s/r}, \quad 0 \leq s \leq r \tag{1.8}$$

provided that $f(\varphi) \in H^r(\mathfrak{T}_m)$.

For each partial derivative $D^\rho f$ of the function $f(\varphi) = \sum_n f^{(n)} e^{i(n,\,\varphi)}$, we have

$$D^\rho f = \sum_n (in)^\rho f^{(n)} e^{i(n,\,\varphi)} \tag{1.9}$$

where $(in)^\rho = (in_1)^{\rho_1} (in_2)^{\rho_2} \ldots (in_m)^{\rho_m}$. It follows from (1.9) that

$$\|D^\rho f\|_r \leq \|f\|_{r+|\rho|}, \tag{1.10}$$

$$\|f\|_r \leq c \sum_{|\rho| \leq r} \|D^\rho f\|_0, \tag{1.11}$$

where c depends on r.

By using the Riesz-Fischer theorem, one can easily show that the space $H^0(\mathfrak{T}_m)$ can be identified (in the sense of isomorphism) with the space $L_2(\mathfrak{T}_m)$ of square integrable functions with the scalar product (1.2) (for $r = 0$), moreover, the Parseval equality

$$\frac{1}{(2\pi)^m} \int\limits_0^{2\pi} \ldots \int\limits_0^{2\pi} \langle f, f \rangle \, d\varphi = \sum_n \|f^{(n)}\|^2$$

holds. A function $f^{(\rho)}(\varphi) \in H^0(\mathfrak{T}_m)$ is called the ρ-th derivative of the function $f(\varphi) \in H^0(\mathfrak{T}_m)$ if

$$(f^{(\rho)}, P)_0 = (-1)^{|\rho|}(f, D^{(\rho)}P)_0, \quad P \in T(\mathfrak{T}_m).$$

Then $H^r(\mathfrak{T}_m)$ is nothing but a subspace of functions from $H^0(\mathfrak{T}_m)$ which have generalized derivatives up to the rth order inclusively, i.e., $f^{(\rho)}(\varphi) \in H^0(\mathfrak{T}_m)$ ($|\rho| \leq r$), where $f^{(\rho)}(\varphi) = D^{(\rho)}f(\varphi)$.

Since the space $H^r(\mathfrak{T}_m)$ has the integral norm, it is convenient to use it in the cases when differential operators or their inverse operators are employed.

Further, we shall need Sobolev's theorems on imbedding and compactness, and some other statements (see Bers, John, Schechter (1966); Sobolev, (1950)).

Theorem 3.1. *If* $f(\varphi) \in H^r(\mathfrak{T}_m)$, *where* $r > m/2 + s$, *then* $f(\varphi) \in C^s(\mathfrak{T}_m)$ *and*

$$|f| \leq c \, \|f\|_r, \tag{1.12}$$

where $c < \sqrt{1 + 2^m \dfrac{2(r-s)+1-m}{2(r-s)-m}}$.

Theorem 3.2 (Sobolev, 1950). *A sequence of functions bounded in* $H^r(\mathfrak{T}_m)$ *is compact in* $H^s(\mathfrak{T}_m)$ *if* $s < r$. *If* $f(\varphi) \in H^r(\mathfrak{T}_m)$, *then*

$$\left\| \sum_{\|n\| \geq N} f^{(n)} e^{i(n,\, \varphi)} \right\|_s \leq N^{-r+s} \|f\|_r, \quad r \geq s. \tag{1.13}$$

If the matrix $A(\varphi) \in C^r(\mathfrak{T}_m)$, *and* $f(\varphi) \in H^r(\mathfrak{T}_m)$, *then*

$$\|A(\varphi) f(\varphi)\|_r \leq c \, |A|_r \, \|f\|_r. \tag{1.14}$$

Lemma 3.1 (Mozer, 1968). *Let* $f(\varphi, y) \in C^r(\mathfrak{T}_m, D_1)$, *where* $D_1 = \{y \mid \|y\| \leq 1\}$, *and let* $g(\varphi) \in H^r(\mathfrak{T}_m) \cap C(\mathfrak{T}_m)$ *be such that* $|g(\varphi)|_0 \leq 1$. *Then*

$$\|f(\varphi, g(\varphi))\|_r \leq c \, |f|_r \, (1 + \|g\|_r), \tag{1.15}$$

where the constant c *does not depend on* f *and* g.

Consider the linear differential operator

$$L = \sum_{v=1}^{m} a_v(\varphi) \frac{\partial}{\partial \varphi_v} + b(\varphi),$$

where $a_v(\varphi)$ and $b(\varphi)$ are $(s \times s)$-dimensional matrices. We have

Lemma 3.2. (Mozer, 1968). *Let the operator* L *be such that*

(i) $a_v(\varphi),\ b(\varphi) \in C^r(\mathfrak{T}_m)$;

(ii) $a_v^T(\varphi) = a_v(\varphi)$;

(iii) *there exists a number* $\gamma_0 > 0$ *such that, for arbitrary vectors* $\xi = (\xi_1, ..., \xi_m)$ *and* $\eta = (\eta_1, ..., \eta_m)$ *for which* $\| \xi \| = \| \eta \| = 1$ *and for arbitrary integer* $l = 0,$ *1,..., r, the following inequality holds*

$$\left\langle \left(l \sum_{v, \mu=1}^{m} \frac{\partial a_v(\varphi)}{\partial \varphi_\mu} \xi_\mu \xi_v + b_0(\varphi) \right) \eta, \eta \right\rangle \geq \gamma_0 > 0, \tag{1.16}$$

where

$$b_0(\varphi) = b(\varphi) - \frac{1}{2} \sum_{v=1}^{m} \frac{\partial a_v(\varphi)}{\partial \varphi_v} .$$

Then we have

$$(Lu, u)_r \geq \gamma_1 \| u \|_0^2 \tag{1.17}$$

for every function $u(\varphi) \in H^0(\mathcal{T}_m)$, *and*

$$(Lu, u)_r \geq \gamma_1 \| u \|_r^2 - \delta_1 \left(1 + \sum_{v=1}^{m} \| a_v \|_r + \| b \|_r \right)^2 , \tag{1.18}$$

for every function $u(\varphi) \in H^r(\mathcal{T}_m) \cap C^2(\mathcal{T}_m)$ *satisfying the inequality* $|u(\varphi)|_2 \leq 1.$ *Here,* γ_1 *and* δ_1 *do not depend on* $u(\varphi)$.

Henceforth, we call processes, described by quasiperiodic functions with frequency basis containing m numbers, m-frequency processes or m-frequency oscillations of systems with lag. These oscillations may appear in real systems described by differential equations of the form

$$\frac{dx(t)}{dt} = X(\omega t, x(t), x(t-\Delta)) \tag{1.19}$$

which are a part of the dynamical system

$$\frac{dx(t)}{dt} = X(\varphi, x(t), x(t-\Delta)),$$

$$\frac{d\varphi(t)}{dt} = \omega.$$

In general, if we say that the dynamical system

$$\frac{dx(t)}{dt} = X(x(t), x(t - \Delta))$$ (1.20)

describes some m-frequency oscillating process $x = x(t)$, this means that $x = f(\omega t)$, where $f \in C(\mathcal{T}_m)$ and $\omega = (\omega_1, \omega_2, ..., \omega_m)$.

Given a quasiperiodic solution of the system (1.20), one can construct an invariant set M corresponding to this solution. The set M is defined by

$$M: \ x = f(\varphi), \quad \varphi \in \mathcal{T}_m,$$ (1.21)

since according to the Weyl theorem, the closure of the quasiperiodic trajectory $f(\omega t)$ coincides with the range of values of the function $f(\varphi)$, $\varphi \in \mathcal{T}_m$, and the closure of an arbitrary trajectory is invariant if it is nonempty and finite-dimensional.

In the case when $f(\varphi) \in C'(\mathcal{T}_m)$ and the rank m of the matrix $\dfrac{\partial f(\varphi)}{\partial \varphi}$ is such that $m < 2n$, where n is the dimensionality of the phase space of (1.20), the equality (1.21) establishes the one-to-one continuous correspondence between points x of the surface M and points of the cube $\mathcal{T}_m: 0 \le \varphi_j \le 2\pi, j = 1, ..., m$, whose opposite sides are identified. Since the cube $\mathcal{T}_m$ is a topological manifold homeomorphic to an m-dimensional torus, the surface M is (in this case) a torus (in the general case, this is a toroidal set). All the trajectories of the system (1.20) which start from this torus are also quasiperiodic with the same frequency basis ω; they are defined by

$$x = f(\omega t + \varphi),$$ (1.22)

where φ is a constant. Let us introduce local coordinates $\varphi = (\varphi_1, \varphi_2, ..., \varphi_m)$ on the invariant torus M. Then the trajectories of (1.20) lying on this torus are given by the formula (1.22) provided that the variable φ is determined by the dynamical system

$$\frac{d\varphi}{dt} = \omega.$$ (1.23)

Let $f(\varphi)$ be a continuously differentiable function. Then, according to the Kronecker lemma (see Levitan (1953)), this function is the classical solution of the equation

$$\sum_{v=1}^{m} \omega_v \frac{\partial f(\varphi)}{\partial \varphi_v} = X(\varphi, f(\varphi), f(\varphi - \omega \Delta)).$$ (1.24)

The problem of existence of quasiperiodic solutions of the system (1.19) is thus reduced to the problem of existence of periodic solutions of (1.24).

Henceforth, we denote by S_N an operator which establishes a correspondence bet-ween a square integrable function $f(\varphi)$ and the segment of its Fourier series of the length N, i.e.,

$$S_N f(\varphi) = \sum_{\|n\| \le N} f^{(n)} e^{i(n,\,\varphi)},$$

$$f^{(n)} = \frac{1}{(2\pi)^m} \int_0^{2\pi} \cdots \int_0^{2\pi} f e^{-i(n,\,\varphi)} d\varphi.$$

In the uniform metric, any quasiperiodic function can be approximated by trigonometric polynomials as accurately as desired. Therefore, one can try to construct a quasiperiodic solution of the system (1.19) (or a periodic solution of the equation (1.24)) in the form of a sequence of trigonometric polynomials

$$f_N(\varphi) = \sum_{\|n\| \le N} u_N^{(n)} e^{i(n,\,\varphi)}, \quad N = 1, 2, \ldots.$$

In order to construct the approximate solution $f_N(\varphi)$ of the equation (1.24), we apply Bubnov-Galerkin's method assuming that the coefficients of this polynomial can be found from the system of nonlinear equations

$$S_N\left(\sum_{v=1}^{m} \omega_v \frac{\partial f_N(\varphi)}{\partial \varphi_v} - X(\varphi, f_N(\varphi), f_N(\varphi - \omega\Delta)) \right) = 0. \tag{1.25}$$

To justify Bubnov-Galerkin's method, it is necessary to prove that the system (1.25) is solvable and that the sequence $f_N(\varphi)$ converges to the exact solution $f(\varphi)$ as $N \to \infty$.

§2. Construction of Quasiperiodic Solutions of Systems with Lag by Bubnov-Galerkin's Method

Consider a system of differential equations (with quasiperiodic right-hand side) of the form

$$\frac{dx(t)}{dt} = f(\omega t, x(t), x(t - \Delta)), \tag{2.1}$$

where $x = (x_1, x_2 \ldots, x_s)$ is an s-dimensional vector, $f = (f_1, f_2 \ldots, f_s)$ is an s-dimensional vector function defined in the region $\mathcal{T}_m \times D \times D$, and $\omega = (\omega_1, \omega_2, \ldots, \omega_m)$ is a frequency basis.

Assume that the system (2.1) possesses a sufficiently smooth quasiperiodic solution $x^0(t)$ with frequency basis ω. Then there exists a function $u^0(t)$ given on the torus $\mathcal{T}_m$ such that $x^0(t) = u^0(\omega t)$; this function is a solution of the equation

$$\sum_{v=1}^{m} \omega_v \frac{\partial u(\varphi)}{\partial \varphi_v} = f(\varphi, u(\varphi), u(\varphi - \omega \Delta)). \tag{2.2}$$

Henceforth, the space of vectors with complex components and the set of all m-dimensional vectors with integer components are denoted by C^m and Z^m, respectively.

Let

$$u_N(\varphi) = \sum_{\|n\| \leq N} u_N^{(n)} e^{i(n, \varphi)} = \sum_{\|n\| \leq N, \, n_i \geq 0} \left(a_N^{(n)} \cos(n, \varphi) + b_N^{(n)} \sin(n, \varphi) \right), \tag{2.3}$$

where

$$u_N^{(n)} \in C^s, \quad u_N^{(-n)} = \bar{u}_N^{(n)}, \quad a_N^{(n)} \in R^s, \quad b_N^{(n)} \in R^s.$$

If vectors of Z^m are ordered, according to a certain law, then one can always establish a correspondence between the polynomial $u_N(\varphi)$ and the real column vector u_N composed of the $2n$-dimensional column vectors $\left(a_N^{(n)}, b_N^{(n)} \right)$ placed one under another and ordered from above downwards according to the law of ordering of the vectors in $n \in Z^m$. Thus, for given polynomial $u_N(\varphi)$, one can always indicate the corresponding $(2sP(N) + s)$-dimensional vector u_N, where $P(N) = \sum_{\|n\| \leq N, \, n_i \geq 0} 1$.

We have

$$\| u_N(\varphi) \|_0 = \| u_N \|; \tag{2.4}$$

and

$$\frac{\partial u_N'(\varphi)}{\partial u_N'} u_N'' = u_N''(\varphi), \tag{2.5}$$

where $\dfrac{\partial u_N'(\varphi)}{\partial u_N'}$ is an $(s \times (sP(N) + s))$-dimensional Jacobian matrix.

We try to find an approximate solution of the equation (2.2) in the form of the polynomial $u_N(\varphi)$. Its coefficients are to be found from the system of nonlinear equations

$$S_N \left(\sum_{v=1}^{m} \omega_v \frac{\partial u_N(\varphi)}{\partial \varphi_v} - f(\varphi, u_N(\varphi), u_N(\varphi - \omega \Delta)) \right) = 0 . \qquad (2.6)$$

Taking the argument presented above into account, we can rewrite (2.6) as follows

$$F_N(u_N) = 0, \qquad (2.7)$$

where F_N is an $(s + 2sP(N))$-dimensional vector. The polynomial (2.3) satisfying the equation (2.6) is called Bubnov-Galerkin's approximation of the N-th order, and the equation (2.7) is called the determining equation.

We now establish what restrictions should be imposed on the right-hand side of (2.1) in order that conditions of Theorem 2.1 hold, and consequently, in order that the determining equation (2.7) be solvable. These conditions are given by the following statements.

Lemma 3.3. (Parasyuk, 1978; Dankanich, 1984). *Suppose that the system of differential equations* (2.1) *satisfies the following conditions:*

(i)   there exists a solution $u^0(\varphi)$ of the equation (2.2) which belongs to the region

$$D_\delta = \{u \mid u \in R^s, \ \| u^0(\varphi) - u(\varphi) \| \le \delta\} \subset D. \quad \delta > 0;$$

moreover, $u^0(\varphi) \in H^r(\mathfrak{T}_m)$, where $r \ge m/2 + 2$;

(ii) $f(\varphi, u, v) \in C^2(\mathfrak{T}_m \times \overline{D} \times \overline{D})$;

(iii) there exists a positive number γ_0 such that, for an arbitrary vector $\eta \in R^s$ with $\|\eta\| = 1$ and an arbitrary $(\varphi, u, v) \in C^2(\mathfrak{T}_m \times D_\delta \times D_\delta)$, the inequalities

$$\left\langle \frac{\partial f(\varphi, u, v)}{\partial v} \eta, \eta \right\rangle \ge 2\gamma_0 \|\eta\|^2, \ \ \gamma_0 > 0,$$

$$\left| \frac{\partial f(\varphi, u, v)}{\partial v} \right|_0 \le \gamma_0$$

hold.

Then there exist numbers $N_0 > 0$ and $c_1 > 0$ such that for $N \ge N_0$, the following relation takes place

$$u_N^0(\varphi) \in D_{\delta/2} = \{u \mid \|u^0(\varphi) - u(\varphi)\| \leq \delta/2\};$$

furthermore, the vector u_N^0 corresponding to the polynomial $u_N^0(\varphi)$ satisfies the inequality

$$\| F_N(u_N^0) \| \leq c_1 N^{-(r-1)}.$$ (2.8)

Proof. By virtue of Theorem 3.2, we have

$$\| u^0(\varphi) - u_N^0(\varphi) \|_0 \leq N^{-r} \| u^0(\varphi) \|_r \leq \delta/2.$$ (2.9)

By using the relations (1.10), (1.13), and (2.4), we obtain

$$\left\| F_N\!\left(u_N^0\right) \right\| = \left\| S_N\!\left(\sum_{v=1}^{m} \omega_v \frac{\partial u_N^0(\varphi)}{\partial \varphi_v} - f(\varphi, u_N^0(\varphi), u_N^0(\varphi - \omega\Delta)) \right) \right\|_0$$

$$\leq \sum_{v=1}^{m} |\omega_v| N^{-(r-1)} \| u^0(\varphi) \|_r + 2s |f(\varphi, u, v)|_1 N^{-r} \| u^0(\varphi) \|_r \leq c_1 N^{-(r-1)},$$

completing the proof.
    Consider a set of vectors

$$D_{\delta_N} = \{u_N \mid \| u_N - u_N^0 \| \leq \delta_N\}, \quad \delta_N = N^{-(m+1)/2}.$$

Let $u_N(\varphi)$ be an arbitrary polynomial, and let $u_N \in D_{\delta_N}$ be the vector which corresponds to this polynomial. Then, taking (2.4) and the inequality

$$|u_{N+1} - u_N|_0 < \sum_{\|n\| \leq N+1} \| u_{N+1}^{(n)} - u_N^{(n)} \|$$

$$\leq \sqrt{\| u_{N+1} - u_N \|_0^2} \sqrt{\sum_{\|n\| \leq N+1} 1} \leq 2^{m/2}(N+1)^{m/2} \| u_{N+1} - u_N \|_0$$

into account, we obtain the estimate

$$\| u_N(\varphi) - u_N^0(\varphi) \| \leq |u_N(\varphi) - u_N^0(\varphi)|_0$$

$$\leq 2^{m/2} N^{m/2} \| u_N(\varphi) - u_N^0(\varphi) \|_0 \leq 2^{m/2} N^{-1/2} \leq \delta/2$$ (2.10)

for all $N \geq N_0$ provided that N_0 is sufficiently large. It follows from the inequality (2.10) that $u_N(\varphi) \in D_\delta$, and that the vector-function $F_N(u_N)$ is defined and twice continuously differentiable on the set D_{δ_N}

Denote by $J_N(u_N)$ the Jacobian matrix of the left-hand side of (2.7), i.e., $J_N(u_N) = \dfrac{\partial F(u_n)}{\partial u_n}$. Let us clarify the main properties of this matrix.

Lemma 3.4. (Parasyuk, 1978; Dankanich, 1984). *Suppose that conditions of Lemma 3.3 hold. Then the matrix $J_N(u_N^0)$ is not degenerate and $J_N^{-1} \| (u_N^0) \| \leq M$, where M is a constant.*

Proof. Let

$$\omega_N(\varphi) = \sum_{\|n\| \leq M} \omega_N^{(n)} e^{i(n,\, \varphi)} = \sum_{\|n\| \leq M} \left(\omega_{1N}^{(n)} \cos(n,\, \varphi) + \omega_{2N}^{(n)} \sin(n,\, \varphi) \right), \quad (2.11)$$

$$v_N(\varphi) = \sum_{\|n\| \leq M} v_N^{(n)} e^{i(n,\, \varphi)} = \sum_{\|n\| \leq M} \left(v_{1N}^{(n)} \cos(n,\, \varphi) + v_{2N}^{(n)} \sin(n,\, \varphi) \right), \quad (2.12)$$

and let w_N and v_N be $(s + 2sP(N))$-dimensional vectors which corresponding to $\omega_N(\varphi)$ and $v_N(\varphi)$, respectively. Consider a linear system

$$J_N(u_N) w_N = v_N. \qquad (2.13)$$

Clearly, in order that the inequality (2.13) hold, it is necessary and sufficient that the following equalities be true:

$$\left(\frac{\partial}{\partial \bar{u}_N} \int_0^{2\pi} \cdots \int_0^{2\pi} \left(\sum_{v=1}^{m} \omega_v \frac{\partial u_N(\varphi)}{\partial \varphi_v} - f(\varphi,\, u_N(\varphi),\, u_N(\varphi - \omega\Delta)) \right) \cos(n,\, \varphi)\, d\varphi \right) w_N = v_{1,N}^{(n)},$$

$$\left(\frac{\partial}{\partial \bar{u}_N} \int_0^{2\pi} \cdots \int_0^{2\pi} \left(\sum_{v=1}^{m} \omega_v \frac{\partial u_N(\varphi)}{\partial \varphi_v} - f(\varphi,\, u_N(\varphi),\, u_N(\varphi - \omega\Delta)) \right) \sin(n,\, \varphi)\, d\varphi \right) w_N = v_{2,N}^{(n)}.$$

for all $\|n\| \leq N$. We can transform these equalities into the following relations

$$\int\limits_0^{2\pi} \cdots \int\limits_0^{2\pi} \left(\sum_{\nu=1}^m \omega_\nu \frac{\partial}{\partial \varphi_\nu} \frac{\partial u_N(\varphi)}{\partial u_N} w_N - \frac{\partial f(\varphi, u_N(\varphi), u_N(\varphi - \omega\Delta))}{\partial u} \frac{\partial u_N(\varphi)}{\partial u_N} w_N \right.$$

$$\left. - \frac{\partial f(\varphi, u_N(\varphi), u_N(\varphi - \omega\Delta))}{\partial y} \frac{\partial u_N(\varphi)}{\partial u_N} w_N \right) \cos(n, \varphi) d\varphi = v_{1,N}^{(n)},$$

$$\int\limits_0^{2\pi} \cdots \int\limits_0^{2\pi} \left(\sum_{\nu=1}^m \omega_\nu \frac{\partial}{\partial \varphi_\nu} \frac{\partial u_N(\varphi)}{\partial u_N} w_N - \frac{\partial f(\varphi, u_N(\varphi), u_N(\varphi - \omega\Delta))}{\partial u} \frac{\partial u_N(\varphi)}{\partial u_N} w_N \right.$$

$$\left. - \frac{\partial f(\varphi, u_N(\varphi), u_N(\varphi - \omega\Delta))}{\partial y} \frac{\partial u_N(\varphi)}{\partial u_N} w_N \right) \sin(n, \varphi) d\varphi = v_{2,N}^{(n)}.$$

Let us introduce that operator

$$L(u) = L_1(u) + L_1^\Delta(u), \tag{2.14}$$

where

$$L_1(u) = \sum_{\nu=1}^m \omega_\nu \frac{\partial u}{\partial \varphi_\nu} - \frac{\partial f(\varphi, u(\varphi), u(\varphi - \omega\Delta))}{\partial u};$$

$$L_1^\Delta(u) = -\frac{\partial f(\varphi, u(\varphi), u(\varphi - \omega\Delta))}{\partial y}, \quad (y = u(\varphi - \omega\Delta));$$

moreover,

$$L_1^\Delta(u)\omega(\varphi) = -\frac{\partial f(\varphi, u(\varphi), u(\varphi - \omega\Delta))}{\partial y} \omega(\varphi - \omega\Delta).$$

By virtue of (2.5), the equality (2.13) can be rewritten in the form

$$S_N(L(u_N^0(\varphi))\omega_N(\varphi) = v_N(\varphi). \tag{2.15}$$

Consider the following equation

$$J_N(u_N)w_N = 0, \tag{2.16}$$

which corresponds to the system (2.13). This equation can be written in the form

$$S_N(L(u_N^0(\varphi))\omega_N(\varphi) = 0. \tag{2.17}$$

Taking condition (iii) of Lemma 3.2 into account, we find

$$0 = (L(u_N^0(\varphi))\omega_N(\varphi), \omega_N(\varphi))_0$$

$$\geq \left(2\gamma_0 - \left|\frac{\partial f(\varphi, u_N^0(\varphi), u_N^0(\varphi - \omega\Delta))}{\partial y}\right|_0\right)\|\omega_N(\varphi)\|^2. \tag{2.18}$$

This estimate implies that for $u_N = u_N^0$, the equation (2.16) possesses the trivial solution only. Therefore, the matrix $J_N(u_N^0)$ is nonsingular. Then for $u_N = u_N^0$ and arbitrary v_N, the non-homogeneous system (2.15) always possesses the unique solution.

In order to find bounds for the matrix $J_N^{-1}(u_N)$, we employ Lemma 3.2. The relation (2.15) implies that

$$(v_N(\varphi), \omega_N(\varphi))_0 = (L(u_N^0(\varphi))\omega_N(\varphi), \omega_N(\varphi))_0$$

$$\geq \left(2\gamma_0 - \left|\frac{\partial f(\varphi, u_N^0(\varphi), u_N^0(\varphi - \omega\Delta))}{\partial y}\right|_0\right)\|\omega_N(\varphi)\|^2. \tag{2.19}$$

This yields

$$\| \omega_N(\varphi)\|_0 \leq \left(2\gamma_0 - \left|\frac{\partial f(\varphi, u_N^0(\varphi), u_N^0(\varphi - \omega\Delta))}{\partial y}\right|_0\right)^{-1}\|v_N(\varphi)\|_0. \tag{2.20}$$

The relation (2.13) gives $w_N = J_N^{-1}(u_N)v_N$. Therefore, the inequality (2.20) can be rewritten in the form

$$\|J_N^{-1}(u_N)v_N\|_0 \leq \left(2\gamma_0 - \left|\frac{\partial f(\varphi, u_N^0(\varphi), u_N^0(\varphi - \omega\Delta))}{\partial y}\right|_0\right)^{-1}\|v_N\|. \tag{2.21}$$

By setting

$$M = \left(2\gamma_0 - \left| \frac{\partial f(\varphi, u_N^0(\varphi), u_N^0(\varphi - \omega\Delta))}{\partial y} \right|_0 \right)^{-1}$$

and inserting this in (2.21), we obtain $\| J_N^{-1}(u_N) \| \le M$.

The estimate for the difference $J_N(u_N) - J_N(u_N^0)$ is given by the following lemma.

Lemma 3.5. (Parasyuk, 1978; Dankanich, 1984). *Suppose that the right-hand side of the system of differential equations* (2.1) *satisfies conditions of Lemma 3.3. Then, for an arbitrary vector* $u_N \in D_{\delta_N}$, $0 < \kappa < 1$ *and sufficiently large* $N \ge N_0$, *we have*

$$\| J_N(u_N) - J_N(u_N^0) \| \le \kappa / M. \tag{2.22}$$

Proof. For an arbitrary polynomial

$$w_N(\varphi) = \sum_{\|n\| \le N} w_N^{(n)} e^{i(n, \varphi)} = w_N(\varphi) = \sum_{\|n\| \le N} w_N^{(n)} e^{i(n, \varphi)} = \,,$$

we construct the corresponding vector w_N. Then it follows from (2.13) that

$$\| J_N(u_N) - J_N(u_N^0) w_N \| = \| S_N(L(u_N(\varphi)) - L(u_N^0(\varphi)) \omega_N(\varphi) \|$$

$$\le \left| \frac{\partial f(\varphi, u_N(\varphi), u_N(\varphi - \omega\Delta))}{\partial u} - \frac{\partial f(\varphi, u_N^0(\varphi), u_N^0(\varphi - \omega\Delta))}{\partial u} \right|_0 \| \omega_N(\varphi) \|_0$$

$$+ \left| \frac{\partial f(\varphi, u_N(\varphi), u_N(\varphi - \omega\Delta))}{\partial y} - \frac{\partial f(\varphi, u_N^0(\varphi), u_N^0(\varphi - \omega\Delta))}{\partial y} \right|_0 \| \omega_N(\varphi - \omega\Delta) \|_0$$

$$\le c_2 |f(\varphi, u, y)|_2 \, |u_N(\varphi) - u_N^0(\varphi)|_0 \| w_N \|. \tag{2.23}$$

This yields

$$\| J_N(u_N) - J_N(u_N^0) \| \le c_2 |f(\varphi, u, y)|_2 \, |u_N(\varphi) - u_N^0(\varphi)|_0. \tag{2.24}$$

Taking into account the definition of the region D_{δ_N} and the inequality (2.10), we finally get

$$\| J_N(u_N) - J_N(u_N^0) \| \leq c_3 N^{m/2} \|u_N(\varphi) - u_N^0(\varphi)\|_0 \leq c_4 N^{-1/2} \leq \kappa / M \qquad (2.25)$$

for all sufficiently large N_0 and $N > N_0$.

By using Lemmas 3.3–3.5 and Theorem 2.1, we now prove the existence of Bubnov-Galerkin's approximations for the system of differential equations (2.1); furthermore, we prove that these approximations converge to the exact quasiperiodic solution.

Theorem 3.3. (Dankanich, 1984). *Suppose that the system of differential equations* (2.1) *satisfies conditions (i)–(iii) of Lemma* 3.3. *Then there exists a sufficiently large number* N_0 *such that, for* $N \geq n_0$, *the equation* (2.2) *possesses a unique solution* $u^*(\varphi)$ *periodic in* φ *for which the following inequality holds*

$$|u_N^0(\varphi) - u_N^*(\varphi)|_0 \leq c_5 N^{-r+m/2+1}. \qquad (2.26)$$

Proof. Assume that the periodic solution $u^0(\varphi)$ of the equation (2.2) is known. Let $u_N^0(\varphi) = S_N u^0(\varphi)$, and let u_N^0 be the corresponding vector. Then, according to Lemma 3.3, the inequality $\| F_N(u_N^0) \| \leq c_1 N^{-(r-1)}$ holds in the region

$$D_{\frac{\delta}{2}} = \{u \mid \| u(\varphi) - u^0(\varphi) \| \leq \delta/2\}.$$

This means that condition (1.8) of Theorem 2.1 is satisfied for $N \geq N_0$, where N_0 is a sufficiently large number, provided that we set $l = l_N = c_1 N^{-(r-1)}$.

We consider the set

$$\Omega_{\delta_N} = \{u_N \mid \| u_N - u_N^0 \| \leq \delta_N\}, \quad \delta_N = N^{-(m+1)/2}.$$

By virtue of Lemmas 3.4 and 3.5, one can always choose sufficiently large N_0 such that for $N \geq N_0$, $F_N(u_N^0)$ satisfy conditions (b) and the inequaly (1.8) in Theorem 2.1. If N_0 is chosen from the inequality

$$\frac{Mc_g N_0^{-(r-1)}}{1-\kappa} \leq N_0^{-(m+1)/2},$$

then condition (c) of Theorem 2.1 is also satisfied. Hence, for any $N \geq N_0$, there exists a solution u_N^* of the equation (2.7) and, consequently, a periodic solution $u_N^*(\varphi)$ of the equation (2.2). The latter satisfies the inequality

$$\|u_N^0(\varphi) - u_N^*(\varphi)\|_0 \leq \frac{Mc_1 N_0^{-(r-1)}}{1-\kappa}. \qquad (2.27)$$

By virtue of (2.10), we can transform the inequality (2.27) as follows

$$|u_N^0(\varphi) - u_N^*(\varphi)|_0 \leq 2^{m/2} N^{m/2} \|u_N^0 - u_N^*\|_0 \leq c N^{-r+m/2+1}.$$

Theorem 3.3 is thus proved.

§3. Construction of Quasiperiodic Solutions of Perturbed Systems with Lag by Bubnov-Galerkin's Method

Consider a system of differential equations with lag

$$\frac{dx(t)}{dt} = f(\omega t, x(t), x(t-\Delta)), \tag{3.1}$$

where $x = (x_1, x_2..., x_s)$ is an s-dimensional vector, $f = (f_1, f_2..., f_s)$ is an s-dimensional vector function, and $\omega = (\omega_1, \omega_2, ..., \omega_m)$ is a frequency basis. Assume that the angle coordinates $\varphi_1, \varphi_2, ..., \varphi_m$ and the normal coordinates $y_1, y_2..., y_s$ are introduced so that this system transforms (see (Mitropolsky, Samoilenko, and Tsydilo, 1977)) into

$$\frac{dy(t)}{dt} = -a_0(\varphi(t), \varphi(t-\Delta), y(t), y(t-\Delta))y(t)$$

$$-b_0(\varphi(t), \varphi(t-\Delta), y(t), y(t-\Delta))y(t-\Delta);$$

$$\frac{d\varphi(t)}{dt} = \omega, \tag{3.2}$$

where a_0 and b_0 are $(s \times s)$-dimensional matrices. Along with (3.2), we shall examine a perturbed system

$$\frac{dy(t)}{dt} = -a(\varphi(t), \varphi(t-\Delta), y(t), y(t-\Delta))y(t)$$

$$-b(\varphi(t), \varphi(t-\Delta), y(t), y(t-\Delta))y(t-\Delta) + c(\varphi(t), \varphi(t-\Delta))$$

$$\frac{d\varphi(t)}{dt} = \omega, \tag{3.3}$$

where $a - a_0$, $b - b_0$, and c are variables which are small in a certain sense. By virtue of the above discussion, it is clear that the investigation of the problem concerning the existence of quasiperiodic solutions for the system (3.3) can be reduced to the investigation of the problem of existence of periodic solutions for the system of partial differential equations

$$L(u)u = \sum_{v=1}^{m} \omega_v \frac{\partial u(\varphi)}{\partial \varphi_v} + a(\varphi, \varphi - \omega\Delta, u(\varphi), u(\varphi - \omega\Delta))u(\varphi)$$

$$+ b(\varphi, \varphi - \omega\Delta, u(\varphi), u(\varphi - \omega\Delta)) u(\varphi - \omega\Delta) = c(\varphi, \varphi - \omega\Delta). \qquad (3.4)$$

Let us show that a periodic solution of (3.4) can be constructed with the help of Bubnov-Galerkin's method.

For the system (3.4), we assume that $(s \times s)$-dimensional matrices $a(\varphi, \psi, y, z)$ and $b(\varphi, \psi, y, z)$ (here, $\psi = \varphi(t - \Delta)$ and $z = y(t - \Delta)$) and an s-dimensional vector function $c(\varphi, \psi)$ are periodic in φ and ψ with period 2π. We also assume that $a(\varphi, \psi, y, z)$, $b(\varphi, \psi, y, z) \in C^r(\mathfrak{M})$, where $\mathfrak{M}$ is the region $\varphi \in \mathfrak{T}_m$, $\psi \in \mathfrak{T}_m$, $\| y \| = \left(\sum_{i=1}^{s} y_i^2 \right)^{1/2} \le d$, $\| z \| \le d$, and that $c(\varphi, \psi) \in H^r(\mathfrak{T}_m \times \mathfrak{T}_m)$, where $r > m/2 + 2$. Clearly, the set $C^r(\mathfrak{M})$ becomes a Banach space if it is equipped with a norm

$$|f(\varphi, \psi, y, z)|_r = \max_{0 \le \rho \le r} |D^\rho f(\varphi, \psi, y, z)|_0,$$

where $D^\rho f$ is an arbitrary derivative of the ρth order, and

$$|f(\varphi, \psi, y, z)|_0 = \max_m \| f(\varphi, \psi, y, z) \|.$$

Consider a linear differential operator

$$L = L_1 + L_1^\Delta, \qquad (3.5)$$

where

$$L_1 u(\varphi) = \sum_{v=1}^{m} \omega_v \frac{\partial u}{\partial \varphi_v} + a(\varphi, \varphi - \omega\Delta) u(\varphi);$$

$$L_1^\Delta u(\varphi) = b(\varphi, \varphi - \omega\Delta) u(\varphi - \omega\Delta).$$

Here, $a(\varphi, \psi)$ and $b(\varphi, \psi)$ are $(s \times s)$-dimensional matrices belonging to the class $H^r(\mathfrak{T}_m \times \mathfrak{T}_m)$. By applying the Schwartz inequality to $(L_1^\Delta u, u)$, we find

$$(L_1^\Delta u, u)_r = (b(\varphi, \varphi - \omega\Delta) u(\varphi - \omega\Delta), u(\varphi))_r$$

$$\geq - \|b(\varphi, \varphi - \omega\Delta)u(\varphi - \omega\Delta)\|_{r-s} \|u(\varphi)\|_{r+s} .$$

For $r = s = 0$, this gives

$$(L_1^\Delta u, u)_0 \geq -\|b(\varphi, \varphi - \omega\Delta)u(\varphi - \omega\Delta)\|_0 \|u(\varphi)\|_0$$

$$\geq - \|b(\varphi, \varphi - \omega\Delta)\|_0 \|u(\varphi - \omega\Delta)\|_0 \|u(\varphi)\|_0 , \qquad (3.6)$$

where

$$|b(\varphi, \varphi - \omega\Delta)|_0 = \max_{\|u\|_0 = 1} (bu, u)_0 .$$

Since

$$\|u(\varphi, \varphi - \omega\Delta)\|_0^2 = \sum_n \left\langle u_\Delta^{(n)}, u_\Delta^{(-n)} \right\rangle = \sum_n \left\langle e^{i(n, \omega)\Delta} u^{(n)}, e^{-i(n, \omega)\Delta} u^{(1-n)} \right\rangle$$

$$= \sum_n \left\langle u^{(n)}, u^{(-n)} \right\rangle = \|u(\varphi)\|_0^2,$$

the inequality (3.6) can be rewritten in the form

$$(L_1^\Delta u, u)_0 \geq -|b(\varphi, \psi)|_0 \|u_0\|^2. \qquad (3.7)$$

By analogy, we can show that

$$(L_1^\Delta u, u)_r \geq -c_r |b(\varphi, \psi)|_r \|u(\varphi)\|_r^2. \qquad (3.8)$$

The following statement (similar to Lemma 3.2) holds for the operator L which is defined by (3.5).

Lemma 3.6. *Suppose that the operator L satisfies the conditions:*

(i) $a(\varphi, \psi), b(\varphi, \psi) \in C^r(\mathfrak{T}_m \times \mathfrak{T}_m);$

(ii) there exist numbers $c_r, \gamma_0,$ and γ_1 such that for an arbitrary vector $\eta = (\eta_1, \eta_2, ..., \eta_s)$ with $\|\eta\| = 1$, the following inequalities hold

$$\langle a(\varphi, \psi)\eta, \eta \rangle \geq 2\gamma_0, \quad \gamma_0 > 0,$$

$$|b(\varphi, \psi)|_0 < \gamma_0, \quad 2c_r |b(\varphi, \psi)|_r < \gamma_1.$$

Then we have

$$(Lu, u)_0 \geq (2\gamma_0 - |b(\varphi, \psi)|_0) \|u\|_0^2, \tag{3.9}$$

for an arbitrary function $u(\varphi) \in H^0(\mathcal{T}_m)$, *and*

$$(Lu, u)_r \geq (2\gamma_1 - c_r |b(\varphi, \psi)|_r) \|u\|_r^2 - \delta_1 (1 + \|a(\varphi, \psi)\|_r)^2, \tag{3.10}$$

for $u(\varphi) \in H^r(\mathcal{T}_m) \cap C^2(\mathcal{T}_m)$ *such that* $|u(\varphi)|_2 < 1$. *Here* γ_1 *and* δ_1 *are constants which depend on* $|a|_r$ *and do not depend on* $u(\varphi)$.

Parallel with (3.4), we consider a system of equations

$$L(\varepsilon u)u = \sum_{v=1}^{m} \omega_v \frac{\partial u(\varphi)}{\partial \varphi_v} + a(\varphi, \varphi - \omega\Delta, \varepsilon u(\varphi), \varepsilon u(\varphi - \omega\Delta))u(\varphi)$$

$$+ b(\varphi, \varphi - \omega\Delta, \varepsilon u(\varphi), \varepsilon u(\varphi - \omega\Delta)) u(\varphi - \omega\Delta) = c(\varphi, \varphi - \omega\Delta), \tag{3.11}$$

which can be easily obtained from (3.4) by substituting εu and εc (where ε is a small positive parameter) for u and c, respectively. We denote $\omega_N(\varphi) = \sum_{\|n\| \leq N} \omega_N^{(n)} e^{i(n, \varphi)}$ and write the system of equations for Galerkin's approximations

$$S_N(L(\varepsilon\omega_N(\varphi)) \omega(\varphi)) = S_N c(\varphi, \psi). \tag{3.12}$$

The existence of periodic solutions of the system of equations (3.11) is established by the following statement.

Theorem 3.4. (Mitropolsky, Martinyuk, and Dankanich, 1980; Martinyuk and Dankanich, 1981; Martinyuk, 1982b). *Suppose that the operator* $L(\varepsilon u)$ *satisfies the conditions:*

(i) $a(\varphi, \psi, y, z)$, $b(\varphi, \psi, y, z) \in C^r(\mathcal{M})$, $c(\varphi, \psi) \in H^r(\mathcal{T}_m \times \mathcal{T}_m)$, *where* $r > m/2 + 2$;

(ii) there exist positive numbers c_r, γ_0, *and* γ_1 *such that for an arbitrary vector* $\eta = (\eta_1, \eta_2, ..., \eta_s)$ *with* $\|\eta\| = 1$, *the inequalities*

$$\langle a(\varphi, \psi, 0, 0) \eta, \eta \rangle \geq 2\gamma_0, \quad \gamma_0 > 0;$$

$$|b(\varphi, \psi, 0, 0)|_0 < \gamma_0, \quad 2c_r |b(\varphi, \psi, 0, 0)|_r < \gamma_1. \tag{3.13}$$

hold.

Then there exist positive numbers ε_0, K, and c_0 such that the system of equations (3.12) possesses a solution for all $N \geq 1$ provided that $\varepsilon \in [0, \varepsilon_0]$ and $\|c(\varphi, \psi)\|_r \leq K$. This solution can be found with the help of the iterative process

$$S_N(L(\varepsilon\omega_N^{j-1}(\varphi))\,\omega_N^{j}(\varphi)) = S_N c(\varphi, \psi), \quad \omega_N^0 = 0, \quad j = 1, 2, 3, \dots. \tag{3.14}$$

As $N \to \infty$, the sequence $\omega_N(\varphi)$ converges in norm of the space $C^k(\mathfrak{T}_m)$, where $k = [r - m/2 - 2] \geq 1$, to the function $u^0(\varphi)$ which is the classical solution of the equation (3.11), and the rate of convergence is determined by the inequality

$$| u^0(\varphi) - \omega_N(\varphi) | \leq c_0 N^{-(r-1)} \| c \|_r. \tag{3.15}$$

Proof. First, we examine the properties of the linear operator $L(\varepsilon u)$ assuming that $\omega(\varphi) \in C^\infty(\mathfrak{T}_m), |\,\omega(\varphi)\,|_2 < 1$ and that $L(0)$ satisfies conditions of Theorem 3.4. Let $|\varepsilon\omega(\varphi)|_0 \leq d$. Then

$$|a(\varphi, \varphi - \omega\Delta, \varepsilon\omega(\varphi), \varepsilon\omega(\varphi - \omega\Delta)) - a(\varphi, \varphi - \omega\Delta, 0, 0)|_0 \leq 2\varepsilon s|a|_1;$$

$$|b(\varphi, \varphi - \omega\Delta, \varepsilon\omega(\varphi), \varepsilon\omega(\varphi - \omega\Delta)) - b(\varphi, \varphi - \omega\Delta, 0, 0)|_0 \leq 2\varepsilon s|b|_1, \tag{3.16}$$

and for sufficiently small ε_0, we have

$$\langle a(\varphi, \varphi - \omega\Delta, \varepsilon\omega(\varphi), \varepsilon\omega(\varphi - \omega\Delta))\eta, \eta \rangle \geq 2\gamma_0, \quad \gamma_0 > 0$$

$$|b(\varphi, \varphi - \omega\Delta, \varepsilon\omega(\varphi), \varepsilon\omega(\varphi - \omega\Delta))|_0 < \gamma_0;$$

$$2c_r |b(\varphi, \varphi - \omega\Delta, \varepsilon\omega(\varphi), \varepsilon\omega(\varphi - \omega\Delta))|_r < \gamma_1. \tag{3.17}$$

This implies that the operator $L(\varepsilon\omega(\varphi))$ satisfies conditions of Lemma 3.6, i.e., for all $u(\varphi) \in H^r(\mathfrak{T}_m) \cap C^2(\mathfrak{T}_m)$ such that $|\,u(\varphi)\,|_2 < 1$, we have

$$(L(\varepsilon\omega(\varphi))u, u)_0 \geq (2\gamma_0 - |b|_0)\|u\|_0^2, \tag{3.18}$$

$$(L(\varepsilon\omega(\varphi))u, u)_r \geq (2\gamma_1 - c_r|b|_r)\|u\|_r^2 - \delta_1(1 + \|a\|_r), \tag{3.19}$$

where γ_0, γ_1, and δ_1 do not depend on $\omega(\varphi)$ and $u(\varphi)$.

According to Lemma 3.1, there exist constants $c_1 > 0$ and $c_2 > 0$ independent of a, b, and ω, such that the inequalities

$$\|a(\varphi, \varphi - \omega\Delta, \varepsilon\omega(\varphi), \varepsilon\omega(\varphi - \omega\Delta))\|_r \leq c_1 |a|_r (1 + \varepsilon\|\omega\|_r);$$

$$\|b(\varphi, \varphi - \omega\Delta, \varepsilon\omega(\varphi), \varepsilon\omega(\varphi - \omega\Delta))\|_r \leq c_2 |b|_r (1 + \varepsilon\|\omega\|_r) \qquad (3.20)$$

hold. By virtue of (3.20), the inequality (3.19) can be rewritten as follows

$$(L(\varepsilon\omega(\varphi))u, u)_r \geq (2\gamma_1 - c_2 c_r |b|_r (1 + \varepsilon\|\omega(\varphi)\|_r))\|u\|_r^2$$

$$- - \delta_2(1 + \varepsilon\|\omega(\varphi)\|_r)^2, \qquad (3.21)$$

where δ_2 does not depend on $\omega(\varphi)$ and $u(\varphi)$.

Assume that the trigonometric polynomial $\omega_N^{j-1}(\varphi)$ satisfies the following inequalities

$$\left\|\omega_N^{j-1}(\varphi)\right\|_0 \leq \frac{\|c(\varphi, \psi)\|_0}{2\gamma_0 - |b|_0};$$

$$\left\|\omega_N^{j-1}(\varphi)\right\|_r \leq \delta\left(1 + \frac{1}{2} + \ldots + \frac{1}{2^{j-1}}\right) < 2\delta; \qquad (3.22)$$

$$\left\|\omega_N^{j-1}(\varphi)\right\|_r < 1,$$

where

$$\delta = 3\sqrt{\frac{\delta_2}{2\gamma_l - |b_0|}}.$$

We now prove that the system (3.14) is always solvable with respect to $\omega_N^j(\varphi)$. For this purpose, we show that for $S_N c(\varphi, \psi) = 0$ it has only the trivial solution. After scalar multiplication of (3.14) by $\omega_N^j(\varphi)$, we obtain

$$S_N\left(L\left(\varepsilon\omega_N^{j-1}\right)\omega_N^j, \omega_N^j\right)_0 = \left(S_N c(\varphi, \psi), \omega_N^j\right)_0.$$

The operator S_N is selfadjoint with respect to the zero scalar product, and thus,

$$\left(L\left(\varepsilon\omega_N^{j-1}\right)\omega_N^j, S_N\omega_N^j\right)_0 = \left(c(\varphi, \psi), S_N\omega_N^j\right)_0;$$

since S_N is an identical operator for trigonometrical polynomials $\omega_N^j(\varphi)$, we find

$$\left(L\left(\varepsilon\omega_N^{j-1}\right)\omega_N^j, \omega_N^j\right)_0 = \left(c(\varphi, \psi), \omega_N^j\right)_0. \tag{3.23}$$

By virtue of (3.22), the operator $L\left(\varepsilon\omega_N^{j-1}\right)$ satisfies the inequalities (3.18) and (3.19). Therefore, for $c(\varphi, \psi) = 0$, the relation (3.23) gives

$$0 = S_N\left(L\left(\varepsilon\omega_N^{j-1}\right)\omega_N^j, \omega_N^j\right)_0 = \left(L\left(\varepsilon\omega_N^{j-1}\right)\omega_N^j, \omega_N^j\right)_0 \geq (2\gamma_0 - |b|_0)\|\omega_N^j\|_0^2,$$

and hence, $\omega_N^j = 0$. Consequently, the system (3.14) is solvable for an arbitrary right-hand side, and thus, by virtue of the Schwartz inequality and (3.18), we have

$$(2\gamma_0 - |b|_0)\|\omega_N^j(\varphi)\|_0^2 < \left(L\left(\varepsilon\omega_N^{j-1}(\varphi)\right)\omega_N^j(\varphi), \omega_N^j(\varphi)\right)_0$$

$$= \left(c(\varphi, \psi), \omega_N^j(\varphi)\right)_0 \leq \|c(\varphi, \psi)\|_0\|\omega_N^j(\varphi)\|_0. \tag{3.24}$$

This yields

$$\left\|\omega_N^j(\varphi)\right\|_0 \leq \frac{\|c(\varphi, \psi)\|_0}{2\gamma_0 - |b|_0}. \tag{3.25}$$

Similarly, by using (3.19) and the Schwartz inequality, we get

$$\left(2\gamma_1 - c_2 c_r |b|_r\left(1 + \varepsilon\|\omega_N^{j-1}(\varphi)\|_r\right)\right)\|\omega_N^j(\varphi)\|_r^2$$

$$-\delta_2\left(1 + \varepsilon\|\omega_N^{j-1}(\varphi)\|_r\right)^2 \leq \|c\|_r\|\omega_N^j(\varphi)\|_r. \tag{3.26}$$

For $\varepsilon \leq \varepsilon_0 \leq 1/2\delta$ with properly chosen δ, (3.26) implies the following estimate

$$\left\|\omega_N^j\right\|_r \leq \frac{\|c\|_r + \sqrt{\|c\|_r^2 + 4\left(2\gamma_1 - c_2 c_r |b|_r\left(1 + \varepsilon\|\omega_N^j\|_r\right)\delta_2\left(1 + \varepsilon\|\omega_N^{j-2}\|_r^2\right)\right)}}{2\left(2\gamma_1 - c_2 c_r |b|_r\left(1 + \varepsilon\|\omega_N^{j-1}\|_r\right)\right)}$$

$$-\delta\left(1 + \varepsilon\|\omega_N^{j-1}\|_r\right) \leq \delta\left(1 + \frac{1}{2} + \ldots + \frac{1}{2\delta}\right) < 2\delta.$$

By virtue of Theorem 3.1 and the inequality (1.8), we obtain

$$\left|\omega_N^j(\varphi)\right|_2 \le c_3 \left\|\omega_N^j(\varphi)\right\|_{r-1} \le 2c_3 \left\|\omega_N^j(\varphi)\right\|_0^{1-(r-1)/r} \left\|\omega_N^j(\varphi)\right\|_r^{(r-1)/r}$$

$$\le 2c_3 \left(\frac{\|c(\varphi,\psi)\|_0}{2\gamma_0 - |b|_0}\right)^{1/r} (2\delta)^{(r-1)/r} \le 1 \qquad (3.27)$$

for sufficiently small $\|c(\varphi,\psi)\|$. Hence, the inequalities (3.22) are always valid for all j = 1, 2, 3, ...

Let us prove that the sequence $\{\omega_N^j(\varphi)\}$ converges as $j \to \infty$. By employing the relation (3.14), we find

$$S_N\left(L\left(\varepsilon\omega_N^j\right)\omega^{j+1}\right) - S_N\left(L\left(\varepsilon\omega_N^j\right)\omega_N^j\right) = S_N\left(L\left(\varepsilon\omega_N^j\right)\omega_N^{j+1}\right)$$

$$- S_N\left(L\left(\varepsilon\omega_N^j\right)\omega_N^j\right) + S_N\left(L\left(\varepsilon\omega_N^j\right)\omega_N^j\right) - S_N\left(L\left(\varepsilon\omega_N^{j-1}\right)\omega_N^j\right)$$

$$= S_N\left(L\left(\varepsilon\omega_N^j\right)\left(\omega_N^{j+1} - \omega_N^j\right)\right) = S_N\left(\left(L\left(\varepsilon\omega_N^{j-1}\right) - L\left(\varepsilon\omega_N^j\right)\right)\omega_N^j\right). \qquad (3.28)$$

By denoting $v_N^{j+1} = \omega_N^{j+1} - \omega_N^j$, we can transform (3.28) as follows

$$S_N\left(L\left(\varepsilon\omega_N^j\right)v_N^{j+1}\right) = S_N\left(\left(L\left(\varepsilon\omega_N^{j-1}\right) - L\left(\varepsilon\omega_N^j\right)\right)\omega_N^j\right). \qquad (3.29)$$

Taking the inequality (3.18) into account, we estimate (3.29) and obtain the following estimate

$$\left\|v_N^{j+1}\right\|_0 \le \frac{\left\|S_N\left(\left(L\left(\varepsilon\omega_N^{j-1}\right) - L\left(\varepsilon\omega_N^j\right)\right)\omega_N^j\right)\right\|_0}{2\gamma_0 - |b|_0}$$

$$\le \varepsilon\left(\frac{|a(\varphi,\psi,0,0)|_0 + |b(\varphi,\psi,0,0)|_0}{2\gamma_0 - |b|_0}\right)\left\|\omega_N^j\right\|_0. \qquad (3.30)$$

It follows from this inequality that for sufficiently small δ_0 the sequence $\{\omega_N^j\}$ is fundamental in $H^0(\mathcal{T}_m)_0$ as $j \to 0$. Therefore, the solution of the equation (3.12) exists for arbitrary $N > 1$, and the inequalities

$$\left\|\omega_N(\varphi)\right\|_r \le 2\delta; \quad \left\|\omega_N(\varphi)\right\|_0 \le \frac{\|c(\varphi, \psi)\|_0}{2\gamma_0 - |b|_0}; \quad \left|\omega_N(\varphi)\right|_2 < 1$$

hold. We now prove that the sequence $\{\omega_N(\varphi)\}$ converges in $H^0(\mathcal{T}_m)$ as $N \to \infty$. By using the equation (3.12), we find

$$S_{N+1}\left(L(\varepsilon\omega_{N+1})\omega_{N+1}\right) - S_N\left(L(\varepsilon\omega_N)\omega_N\right) \equiv S_{N+1}\left(L(\varepsilon\omega_{N+1})\omega_{N+1}\right)$$

$$- S_{N+1}\left(L(\varepsilon\omega_{N+1})\omega_N\right) + S_{N+1}\left(L(\varepsilon\omega_{N+1})\omega_N\right) - S_{N+1}\left(L(\varepsilon\omega_N)\omega_N\right)$$

$$+ S_{N+1}\left(L(\varepsilon\omega_N)\omega_N\right) - S_N\left(L(\varepsilon\omega_N)\omega_N\right) \equiv S_{N+1}\left(L(\varepsilon\omega_{N+1})(\omega_{N+1} - \omega_N)\right)$$

$$+ S_{N+1}\left((L(\varepsilon\omega_{N+1}) - L(\varepsilon\omega_N))\omega_N\right) + (S_{N+1} - S_N)\left(L(\varepsilon\omega_N)\omega_N\right)$$

$$= (S_{N+1} - S_N)\,c(\varphi, \psi) \ . \tag{3.31}$$

This yields

$$S_{N+1}\left(L(\varepsilon\omega_{N+1})(\omega_{N+1} - \omega_N)\right) = - S_{N+1}\left(L(\varepsilon\omega_{N+1}) - L(\varepsilon\omega_N)\right)\omega_N$$

$$- (S_{N+1} - S_N)\left(c(\varphi, \psi) - L(\varepsilon\omega_N)\omega_N\right). \tag{3.32}$$

It follows from (3.32) that $\omega_{N+1} - \omega_N$ is Galerkin's approximation to the solution of the non-linear equation with the right-hand side α equal to the right-hand side of (3.32). Thus, taking (3.22) into account, we obtain

$$\|\omega_{N+1} - \omega_N\| \le \frac{\|\alpha\|_0}{2\gamma_0 - |b|_0}. \tag{3.33}$$

We now estimate $\|\alpha\|_0$:

$$\left\| S_{N+1}\left(L(\varepsilon\omega_{N+1}) - L(\varepsilon\omega_N)\right)\omega_N \right\|_0$$

$$\le c\left\|(L(\varepsilon\omega_{N+1}) - L(\varepsilon\omega_N))\omega_N\right\|_0 \le \varepsilon c_4\left\|\omega_{N+1} - \omega_N\right\|_0. \tag{3.34}$$

According to the inequalities (1.11), (1.13), and (1.14), we find

$$\left\|(S_{N+1} - S_N)\left(c(\varphi, \psi) - L(\varepsilon\omega_N)\omega_N\right)\right\|_0$$

$$\leq c_5 N^{-(r-1)} \left\| c(\varphi, \psi) - L(\varepsilon\omega_N)\omega_N \right\|_{r-1}$$

$$\leq c_5 N^{-(r-1)} \left[\| c(\varphi, \psi) \|_{r-1} + c_6 \| c(\varphi, \psi) \|_r \right] \leq c_7 N^{-(r-1)}. \tag{3.35}$$

By employing (3.34) and (3.35), we can transform the inequality (3.33) as follows

$$\| \omega_{N+1}(\varphi) - \omega_N(\varphi) \|_0 \leq \varepsilon c_8 \| \omega_{N+1}(\varphi) - \omega_N(\varphi) \|_0 + c_9 N^{-(r-1)},$$

and, for sufficiently small ε, we finally obtain

$$\| \omega_{N+1}(\varphi) - \omega_N(\varphi) \|_0 \leq 2c_9 N^{-(r-1)}. \tag{3.36}$$

The inequality (3.36) indicates that the sequence $\{\omega_N(\varphi)\}$ converges in $H^0(\mathfrak{T}_m)$ as $N \to \infty$, since $r > m/2 + 2$. According to Theorem 3.2, the sequence $\{\omega_N(\varphi)\}$ is compact in $H^{r-1}(\mathfrak{T}_m)$. Furthermore, it is convergent in $H^{r-1}(\mathfrak{T}_m)$, because otherwise one can find on $H^{r-1}(\mathfrak{T}_m)$ two subsequences $\omega_{N_j}(\varphi)$ and $\omega_{N_{j_1}}(\varphi)$ which converge to two different functions from $H^{r-1}(\mathfrak{T}_m)$, namely, to $\omega^0(\varphi)$ and $\omega_1^0(\varphi)$, respectively. But it follows from the statements proved above that the Fourier coefficients of these functions coincide, and this means that the sequence $\{\omega_N(\varphi)\}$ is convergent on $H^{r-1}(\mathfrak{T}_m)$. By employing Theorem 3.1, we find that this sequence is convergent in $C^k(\mathfrak{T}_m)$ for $k = [r -2 - m/2] \geq 1$.

We now estimate the rate of convergence of Bubnov-Galerkin's approximations. Assume that the approximations $\omega_N(\varphi)$ are defined by the relations

$$S_N(L(\varepsilon\omega_N(\varphi))\,\omega(\varphi)) = S_N c(\varphi, \psi) \tag{3.37}$$

and the sequence $\{\omega_N(\varphi)\}$ converges to a function $\omega^0(\varphi) \in H^{r-1}(\mathfrak{T}_m)$. Let us estimate the difference $\omega_N(\varphi) - \omega^0(\varphi)$. For a function $f(\varphi) = \sum_n f^{(n)} e^{i(n,\,\varphi)}$, we denote $R_N f = \sum_{\|n\| \leq N} f^{(n)} e^{i(n,\,\varphi)}$. Then

$$R_n = E - S_N, \tag{3.38}$$

and the following estimate is true

$$\| R_n f \|_s \leq N^{s-r} \| f \|_r, \quad s \leq r. \tag{3.39}$$

Since $\omega^0(\varphi)$ is a solution of the equation (3.11), we have

$$L(\varepsilon\omega^0(\varphi))\,\omega^0(\varphi) = c(\varphi, \psi), \tag{3.40}$$

and employing (3.37), (3.38), and (3.40), we obtain

$$L(\varepsilon\omega^0(\varphi))\,\omega^0(\varphi) - L(\varepsilon\omega_N(\varphi))\,\omega_N(\varphi) = R_N\,c(\varphi, \psi)$$

$$- R_n L(\varepsilon\omega_N(\varphi))\,\omega_N(\varphi),$$

or

$$L(\varepsilon\omega^0(\varphi))\,(\omega^0(\varphi) - \omega_N(\varphi)) = R_N\,c(\varphi, \psi) - R_n L(\varepsilon\omega_N(\varphi))\,\omega_N(\varphi)$$

$$+ (L(\varepsilon\omega_N(\varphi)) - L(\varepsilon\omega^0(\varphi)))\,\omega_N(\varphi). \tag{3.41}$$

Multiplying scalarly (3.41) by $\omega^0(\varphi) - \omega_N(\varphi)$, we get

$$(L(\varepsilon\omega^0(\varphi))\,(\omega^0 - \omega_N),\,\omega^0 - \omega_N)_0 = R_n\,((c(\varphi, \psi) - L(\varepsilon\omega_N(\varphi))\,\omega_N),\,\omega^0 - \omega_N)_0$$

$$+ ((L(\varepsilon\omega_N(\varphi)) - L(\varepsilon\omega^0(\varphi)))\,\omega_N),\,\omega^0 - \omega_N)_0 \tag{3.42}$$

Applying the inequality (3.18) and the Schwartz inequality to the relation (3.42) and taking into account the estimate (3.39), we find

$$(2\gamma_0 - |b|_0)\|\omega^0(\varphi) - \omega_N(\varphi)\|_0^2 \leq N^{-(r-1)}\|c(\varphi, \psi)\|_{r-1}$$

$$+ \|\,L(\varepsilon\omega_N(\varphi)\omega_N)\|\,\|\,\omega^0(\varphi) - \omega_N(\varphi)\|_0 + \|\,\|\,L(\varepsilon\omega_N(\varphi))$$

$$- L(\varepsilon\omega^0(\varphi))\|_0\,\|\,\omega_N(\varphi)\|_0\,\|\,\omega^0(\varphi) - \omega_N(\varphi)\|_0. \tag{3.43}$$

Taking into account the boundedness of $\omega_N(\varphi)$, we obtain

$$\|\,L(\varepsilon\omega_N(\varphi)) - L(\varepsilon\omega^0(\varphi))\|_0\,\|\,\omega_N(\varphi)\|_0 \leq \varepsilon c_4\,\|\,\omega^0(\varphi) - \omega_N(\varphi)\|_0. \tag{3.44}$$

Since

$$\|\,L(\varepsilon\omega_N(\varphi))\omega_N\,\|_{r-1} \leq c'\,\|\,\omega_N\,\|_r,$$

and

$$\|\,\omega_N(\varphi)\|_r \leq c''\,\|\,c(\varphi, \psi)\|_r, \tag{3.45}$$

the inequality (3.43) takes the form

$$\| \omega^0(\varphi) - \omega_N(\varphi)\|_0 \leq \varepsilon c_8 \| \omega^0(\varphi) - \omega_N(\varphi)\|_0 + c_9 N^{-(r-1)} \| c \|_r. \tag{3.46}$$

For sufficiently small ε, we have

$$\| \omega^0(\varphi) - \omega_N(\varphi)\|_0 \leq c_0 N^{-(r-1)} \| c(\varphi, \psi)\|_r. \tag{3.47}$$

Similarly, one can easily prove that

$$\| \omega^0(\varphi) - \omega_N(\varphi)\|_s \leq c_0 N^{s-(r-1)} \| c \|_r, \quad s > m/2 + 2.$$

4. EXISTENCE OF INVARIANT TOROIDAL MANIFOLDS FOR SYSTEMS WITH LAG. INVESTIGATION OF THE BEHAVIOR OF TRAJECTORIES IN THEIR VICINITIES

In this chapter, we present results obtained in the investigation of invariant manifolds for different types of nonlinear systems with lag. The results concerning the behavior of these systems in the vicinity of these manifolds are also presented here.

Note that invariant manifolds for systems with lag have been considered by different authors (Zverkin, (1970); Mitropolsky and Lykova, (1974); Neimark, (1975); Fodchuk, (1965, 1970); Hale, (1966b); Halanay, (1967)). In the works of Samoilenko (1970a,b; 1975), a new method has been proposed for the investigation of invariant toroidal manifolds; namely, Green's function method for problems of invariant tori. This method enables new theorems to be proved on the existence of toroidal manifolds for systems with lag and, furthermore, it gives the algorithm for constructing these manifolds.

§1. Existence of Invariant Toroidal Manifolds With Loss of Smoothness

Consider a system of differential equations

$$\frac{dy(t)}{dt} = -b(\varphi(t), \varphi(t-\Delta), y(t), y(t-\Delta), \mu)\, y(t)$$

$$-\, b_1(\varphi(t), \varphi(t-\Delta), y(t), y(t-\Delta), \mu)\, y(t-\Delta) + c(\varphi(t), \varphi(t-\Delta), \mu);$$

$$\frac{dy(t)}{dt} = a(\varphi(t), y(t)),$$

(1.1)

135

where the functions $y = (y_1, ..., y_n)$, $\varphi = (\varphi_1, ..., \varphi_m)$, a, b, b_1, and c are periodic in φ and $\psi = \varphi(t - \Delta)$ with period 2π; they are defined for all y, $z = y(t - \Delta)$, and μ belonging to the region

$$|y| = \left(\sum_{i=1}^{n} y_i^2 \right)^{1/2} \leq d, \ \| z \| \leq d, \ \mu \in [0, \mu_0]. \tag{1.2}$$

Here, Δ is a constant which characterizes a lag in the system, and μ is a small positive parameter.

Let us consider the problem of the existence of invariant tori for the system (1.1), under the assumption that $c(\varphi, \psi, 0) = 0$, i. e., we assume that for $\mu = 0$ the system (1.1) possesses an invariant torus $y = 0$. The invariant manifold $\mathcal{T}(\mu)$ of (1.1) will be sought for in the form

$$y = u(\varphi, \mu) \tag{1.3}$$

where $u(\varphi, \mu)$ is a continuous function periodic in φ with period 2π.

Let $\varphi_t = \varphi_t(\varphi)$ and $\varphi_\tau(\varphi) = \varphi$ be a solution of the system of equations

$$\frac{d\varphi}{dt} = a(\varphi, u(\varphi, \mu), \mu), \tag{1.4}$$

where τ and φ are arbitrary constants and $u(\varphi, \mu)$ is some function periodic in φ with period 2π. The function $u(\varphi, \mu)$ defines an invariant manifold of (1.1) if for all $-\infty < t < \infty$ the identity

$$\begin{aligned} \frac{du(\varphi_t, \mu)}{dt} &\equiv - b(\varphi_t, \varphi_{t\Delta}, u(\varphi_t, \mu), u(\varphi_{t\Delta}, \mu), \mu) u(\varphi_t, \mu) \\ &- b_1(\varphi_t, \varphi_{t\Delta}, u(\varphi_t, \mu), u(\varphi_{t\Delta}, \mu), \mu) u(\varphi_{t-\Delta}, \mu) + c(\varphi_t, \varphi_{t\Delta}, \mu) \end{aligned} \tag{1.5}$$

holds.

In order to find the torus $\mathcal{T}(\mu)$, we use the iteration method which enables us to find $\mathcal{T}(\mu)$ as the limit of a sequence of tori $\mathcal{T}^0(\mu) = \mathcal{T}(0)$, $\mathcal{T}^2(\mu)$, ... , $\mathcal{T}^i(\mu)$, each is an invariant torus

$$\mathcal{T}^{i+1}(\mu): y = u^{i+1}(\varphi, \mu), \ i = 0, 1, 2, ... \tag{1.6}$$

of the corresponding system of differential equations

$$\frac{dy(t)}{dt} = -b(\varphi(t), \varphi(t-\Delta), u^i(\varphi(t), \mu), u^i(\varphi(t-\Delta), \mu), \mu)\, y(t)$$

$$-b_1(\varphi(t), \varphi(t-\Delta), u^i(\varphi(t), \mu), u^i(\varphi(t-\Delta), \mu), \mu)\, u^i(\varphi(t-\Delta), \mu) \qquad (1.7)$$

$$+ c(\varphi(t), \varphi(t-\Delta), \mu).$$

This method of finding $\mathcal{T}(\mu)$ was justified by Martinyuk and Samoilenko (1974a).

Lemma 4.1. *Suppose that functions* $a(\varphi, y, \mu)$, $b(\varphi, \psi, y, z, \mu)$, $b_1(\varphi, \psi, y, z, \mu)$, *and* $c(\varphi, \psi, \mu)$ *are continuous in* φ, ψ, y, *and* z *for* $\| y \| \le d, \| z \| \le d, \mu \in [0, \mu_0]$ *and periodic in* φ, ψ *with period* 2π. *Then a limiting function* $u^i = \lim\limits_{i \to \infty} u^i(\varphi, \mu)$ *defines the invariant torus* $\mathcal{T}(\mu) : y = u(\varphi, \mu)$ *of the system* (1.1) *provided that the sequence* (1.6) *converges uniformly for every* $\mu \in [0, \mu_0]$.

Proof. Since $y = u^{i+1}(\varphi, \mu)$ is the invariant torus of the system (1.7), the relations

$$\varphi_t^i = \varphi + \int\limits_\tau^t a\Big(\varphi_t^i, u_t^i \big(\varphi_t^i, \mu \big), \mu \Big) dt;$$

$$(1.8)$$

$$u^{i+1}\big(\varphi_t^i, \mu\big) = u^{i+1}(\varphi, \mu) + \int\limits_\tau^t \Big[-b\big(\varphi_t^i, \varphi_{t-\Delta}^i, u^i\big(\varphi_t^i, \mu\big), u^i\big(\varphi_{t-\Delta}^i, \mu\big), \mu\big) u^{i+1}\big(\varphi_t^i, \mu\big)$$

$$- b_1\big(\varphi_t^i, \varphi_{t-\Delta}^i, u^i\big(\varphi_t^i, \mu\big), u^i\big(\varphi_{t-\Delta}^i, \mu\big), \mu\big) u^i\big(\varphi_{t-\Delta}^i, \mu\big) + c\big(\varphi^i, \varphi_{t-\Delta}^i, \mu\big) \Big] dt$$

hold for the trajectories φ_t^i, y_t^i which are situated on it.

The function $a(\varphi, y, \mu)$ is periodic and continuous for $\| y \| \le d$, and thus, by virtue of the inequality $\| u^i \| \le d$, the relations (1.8) imply that the sequence $\varphi_t^i, i = 0, 1, \ldots$ is uniformly bounded and equicontinuous for t from an arbitrary bounded interval T of the real axis $R : -\infty < t < \infty$. Thus, the sequence φ_t^i contains a subsequence $\varphi_t^{i_k}, k = 1, 2, \ldots$ uniformly convergent on T. Let us denote the limit of this subsequence by φ_t, i.e.,

$$\varphi_t = \lim\limits_{k \to \infty} \varphi_t^{i_k} \qquad (1.9)$$

and pass in (1.8) to the limit for $i = i_k$ (under the assumption that the functions a, b, b_1, c, and u^i are continuous we can change all the positions of the limit signs). As a result, we obtain the identities

$$\varphi_t = \varphi + \int_\tau^t a\big(\varphi_t, u(\varphi_t, \mu), \mu\big)\, dt, \quad (t \in T),$$

(1.10)

$$u(\varphi_t, \mu) \equiv u(\varphi, \mu) + \int_\tau^t \Big[-b\big(\varphi_t, \varphi_{t-\Delta}, u(\varphi_t, \mu), u(\varphi_{t-\Delta}, \mu), \mu\big) u(\varphi_t, \mu)$$

$$-b_1\big(\varphi_t, \varphi_{t-\Delta}, u(\varphi_t, \mu), u(\varphi_{t-\Delta}, \mu), \mu\big) u(\varphi_{t-\Delta}, \mu) + c(\varphi_t, \varphi_{t-\Delta}, \mu)\Big]\, dt$$

which imply that on every bounded interval T (and therefore, for all $t \in (-\infty, \infty)$) the continuous periodic function $u(\varphi, \mu)$ satisfies the system of equations (1.4) and (1.5). The manifold $y = u(\varphi, \mu)$ is thus the invariant torus of the system (1.1).

Hence, the problem of the existence of an invariant torus for (1.1) is connected with the problem of existence of an invariant torus for (1.7).

Let us now find under what conditions (1.7) possesses an invariant torus. Consider the system of equations

$$\frac{dy(t)}{dt} = -b(\varphi_t(\varphi), \varphi_{t-\Delta}(\varphi), u^i(\varphi_t(\varphi), \mu), u^i(\varphi_{t-\Delta}(\varphi), \mu), \mu)\, y(t)$$

$$- b_1(\varphi_t(\varphi), \varphi_{t-\Delta}(\varphi), u^i(\varphi_t(\varphi), \mu), u^i(\varphi_{t-\Delta}(\varphi), \mu), \mu)\, u^i(\varphi_{t-\Delta}(\varphi), \mu) \qquad (1.11)$$

$$+ c(\varphi_t(\varphi), \varphi_{t-\Delta}(\varphi), \mu),$$

where $u^i(\varphi, \mu)$ is a given function periodic in φ with a period 2π and $\varphi_t(t) = \varphi_t^i(\varphi)$ $(\varphi_0(\varphi) = \varphi)$ is the general solution of the first equation of (1.7).

Denote by $G^i(t, \tau, \varphi, \mu)$ a Green's function for the problem concerning the bounded solutions of the system (1.11) and assume that this function exists. Then the relation

$$y_t(\varphi, \mu) = \int_{-\infty}^{\infty} G^i(t, \tau, \varphi, \mu)\, c_1(\varphi_\tau(\varphi), \varphi_{\tau-\Delta}(\varphi), \mu)\, d\tau, \qquad (1.12)$$

where

$$c_1(\varphi_t(\varphi), \varphi_{t-\Delta}(\varphi), \mu) = c(\varphi_t(\varphi), \varphi_{t-\Delta}(\varphi), \mu)$$

$$- b_1(\varphi_t(\varphi), \varphi_{t-\Delta}(\varphi), u^i(\varphi_t(\varphi), \mu), u^i(\varphi_{t-\Delta}(\varphi), \mu), \mu)\, u^i(\varphi_{t-\Delta}(\varphi), \mu),$$

defines a family of bounded solutions of (1.11), which depends on φ and μ (these are the parameters). These solutions fill in the invariant set

$$\mathcal{T}^{i+1}(\mu) : y = u^i(\varphi, \mu) \equiv \int_{-\infty}^{\infty} G^i(0, \tau, \varphi, \mu)\, c_1(\varphi_\tau(\varphi), \varphi_{\tau-\Delta}(\varphi), \mu)\, d\tau, \qquad (1.13)$$

which is an invariant toroidal set of the system of equations (1.7).

The function $G^i(0, \tau, \varphi, \mu)$ which defines the invariant set $y = u^{i+1}(\varphi, \mu)$ of the system (1.11) according to (1.13) is called the Green's function for the problem of the invariant tori of the system (1.7).

The smoothness of this function with respect to φ and μ ensures the smoothness of the invariant tori $\mathcal{T}^i(\mu)$, whereas the small difference between the Green's function for the system (1.10) with $i = k+1$ and $i = k$ involves the convergence of the sequence of tori $\mathcal{T}^i(\mu)$, as $i \to \infty$. Let us determine the general form of this function.

Consider a system of equations

$$\frac{d\varphi}{dt} = a(\varphi, \mu), \quad \frac{dy}{dt} = -b(\varphi, \mu)y.$$

Let us denote by $\Omega_\tau^t(\varphi, \mu)$ a fundamental matrix of solutions of the linear system

$$\frac{dy}{dt} = -b(\varphi_t(\varphi, \mu), \mu)\, y,$$

where $\varphi_t(\varphi, \mu)$ is the solution of the first equation of this system, $\varphi_0(\varphi, \mu) = \varphi$, $\Omega_\tau^\tau(\varphi, \mu)$ $= E$, where E is the unit matrix. Let $\varphi_t(\varphi, \mu)$ be periodic in each $\varphi_1, \ldots, \varphi_m$ with period 2π.

Consider the function

$$G_0(\tau, \varphi, \mu) = \begin{cases} \Omega_\tau^0(\varphi,\mu)\, c_1\big(\varphi_\tau(\varphi,\mu), \varphi_{\tau-\Delta}(\varphi,\mu), \mu\big) & \text{for } \tau < 0, \\ \Omega_\tau^0(\varphi,\mu)\big(c_1\big(\varphi_\tau(\varphi,\mu), \varphi_{\tau-\Delta}(\varphi,\mu), \mu\big) - E\big) & \text{for } \tau > 0, \end{cases}$$

where $c_1(\varphi_t, \varphi_{t-\Delta}, \mu)$ is a matrix periodic in φ_t and $\varphi_{t-\Delta}$ with period 2π. Assume that the matrix $G_0(\tau, \varphi, \mu)$ is such that

$$\int_{-\infty}^{\infty} \|G_0(\tau, \varphi, \mu)\|\, d\tau < K < \infty.$$

Then the function

$$G(t, \tau, \varphi, \mu) = \begin{cases} \Omega_\tau^t(\varphi,\mu)c_1\big(\varphi_\tau(\varphi,\mu),\varphi_{\tau-\Delta}(\varphi,\mu),\mu\big) & \text{for } \tau<0, \\ \Omega_\tau^t(\varphi,\mu)\big(c_1\big(\varphi_\tau(\varphi,\mu),\varphi_{\tau-\Delta}(\varphi,\mu),\mu\big)-E\big) & \text{for } \tau>0, \end{cases}$$

is a Green's function for the problem concerning the bounded solutions of the system

$$\frac{dy}{dt} = - b(\varphi_t(\varphi, \mu), \mu)\, y + c_1(\varphi_t(\varphi, \mu), \varphi_{t-\Delta}(\varphi, \mu), \mu).$$

Therefore, $G_0(\tau, \varphi, \mu) = G(0, \tau, \varphi, \mu)$ is the Green's function for the problem of the invariant tori; moreover $G_0(\tau, \varphi, \mu)$ is the Green's function for the problem under consideration provided that it is periodic in φ with period 2π and

$$\int_{-\infty}^{\infty} \| G_0(\tau,\varphi,\mu) \|\, d\tau \le K < \infty.$$

Let us now consider the problem of the existence of the sequence of tori (1.6). In the system of equations

$$\frac{d\varphi(t)}{dt} = a_0(\varphi(t)) + a_1(\varphi(t)),$$

$$(1.14)$$

$$\frac{dy(t)}{dt} = -[b_0(\varphi(t), \varphi(t-\Delta)) + b_1^0(\varphi(t), \varphi(t-\Delta))]\, y(t) + c_0(\varphi(t), \varphi(t-\Delta)),$$

$$a_0, b_0, a_1, b_1^0, c_0 \in C^r\left(\mathbb{T}_m \times \mathbb{T}_m : 0 \le \varphi_j \le 2\pi, 0 < \psi_j \le 2\pi, i = 1, \ldots, m\right);$$

a_0 and b_0 are fixed functions, whereas a_1, b_1^0, and c_0 are arbitrary (but small in the sense of the norm in $C^r(\mathbb{T}_m \times \mathbb{T}_m)$) functions.

Lemma 4.2. (Martinyuk and Samoilenko, 1974). *Suppose that one can find an integer r and positive M, γ, and K such that for all $a_1(\varphi)$ and $b_1^0(\varphi, \psi)$ satisfying the inequality*

$$\max\left\{|a_1|_r, |b_1^0|_r\right\} \le M, \qquad (1.15)$$

where $|b_1^0|_r = \max_{|u|_r=1} |b_1^0 u|_r$, the system of equations (1.14) possesses the Green's function for the problem of invariant tori $G_0(\tau,\varphi)$ which satisfies the inequality

$$| G_0(\tau, \varphi)\, c_0(\varphi_\tau(\varphi),\, \varphi_{\tau-\Delta}(\varphi))\, |_r \; \leq K\, e^{-\gamma|\tau|}\, |c_0(\varphi, \psi)|_r \qquad (1.16)$$

for every $\tau \in (-\infty, \infty)$. *Then the system* (1.14) *has the invariant torus* $\mathfrak{T} : u = u(\varphi)$ *for which the following estimate holds*

$$| u(\varphi) |_r \; \leq \; \frac{2K}{\gamma}\, | c_0(\varphi, \psi)\, |_r. \qquad (1.17)$$

We now employ this lemma to construct the sequence of the tori (1.6). For this purpose, we set

$$a_0(\varphi) \; = \; a(\varphi, 0, 0);\; b_0(\varphi, \psi) = b(\varphi, \psi, 0, 0, 0);$$

$$a^1(\varphi, y, \mu) \; = \; a(\varphi, y, \mu) - a(\varphi, 0, 0); \qquad (1.18)$$

$$b^1(\varphi, \psi, y, z, \mu) \; = \; b(\varphi, \psi, y, z, \mu) - b(\varphi, \psi, 0, 0, 0),$$

and assume that $c(\varphi, \psi, \mu)$, $a(\varphi, y, \mu)$, $b(\varphi, \psi, y, z, \mu)$, $b_1(\varphi, \psi, y, z, \mu) \in C^r(\mathfrak{M})$ for all $0 \leq \mu \leq \mu_0$, where $\mathfrak{M}$ is the region $\| y \| \leq d,\; \| z \| \leq d,\; (\varphi, \psi) \in \mathfrak{T}_m \times \mathfrak{T}_m$. Assume also that the inequalities

$$\max_\mu \left\{ \left| a^1(\varphi, y, \mu) \right|_r, \left| b^1(\varphi, \psi, y, z, \mu) \right|_r, \left| c(\varphi, \psi, \mu) \right|_r \right\} \leq L_r(d, \mu_0), \qquad (1.19)$$

$$\frac{2K}{\gamma} | b_1 |_r \leq R$$

hold. Here, $L_r\, (d,\, \mu_0)$ is a positive monotonically decreasing function of one of the arguments (when the value of the second argument is fixed) such that $L_r(d,\, \mu_0) \to 0$ as $d \to 0,\, \mu_0 \to 0, R < 1$.

The existence of the sequence of tori (1.6) is established by:

Theorem 4.1. (Martinyuk and Samoilenko, 1974, 1976). *Suppose that the functions* $a_0(\varphi)$ *and* $b_0(\varphi,\, \mu)$ *are such that the system* (1.14) *satisfies conditions of Lemma 4.2. Then one can find* $\mu^0 (0 \leq \mu^0 \leq \mu^0)$ *such that for* $\mu \leq \mu^0$ *the sequence of the systems* (1.7) *defines the sequence of the invariant tori* (1.6), *each belonging to the space* $C^r(\mathfrak{T}_m)$ *and satisfying the inequality*

$$| u^i\, (\varphi, \mu)\, |_r \; \leq \; \frac{2L_r(0, \mu^0)K}{\gamma(1 - R)} \leq d. \qquad (1.20)$$

Proof. Since $u^0 = 0$, Theorem 4.1 holds for $i = 0$. Assume that it is valid for $i = k$; let us show that the invariant torus $\mathcal{T}^{i+1}(\mu) : y = u^{k+1}(\mu)$ exists and satisfies the inequality (1.20). For this purpose, we put $i = k$ in (1.8) and obtain the set of equations for the torus $y = u^{k+1}(\varphi, \mu)$

$$\frac{d\varphi_t^k}{dt} = a\big(\varphi_t^k, u^k(\varphi_t^k, \mu), \mu\big);$$

$$\frac{dy(t)}{dt} = -b\big(\varphi_t^k, \varphi_{t-\Delta}^k, u^k(\varphi_t^k, \mu), u^k(\varphi_{t-\Delta}^k, \mu), \mu\big)y(t)$$

$$-b_1\big(\varphi_t^k, \varphi_{t-\Delta}^k, u^k(\varphi_t^k, \mu), u^k(\varphi_{t-\Delta}^k, \mu), \mu\big)u^k(\varphi_{t-\Delta}^k, \mu) + c\big(\varphi_t^k, \varphi_{t-\Delta}^k, \mu\big).$$

(1.21)

These relations can be rewritten as follows

$$\frac{d\varphi_t^k}{dt} = a_0\big(\varphi_t^k\big) + a^1\big(\varphi_t^k, u^k(\varphi_t^k, \mu), \mu\big);$$

$$\frac{dy(t)}{dt} = -[b_0\big(\varphi_t^k, \varphi_{t-\Delta}^k\big) + b^1\big(\varphi_t^k, \varphi_{t-\Delta}^k, u^k(\varphi_t^k, \mu), u^k(\varphi_{t-\Delta}^k, \mu), \mu\big)]y(t)$$

$$+ c_1\big(\varphi_t^k, \varphi_{t-\Delta}^k, u^k(\varphi_t^k, \mu), u^k(\varphi_{t-\Delta}^k, \mu), \mu\big)$$

(1.22)

Using the fact that the functions a^1, b^1 and $u^k(\varphi, \mu)$ satisfy the inequalities (1.19) and (1.20), respectively, one can prove that the inequality

$$\max_{\mu} \left\{ \left| a^1\big(\varphi_t^k, u^k(\varphi_t^k, \mu), \mu\big) \right|_r, \left| b^1\big(\varphi_t^k, \varphi_{t-\Delta}^k, u^k(\varphi_t^k, \mu), u^k(\varphi_{t-\Delta}^k, \mu), \mu\big) \right|_r \right\}$$

$$\leq 2L_r \left(\frac{2L_r(0, \mu^0)K}{\gamma(1-R)}, \mu_0 \right).$$

(1.23)

is valid for sufficiently small μ^0 and all $\mu \leq \mu^0$. Since $L_r(0, \mu_0) \to 0$ as $\mu_0 \to 0$, one can always consider μ^0 to be so small that the inequalities

$$2L_r \left(\frac{2L_r(0, \mu^0)K}{\gamma(1-R)}, \mu_0 \right) \leq M,$$

$$\frac{2L_r\left(0, \mu^0\right)K}{\gamma(1-R)} \le d, \tag{1.24}$$

hold together with (1.23). For this choice of μ^0 and all $\mu \le \mu^0$, the lemma which establishes the existence of the torus $\mathcal{C}^{k+1}(\mu) : y = u^{k+1}(\varphi, \mu)$ such that

$$|u^{k+1}(\varphi, \mu)|_r \le \frac{2K}{\gamma}|c_1(\varphi, \psi, z, \mu)|_r \le \frac{2K}{\gamma}[|c|_r + |b_1|_r |u^k_r], \quad \mu \le \mu^0, \tag{1.25}$$

can be applied to the system (1.22).

Since $|c(\varphi, \psi, \mu)|_r \le L_r(0, \mu_0)$, the estimate (1.20) follows from the inequality (1.25).

Let us illustrate the choice of μ^0 by taking the case $r = 0$ as an example. We assume that the functions $a(\varphi, y, \mu)$ and $b(\varphi, \psi, y, z, \mu)$ are continuously differentiable with respect to the arguments φ, ψ, y, z, and μ in the region $(\varphi, \psi) \in \mathcal{C}_m \times \mathcal{C}_m$, $\|y\| \le d, \|z\| \le d, 0 \le \mu \le \mu^0$. We have

$$\max_{\mu \in [0, \mu_0]}\left\{\left|a^1(\varphi, y, \mu)\right|_0, \left|b^1(\varphi, \psi, y, z, \mu)\right|_0\right\} \le \max_{\mu \in [0, \mu_0]}\left\{\left|a(\varphi, y, \mu) - a(\varphi, 0, \mu)\right|_0\right.$$

$$+ \left|a(\varphi, 0, \mu) - a(\varphi, 0, 0)\right|_0; \left|b(\varphi, \psi, y, z, \mu) - b(\varphi, \psi, 0, z, \mu)\right|_0 + |b(\varphi, \psi, 0, z, \mu)$$

$$- b(\varphi, \psi, 0, 0, \mu)|_0 + \left|b(\varphi, \psi, 0, 0, \mu) - b(\varphi, \psi, 0, 0, 0)\right|_0\right\} \le \max\left\{l|a|_1 d + \right.$$

$$+ \max_{\mu \in [0, \mu_0]}\left|\frac{\partial a}{\partial \mu}\right|\mu; 2l|b|_1 d + \max_{\mu \in [0, \mu_0]}\left|\frac{\partial b}{\partial \mu}\right|\mu\right\} \le l_1(d + \mu_0), \tag{1.26}$$

where l_1 is a positive constant. Consequently,

$$L_0(d, \mu_0) = l_1(d + \mu_0). \tag{1.27}$$

The inequality (1.23) thus takes the form

$$\max_{\mu}\left\{\left|a^1\left(\varphi^k_t, u^k\left(\varphi^k_t, \mu\right), \mu\right)\right|_0; \left|b^1\left(\varphi^k_t, \varphi^k_{t-\Delta}, u^k\left(\varphi^k_t, \mu\right), u^k\left(\varphi^k_{t-\Delta}, \mu\right), \mu\right)\right|_0\right\}$$

$$\leq l_1 \left(\frac{2K l_1 \mu^0}{\gamma(1-R)} + \mu^0 \right). \tag{1.28}$$

By virtue of (1.20) and (1.26), it is valid for all μ^0. Hence, the value of μ^0 should be chosen from the inequalities (1.24). In the case $r = 0$, these have the simple form

$$l_1 \left(\frac{2K l_1 \mu^0}{\gamma(1-R)} + \mu^0 \right) = c_1 \mu^0 \leq M,$$

$$\frac{2K l_1 \mu^0}{\gamma(1-R)} = c_2 \mu^0 \leq d. \tag{1.29}$$

And one can easily show that for

$$\mu^0 \leq \min \left\{ \frac{M}{c_1} ; \frac{d}{c_2} \right\} \tag{1.30}$$

these inequalities hold.

Let us now prove the convergence of the sequence of the invariant tori (1.7) in the norm of the space $C^{r-1}(\mathbb{C}_m)$ for $r \geq 1$. Since the function $u^{k+1}(\varphi, \mu)$ defines the smooth manifold of the system (1.8) with $i = k$, the following identity holds:

$$\frac{du^{k+1}\left(\varphi_t^k, \mu\right)}{dt} \equiv \frac{\partial u^{k+1}\left(\varphi_t^k, \mu\right)}{\partial \varphi_t^k} a\left(\varphi_t^k, u^k\left(\varphi_t^k, \mu\right), \mu\right)$$

$$\equiv -b\left(\varphi_t^k, \varphi_{t-\Delta}^k, u^k\left(\varphi_t^k, \mu\right), u^k\left(\varphi_{t-\Delta}^k, \mu\right), \mu\right) u^{k+1}\left(\varphi_t^k, \mu\right)$$

$$-b_1\left(\varphi_t^k, \varphi_{t-\Delta}^k, u^k\left(\varphi_t^k, \mu\right), u^k\left(\varphi_{t-\Delta}^k, \mu\right), \mu\right) u^k\left(\varphi_{t-\Delta}^k, \mu\right) + c\left(\varphi_t^k, \varphi_{t-\Delta}^k, \mu\right). \tag{1.31}$$

For $t = 0$, this identity takes the form

$$\frac{\partial u^{k+1}(\varphi, \mu)}{\partial \varphi} a\left(\varphi, u^k(\varphi, \mu), \mu\right) \equiv -b\left(\varphi, \varphi_{-\Delta}^k(\varphi), u^k(\varphi, \mu), u^k\left(\varphi_{-\Delta}^k(\varphi), \mu\right), \mu\right) u^{k+1}(\varphi, \mu)$$

$$-b_1\left(\varphi, \varphi_{-\Delta}^k(\varphi), u^k(\varphi, \mu), u^k\left(\varphi_{-\Delta}^k(\varphi), \mu\right), \mu\right) u^k\left(\varphi_{-\Delta}^k(\varphi), \mu\right) + c\left(\varphi, \varphi_{-\Delta}^k(\varphi), \mu\right),$$

where $\varphi_{-\Delta}^k(\varphi) = \varphi_t^k(\varphi)\big|_{t=-\Delta}$ and $\varphi_0^k(\varphi) = \varphi$. We set $u^{k+1}(\varphi, \mu) - u^k(\varphi, \mu) = w^{k+1}(\varphi, \mu)$.

For $w^{k+1}(\varphi, \mu)$, we obtain the identity

$$\frac{\partial w^{k+1}(\varphi,\mu)}{\partial \varphi}\, a\big(\varphi, u^k(\varphi,\mu),\mu\big) + \frac{\partial u^k(\varphi,\mu)}{\partial \varphi}\Big[a\big(\varphi, u^k(\varphi,\mu),\mu\big) - a\big(\varphi, u^{k-1}(\varphi,\mu),\mu\big)\Big]$$

$$\equiv - b\Big(\varphi,\, \varphi^k_{-\Delta}(\varphi), u^k(\varphi,\mu), u^k\big(\varphi^k_{-\Delta}(\varphi),\mu\big),\mu\Big)\, w^{k+1}(\varphi,\mu)$$

$$+\, [\, b\Big(\varphi,\, \varphi^k_{-\Delta}(\varphi),\, u^k(\varphi,\mu),\, u^k\big(\varphi^k_{-\Delta}(\varphi),\mu\big),\mu\Big)$$

$$-\, b\Big(\varphi,\, \varphi^{k-1}_{-\Delta}(\varphi),\, u^{k-1}(\varphi,\mu),\, u^{k-1}\big(\varphi^{k-1}_{-\Delta}(\varphi),\mu\big),\mu\Big)\,]\, u^k(\varphi,\mu)$$

$$-\, b_1\Big(\varphi,\, \varphi^k_{-\Delta}(\varphi), u^k(\varphi,\mu), u^k\big(\varphi^k_{-\Delta}(\varphi),\mu\big),\mu\Big)\, u^k\big(\varphi^k_{-\Delta}(\varphi),\mu\big)$$

$$+\, b_1\Big(\varphi,\, \varphi^{k-1}_{-\Delta}(\varphi), u^{k-1}(\varphi,\mu), u^{k-1}\big(\varphi^{k-1}_{-\Delta}(\varphi),\mu\big),\mu\Big)\, u^{k-1}\big(\varphi^{k-1}_{-\Delta}(\varphi),\mu\big)$$

$$+\, c\Big(\varphi,\, \varphi^k_{-\Delta}(\varphi),\, \mu\Big) - c\Big(\varphi, \varphi^{k-1}_{-\Delta}(\varphi), \mu\Big),. \tag{1.32}$$

which can be rewritten as follows

$$\frac{\partial w^{k+1}(\varphi,\mu)}{\partial \varphi}\, a\big(\varphi, u^k(\varphi,\mu),\mu\big) + b\Big(\varphi,\, \varphi^k_{-\Delta}(\varphi), u^k(\varphi,\mu), u^k\big(\varphi^k_{-\Delta}(\varphi),\mu\big),\mu\Big)\, w^{k+1}(\varphi,\mu)$$

$$=\frac{\partial u^k(\varphi,\mu)}{\partial \varphi}\Big[a\big(\varphi, u^{k-1}(\varphi,\mu),\mu\big) - a\big(\varphi, u^k(\varphi,\mu),\mu\big)\Big]$$

$$+\, [\, b\Big(\varphi,\, \varphi^k_{-\Delta}(\varphi),\, u^k(\varphi,\mu),\, u^k\big(\varphi^k_{-\Delta}(\varphi),\mu\big),\mu\Big)$$

$$-\, b\Big(\varphi,\, \varphi^{k-1}_{-\Delta}(\varphi), u^k(\varphi,\mu), u^k\big(\varphi^k_{-\Delta}(\varphi),\mu\big),\mu\Big)\Big]\, u^k(\varphi,\mu)$$

$$+\, [\, b\Big(\varphi,\, \varphi^{k-1}_{-\Delta}(\varphi),\, u^k(\varphi,\mu),\, u^k\big(\varphi^k_{-\Delta}(\varphi),\mu\big),\mu\Big)$$

$$-\, b\Big(\varphi,\, \varphi^{k-1}_{-\Delta}(\varphi), u^{k-1}(\varphi,\mu), u^k\big(\varphi^k_{-\Delta}(\varphi),\mu\big),\mu\Big)\Big]\, u^k(\varphi,\mu)$$

$$+\, [\, b\Big(\varphi,\, \varphi^{k-1}_{-\Delta}(\varphi),\, u^{k-1}(\varphi,\mu),\, u^k\big(\varphi^k_{-\Delta}(\varphi),\mu\big),\mu\Big)$$

$$-\, b\Big(\varphi,\, \varphi^{k-1}_{-\Delta}(\varphi), u^{k-1}(\varphi,\mu), u^{k-1}\big(\varphi^k_{-\Delta}(\varphi),\mu\big),\mu\Big)\Big]\, u^k(\varphi,\mu)$$

$$+ \left[b\left(\varphi, \varphi_{-\Delta}^{k-1}(\varphi), u^{k-1}(\varphi, \mu), u^{k-1}\left(\varphi_{-\Delta}^{k}(\varphi), \mu\right), \mu \right) \right.$$

$$\left. - b\left(\varphi, \varphi_{-\Delta}^{k-1}(\varphi), u^{k-1}(\varphi,\mu), u^{k-1}\left(\varphi_{-\Delta}^{k-1}(\varphi),\mu\right),\mu \right) \right] u^{k}(\varphi,\mu)$$

$$+ b_1\left(\varphi, \varphi_{-\Delta}^{k}(\varphi), u^{k}(\varphi,\mu), u^{k}\left(\varphi_{-\Delta}^{k}(\varphi), \mu\right), \mu \right) (u^{k-1}\left(\varphi_{-\Delta}^{k}(\varphi), \mu\right) - u^{k}(\varphi_{-\Delta}^{k}(\varphi), \mu))$$

$$+ b_1\left(\varphi, \varphi_{-\Delta}^{k-1}(\varphi), u^{k-1}(\varphi, \mu), u^{k-1}\left(\varphi_{-\Delta}^{k-1}(\varphi), \mu\right), \mu \right) (u^{k-1}\left(\varphi_{-\Delta}^{k-1}(\varphi), \mu\right)$$

$$- u^{k-1}(\varphi_{-\Delta}^{k}(\varphi), \mu)) + \left[b_1\left(\varphi, \varphi_{-\Delta}^{k-1}(\varphi), u^{k-1}(\varphi, \mu), u^{k-1}\left(\varphi_{-\Delta}^{k-1}(\varphi), \mu\right), \mu \right) \right.$$

$$\left. - b_1\left(\varphi, \varphi_{-\Delta}^{k}(\varphi), u^{k-1}(\varphi,\mu), u^{k-1}\left(\varphi_{-\Delta}^{k-1}(\varphi),\mu\right),\mu \right) \right] u^{k-1}\left(\varphi_{-\Delta}^{k}(\varphi),\mu\right)$$

$$+ \left[b_1\left(\varphi, \varphi_{-\Delta}^{k}(\varphi), u^{k-1}(\varphi, \mu), u^{k-1}\left(\varphi_{-\Delta}^{k-1}(\varphi), \mu\right), \mu \right) \right.$$

$$\left. - b_1\left(\varphi, \varphi_{-\Delta}^{k}(\varphi), u^{k}(\varphi,\mu), u^{k-1}\left(\varphi_{-\Delta}^{k-1}(\varphi),\mu\right),\mu \right) \right] u^{k-1}\left(\varphi_{-\Delta}^{k}(\varphi),\mu\right)$$

$$+ \left[b_1\left(\varphi, \varphi_{-\Delta}^{k}(\varphi), u^{k}(\varphi, \mu), u^{k}\left(\varphi_{-\Delta}^{k-1}(\varphi), \mu\right), \mu \right) \right.$$

$$\left. - b_1\left(\varphi, \varphi_{-\Delta}^{k}(\varphi), u^{k}(\varphi, \mu), u^{k}\left(\varphi_{-\Delta}^{k}(\varphi), \mu\right), \mu \right) \right] u^{k-1}\left(\varphi_{-\Delta}^{k}(\varphi), \mu\right)$$

$$+ c\left(\varphi, \varphi_{-\Delta}^{k}(\varphi), \mu \right) - c\left(\varphi, \varphi_{-\Delta}^{k-1}(\varphi), \mu \right).$$

Denote the right-hand side of the last identity by $c_k(\varphi, \mu)$. Then

$$\frac{\partial w^{k+1}(\varphi,\mu)}{\partial \varphi} \left[a(\varphi, 0, 0) + a(\varphi, u^{k}(\varphi, \mu), \mu) - a(\varphi, 0, 0) \right]$$

$$+ \left[b(\varphi, \varphi_{-\Delta}^{k}(\varphi), 0, 0, 0) + \left(b\left(\varphi, \varphi_{-\Delta}^{k}(\varphi), u^{k}(\varphi, \mu), u^{k}\left(\varphi_{-\Delta}^{k}(\varphi), \mu\right), \mu \right) \right. \right.$$

$$- b(\varphi, \varphi_{-\Delta}^{k}(\varphi), 0, 0, 0)) \right] w^{k+1}(\varphi, \mu) = c_k(\varphi, \mu). \tag{1.33}$$

Equation (1.33) implies that $y = w^{k+1}(\varphi, \mu)$ is the invariant torus of the system

$$\frac{d\varphi}{dt} = a(\varphi, \theta, 0) + (a(\varphi, u^{k}(\varphi, \mu), \mu) - a(\varphi, 0, 0));$$

$$\frac{dy(t)}{dt} = - \left[b(\varphi, \varphi_{-\Delta}^{k}(\varphi), 0, 0, 0) + \left(b\left(\varphi, \varphi_{-\Delta}^{k}(\varphi), u^{k}(\varphi,\mu), u^{k}\left(\varphi_{-\Delta}^{k}(\varphi),\mu\right),\mu \right) \right. \right.$$

$$- b(\varphi, \varphi_{-\Delta}^{k}(\varphi), 0, 0, 0)) \right] y(t) + c_k(\varphi, \mu). \tag{1.34}$$

This system has the same form as (1.22) and we can thus write $y = w^{k+1}(\varphi, \mu)$ in terms of the Green's function of the problem of the invariant tori for the system (1.34)

$$y = w^{k+1}(\varphi, \mu) = \int_{-\infty}^{\infty} G_0^k(\tau, \varphi) \, c_k(\varphi_\tau(\varphi), \mu) \, d\tau.$$

Moreover, by virtue of (1.17), the estimate

$$w^{k+1}(\varphi, \mu)|_0 \leq \frac{2K}{\gamma} |c_k(\varphi, \mu)|_0$$

holds for $\mu \leq \mu_0$.

Using the definition of the function $c_k(\varphi, \mu)$, we get, for $w^{k+1}(\varphi, \mu)$,

$$|w^{k+1}(\varphi, \mu)|_0 \leq \frac{2K}{\gamma} \Big\{ l |u^k|_1 |a|_1 |w^k(\varphi, \mu)|_0 + l |u^k|_0 |b|_1 |\varphi_{-\Delta}^k(\varphi) - \varphi_{-\Delta}^{k-1}(\varphi)|_0$$

$$+ 2l |u^k|_0 |b|_1 |w^k(\varphi, \mu)|_0 + l |b|_1 |u^{k-1}|_1 |\varphi_{-\Delta}^k(\varphi) - \varphi_{-\Delta}^{k-1}(\varphi)|_0 |u^k|_0$$

$$+ |b_1|_0 |w^k(\varphi, \mu)|_0 + l l |b_1|_0 + |u^{k-1}|_1 |\varphi_{-\Delta}^k(\varphi) - \varphi_{-\Delta}^{k-1}(\varphi)|_0$$

$$+ l |b_1|_1 |u^{k-1}|_0 |\varphi_{-\Delta}^k(\varphi) - \varphi_{-\Delta}^{k-1}(\varphi)|_0 + 2l |b_1|_1 |u^{k-1}|_0 |w^k(\varphi, \mu)|_0$$

$$+ l^2 |b_1|_1 |u^{k-1}|_0 |\varphi_{-\Delta}^k(\varphi) - \varphi_{-\Delta}^{k-1}(\varphi)|_0 |u^k|_1 + l |c|_1 |\varphi_{-\Delta}^k(\varphi) - \varphi_{-\Delta}^{k-1}(\varphi)|_0 \Big\}. \quad (1.35)$$

We now estimate the difference $\varphi_{-\Delta}^k(\varphi) - \varphi_{-\Delta}^{k-1}(\varphi)$. Taking into account the first relation of (1.8), we find

$$\varphi_t^k - \varphi_t^{k-1} = \int_0^t \Big[a\big(\varphi_t^k, u^k(\varphi_t^k, \mu), \mu\big) - a\big(\varphi_t^{k-1}, u^{k-1}(\varphi_t^{k-1}, \mu), \mu\big) \Big] \, dt. \quad (1.36)$$

This yields

$$|\varphi_t^k - \varphi_t^{k-1}|_0 \leq \int_0^{|t|} \Big| a\big(\varphi_t^k, u^k(\varphi_t^k, \mu), \mu\big) - a\big(\varphi_t^{k-1}, u^{k-1}(\varphi_t^{k-1}, \mu), \mu\big) \Big| \, dt$$

$$\leq \int_0^{|t|} \Big[\Big| a\big(\varphi_t^k, u^k(\varphi_t^k, \mu), \mu\big) - a\big(\varphi_t^{k-1}, u^k(\varphi_t^k, \mu), \mu\big) \Big|$$

$$+\left|a\left(\varphi_t^{k-1}, u^k\left(\varphi_t^k, \mu\right), \mu\right) - a\left(\varphi_t^{k-1}, u^{k-1}\left(\varphi_t^k, \mu\right), \mu\right)\right|$$

$$+\left|a\left(\varphi_t^{k-1}, u^{k-1}\left(\varphi_t^k, \mu\right), \mu\right) - a\left(\varphi_t^{k-1}, u^{k-1}\left(\varphi_t^{k-1}, \mu\right), \mu\right)\right|\right]dt$$

$$\leq \int_0^{|t|}\left[l|a|_1\left|\varphi_t^k - \varphi_t^{k-1}\right|_0 + l|a|_1\left|u^k - u^{k-1}\right|_0 + l^2|a|_1\left|u^{k-1}\right|_1\left|\varphi_t^k - \varphi_t^{k-1}\right|_0\right]dt$$

$$\leq l\,|a|_1|w^k(\varphi, \mu)|_0\,|t| + k_1\int_0^{|t|}\left|\varphi_t^k - \varphi_t^{k-1}\right|_0 dt. \qquad (1.37)$$

where $k_1 = l\,|a|_1(1 + ld)$. The inequality (1.37) implies

$$\left|\varphi_{-\Delta}^k - \varphi_{-\Delta}^{k-1}\right| \leq \frac{1}{1+ld}\left|w^k(\varphi, \mu)\right|_0\left(e^{k_1\Delta} - 1\right). \qquad (1.38)$$

Taking into account (1.38), we can rewrite (1.35) as follows:

$$\left|w^k(\varphi, \mu)\right|_0 \leq \frac{2K}{\gamma}\left\{ld|a|_1 + \frac{l|b|_1 d}{1+ld}\left(e^{k_1\Delta} - 1\right) + 2l|b|_1 d + \frac{l^2|b|_1 d^2}{1+ld}\left(e^{k_1\Delta} - 1\right)\right.$$

$$+ l|b|_0 + \frac{l|b|_1 d}{1+ld}\left(e^{k_1\Delta} - 1\right) + 2l|b_1|_1 d + \frac{l^2|b_1|_1 d^2}{1+ld}\left(e^{k_1\Delta} - 1\right)$$

$$\left.+ \frac{l|c|}{1+ld}\left(e^{k_1\Delta} - 1\right)\right\}\left|w^k(\varphi, \mu)\right|_0. \qquad (1.39)$$

Using the estimates (1.23) and (1.25), we find from (1.39) that

$$\left|w^{k+1}(\varphi, \mu)\right|_0 \leq \rho_0\left|w^k(\varphi, \mu)\right|_0 \leq \rho_0^{k-1}\left|u^1(\varphi, \mu)\right|_0 \leq \rho_0^{k-1}\frac{2K}{\gamma}\left|c(\varphi, \psi, \mu)\right|_0, \qquad (1.40)$$

where ρ_0 is a positive constant, less than unity for small μ. The last inequality implies the convergence of the sequence of the invariant tori (1.6) in the space $C_0(\mathcal{C}_m)$ and the continuity of the limiting function with respect to μ at the point $\mu = 0$.

Let

$$\lim_{k \to \infty} u^k(\varphi, \mu) = u(\varphi, \mu). \qquad (1.41)$$

We now show that $u(\varphi, \mu) \in C^{r-1}(\mathcal{T}_m)$. Taking into account the uniform boundedness and the uniform continuity of the sequence of functions $\{D^\rho u^k(\varphi, \mu)\}$ which consists of the partial derivatives of the functions $u^k(\varphi, \mu)$ (of the order $\rho \leq r-1$), we conclude, using Arzela's lemma, that any infinite subsequence of $\{D^\rho u^k(\varphi, \mu)\}$ converges uniformly to some function $D^\rho u^k(\varphi, \mu)$. The last statement and (1.41) together imply that $u(\varphi, \mu)$ has continuous derivatives of the order $\rho \leq r-1$.

By using Lemma 4.1 and the convergence of the sequence $\{u_k(\varphi, \mu)\}$ in $C^{r-1}(\mathcal{T}_m)$, one can prove the following theorem on the existence of an invariant torus for the perturbed system (1.1) (Martinyuk and Samoilenko, 1974).

Theorem 4.2. *Suppose that the right-hand side of the system*

$$\frac{dy(t)}{dt} = -b(\varphi(t), \varphi(t-\Delta), y(t), y(t-\Delta), \mu)\, y(t)$$

$$-b_1(\varphi(t), \varphi(t-\Delta), y(t), y(t-\Delta), \mu)\, y(t-\Delta) + c(\varphi(t), \varphi(t-\Delta), \mu);$$

$$\frac{d\varphi(t)}{dt} = a(\varphi(t), y(t), \mu)$$

(1.42)

satisfies the conditions :

(i) for all $0 \leq \mu \leq \mu_0$,

$$a(\varphi, y, \mu), b(\varphi, \psi, y, z, \mu), b_1(\varphi, \psi, y, z, \mu), c(\varphi, \psi, \mu) \in C^r(\mathcal{M}),$$

where $\mathcal{M}$ is the region $\| y \| \leq d, \| z \| \leq d, \varphi \in \mathcal{T}_m, \psi \in \mathcal{T}_m$;

(ii)

$$\max_{\mu} \{ |\, a(\varphi, y, \mu) - a(\varphi, 0, 0) \,|_r \,;\, |\, b(\varphi, \psi, y, z, \mu)$$

$$-b(\varphi, \psi, 0, 0, 0) \,|_r \,;\, |\, c(\varphi, \psi, \mu) \,|_r \} \leq L_r(d, \mu_0);$$

$$\frac{2K}{\gamma} |\, b_1 |_r \leq R,$$

where $L_r(d, \mu_0)$ is a positive monotonically decreasing function of one of its arguments (when the value of the second argument is fixed); moreover $L_r(d, \mu_0) \to 0$ when

$d \to 0, \mu_0 \to 0, R < 1$;

   (iii) for arbitrary functions $a_1(\varphi)$, $b_1^0(\varphi)$, and $c_0(\varphi, \psi)$ sufficiently small in the norm of $C^r(\mathfrak{T} \times \mathfrak{T})$, a system of equations

$$\frac{d\varphi(t)}{dt} = a(\varphi(t), 0, 0) + a_1(\varphi(t));$$

$$(1.43)$$

$$\frac{dy(t)}{dt} = -[b(\varphi(t), \varphi(t-\Delta), 0, 0, 0) + b_1^0(\varphi(t), \varphi(t-\Delta))]\, y(t) + c_0(\varphi(t), \varphi(t-\Delta))$$

possesses the Green's function $G_0(\tau, \varphi)$ of the problem on the invariant tori, satisfying, for all $\tau \in (-\infty, \infty)$, the inequality

$$|\, G_0(\tau, \varphi)\, c_0(\varphi_\tau(\varphi), \psi_\tau(\varphi))\, |_r \le K\, e^{-\gamma|\tau|}\, |\, c_0(\varphi, \psi)\, |_r\, ,$$

$$(1.44)$$

where K and γ are some positive constants. Then one can find $0 \le \mu^0 \le \mu_0$ such that for all $\mu \in [0, \mu^0]$ the system of equations (1.42) has an invariant torus $\mathfrak{T}(\mu)$: $y = u(\varphi, \mu), u(\varphi, \mu) \in C^{r-1}(\mathfrak{T}_m)$ for which

$$\lim_{\mu \to 0} |\, u(\varphi, \mu)\, |_{r-1} = 0.$$

$$(1.45)$$

§2. Existence of Lipschitz Tori for Nonlinear Systems with Lag

Consider a system of differential equations

$$\frac{dy(t)}{dt} = -b(\varphi(t), \varphi(t-\Delta), y(t), y(t-\Delta))\, y(t)$$

$$-b_1(\varphi(t), \varphi(t-\Delta), y(t), y(t-\Delta))\, y(t-\Delta) + c(\varphi(t), \varphi(t-\Delta));$$

$$(2.1)$$

$$\frac{d\varphi(t)}{dt} = a(\varphi(t), y(t)),$$

where the functions $\varphi = (\varphi_1, ..., \varphi_m), y = (y_1, ..., y_n), a, b, b_1$, and c are periodic in φ

and $\psi = \varphi\,(t - \Delta)$ with period 2π; they are defined in the region

$$(\varphi, \psi, y, z) \in (\mathfrak{C} \times \mathfrak{C} \times d \times d). \tag{2.2}$$

Assume that the right-hand side of (2.1) satisfies the following conditions in the region (2.2):

(i) $\qquad\qquad a(\varphi, y), b(\varphi, \psi, y, z), b_1(\varphi, \psi, y, z), c(\varphi, \psi) \in C^1(\mathfrak{M});$

(ii)   the derivatives of the vector function $a(\varphi, y)$ and the matrices $b(\varphi, \psi, y, z)$ and $b_1(\varphi, \psi, y, z)$ satisfy the Lipschitz conditions

$$\left\| \frac{\partial a(\varphi', y')}{\partial \varphi} - \frac{\partial a(\varphi'', y'')}{\partial \varphi} \right\| \le K_1 \left\| \varphi' - \varphi'' \right\| + K_2 \left\| y' - y'' \right\|;$$

$$\left\| \frac{\partial a(\varphi', y')}{\partial \varphi} - \frac{\partial a(\varphi'', y'')}{\partial \varphi} \right\| \le K_1 \left\| \varphi' - \varphi'' \right\| + K_2 \left\| y' - y'' \right\|;$$

$$\left\| \frac{\partial b(\varphi', \psi', y', z')}{\partial \varphi_r} - \frac{\partial b(\varphi'', \psi'', y'', z'')}{\partial \varphi_r} \right\| \le K_{5r} \left\| \varphi' - \varphi'' \right\| + K_{5r}^\Delta \left\| \psi' - \psi'' \right\|$$
$$+ K_{6r} \left\| y' - y'' \right\| + K_{6r}^\Delta \left\| z' - z'' \right\|;$$

$$\left\| \frac{\partial b(\varphi', \psi', y', z')}{\partial \psi_r} - \frac{\partial b(\varphi'', \psi'', y'', z'')}{\partial \psi_r} \right\| \le L_{7r} \left\| \varphi' - \varphi'' \right\| + L_{7r}^\Delta \left\| \psi' - \psi'' \right\|$$
$$+ L_{8r} \left\| y' - y'' \right\| + L_{8r}^\Delta \left\| z' - z'' \right\|;$$

$$\left\| \frac{\partial b(\varphi', \psi', y', z')}{\partial y_k} - \frac{\partial b(\varphi'', \psi'', y'', z'')}{\partial y_k} \right\| \le K_{9k} \left\| \varphi' - \varphi'' \right\| + K_{9k}^\Delta \left\| \psi' - \psi'' \right\|$$
$$+ K_{10k} \left\| y' - y'' \right\| + K_{10k}^\Delta \left\| z' - z'' \right\|;$$

$$\left\| \frac{\partial b(\varphi', \psi', y', z')}{\partial z_k} - \frac{\partial b(\varphi'', \psi'', y'', z'')}{\partial z_k} \right\| \le K_{11k} \left\| \varphi' - \varphi'' \right\| + K_{11k}^\Delta \left\| \psi' - \psi'' \right\|$$
$$+ K_{12k} \left\| y' - y'' \right\| + K_{12k}^\Delta \left\| z' - z'' \right\|;$$

$$\left\| \frac{\partial b_1(\varphi',\psi',y',z')}{\partial \varphi_r} - \frac{\partial b_1(\varphi'',\psi'',y'',z'')}{\partial \varphi_r} \right\| \le K_{13r}\|\varphi'-\varphi''\| + K_{13r}^\Delta\|\psi'-\psi''\|$$

$$+ K_{14r}\|y'-y''\| + K_{14r}^\Delta\|z'-z''\|;$$

$$\left\| \frac{\partial b_1(\varphi',\psi',y',z')}{\partial \psi_r} - \frac{\partial b_1(\varphi'',\psi'',y'',z'')}{\partial \psi_r} \right\| \le K_{15r}\|\varphi'-\varphi''\| + K_{15r}^\Delta\|\psi'-\psi''\|$$

$$+ K_{16r}\|y'-y''\| + K_{16r}^\Delta\|z'-z''\|;$$

$$\left\| \frac{\partial b_1(\varphi',\psi',y',z')}{\partial y_k} - \frac{\partial b_1(\varphi'',\psi'',y'',z'')}{\partial y_k} \right\| \le K_{17k}\|\varphi'-\varphi''\| + K_{17k}^\Delta\|\psi'-\psi''\|$$

$$+ K_{18k}\|y'-y''\| + K_{18k}^\Delta\|z'-z''\|;$$

$$\left\| \frac{\partial b_1(\varphi',\psi',y',z')}{\partial z_k} - \frac{\partial b_1(\varphi'',\psi'',y'',z'')}{\partial z_k} \right\| \le K_{19k}\|\varphi'-\varphi''\| + K_{19k}^\Delta\|\psi'-\psi''\|$$

$$+ K_{20k}\|y'-y''\| + K_{20k}^\Delta\|z'-z''\|;$$

$$\left\| \frac{\partial c(\varphi',\psi')}{\partial \varphi_r} - \frac{\partial c(\varphi'',\psi'')}{\partial \varphi_r} \right\| \le K_{21}\|\varphi'-\varphi''\| + K_{21}\|\psi'-\psi''\|;$$

$$\left\| \frac{\partial c(\varphi',\psi')}{\partial \psi_r} - \frac{\partial c(\varphi'',\psi'')}{\partial \psi_r} \right\| \le K_{22}\|\varphi'-\varphi''\| + K_{22}^\Delta\|\psi'-\psi''\|;$$

$$r = 1, 2, \ldots, m; \quad k = 1, 2, \ldots, n;$$

(iii) $$\min_{\|\eta=1\|} \langle b(\varphi,\psi,0,0)\eta, \eta \rangle \le \beta, \quad \min_{\|\eta=1\|} \left\langle \frac{\partial a(\varphi,0)}{\partial \varphi}\eta, \eta \right\rangle \le \alpha,$$

$$\alpha > 0, \quad \beta + 2\alpha > 0.$$

Under these assumptions, we now find the conditions which ensure the existence of an invariant toroidal manifold $\mathcal{T} : y = u(\varphi)$ of the system (2.1) whose derivatives satisfy the Lipschitz conditions. To find this torus, we use, as before, an iteration process,

according to which $\mathcal{T}$ is constructed as the limit of a sequence of tori $\mathcal{T}^0$, $\mathcal{T}^1$, ...,
$\mathcal{T}^i$,..., each being an invariant torus

$$\mathcal{T}^{i+1} : y = u^{i+1}(\varphi), \ u^0(\varphi) \equiv 0, \ i = 0, 1, ... \tag{2.3}$$

of the system of equations

$$\frac{dy}{dt} = -b(\varphi(t), \ \varphi(t-\Delta), \ u^i(\varphi(t)), \ u^i(\varphi(t-\Delta))) \, y(t)$$
$$- b_1(\varphi(t), \ \varphi(t-\Delta), \ u^i(\varphi(t)), \ u^i(\varphi(t-\Delta))) \, u^i(\varphi(t-\Delta)) + c(\varphi(t), \ \varphi(t-\Delta)); \tag{2.4}$$

$$\frac{d\varphi}{dt} = a(\varphi(t), \ u^i(\varphi(t))).$$

Denote by $\varphi_t^i(\varphi)$ $\left(\varphi_0^i(\varphi) = \varphi\right)$ a solution of the second equation of (2.4) and by
$\varphi_t^i(\tau, \varphi, y_0)$ $\left(\varphi_\tau^i(\tau, \varphi, y_0) = y_0\right)$ a solution of the equation

$$\frac{dy}{dt} = -b\left(\varphi_t^i(\varphi), \ \varphi_{t-\Delta}^i(\varphi), u^i\left(\varphi_t^i(\varphi)\right), u^i\left(\varphi_{t-\Delta}^i(\varphi)\right)\right) y. \tag{2.5}$$

Since the matrix $b(\varphi, \psi, y, z)$ is continuous, one can always find $0 < d_1 < d$ such that,
for given $0 < \varepsilon_1 < \beta$, the inequality

$$\left\| b\left(\varphi_t^i(\varphi), \ \varphi_{t-\Delta}^i(\varphi), \ u^i\left(\varphi_t^i(\varphi)\right), \ u^i\left(\varphi_{t-\Delta}^i(\varphi)\right)\right) - b\left(\varphi_t^i(\varphi), \ \varphi_{t-\Delta}^i(\varphi), 0, 0\right) \right\| \leq \varepsilon_1, \tag{2.6}$$

holdsprovided that $\| u^i(\varphi) \| \leq d_1$. Using (2.5), (iii), (2.6), and the Cauchy inequality,
we obtain

$$\frac{d}{dt}\left\langle y_t^i(\tau, \varphi, y_0), \ y_t^i(\tau, \varphi, y_0)\right\rangle = 2\left\langle \frac{dy_t^i(\tau, \varphi, y_0)}{dt}, \ y_t^i(\tau, \varphi, y_0)\right\rangle$$

$$= -2b \left\langle \left\| (\varphi_t^i(\varphi), \ \varphi_{t-\Delta}^i(\varphi), 0, 0)\right. \right.$$

$$+ b(\varphi_t^i(\varphi), \ \varphi_{t-\Delta}^i(\varphi), u^i\left(\varphi_t^i(\varphi)\right), u^i\left(\varphi_{t-\Delta}^i(\varphi)\right))$$

$$\left. - b(\varphi_t^i(\varphi), \ \varphi_{t-\Delta}^i(\varphi), 0, 0)\right| y_t^i(\tau, \varphi, y_0), \ y_t^i(\tau, \varphi, y_0)\right\rangle$$

$$\leq -2\left(\beta - \varepsilon_1\right)\left\langle y_t^i(\tau, \varphi, y_0), \ y_t^i(\tau, \varphi, y_0)\right\rangle$$

This yields

$$\| y^i(\tau, \varphi, y_0) \| \le e^{(\beta - \varepsilon_1)(\tau - t)} \| y_0 \|, \quad t \ge \tau. \tag{2.7}$$

If (2.7) holds, then there exists a Green's function of the problem concerning the invariant tori of the system (2.4) which has the form (Samoilenko, 1970a)

$$G_0^i(\tau, \varphi) = \begin{cases} \Omega_\tau^0(\varphi, i), & \tau < 0, \\ 0, & \tau > 0, \end{cases} \tag{2.8}$$

where $\Omega_\tau^t(\varphi, i)$ is the fundamental matrix of the system (2.5), $\Omega_\tau^\tau(\varphi, i) = E$, and E is the unit matrix. Then the invariant manifold (2.3) of the system of equations (2.4) is defined by

$$u^{i+1}(\varphi) = \int_{-\infty}^{0} \Omega_{t_1}^0(\varphi, i) \left[c\left(\varphi_{t_1}^i(\varphi), \varphi_{t_1 - \Delta}^i(\varphi)\right) \right.$$

$$\left. - b_1\left(\varphi_{t_1}^i(\varphi), \varphi_{t_1 - \Delta}^i(\varphi), u^i\left(\varphi_{t_1}^i(\varphi)\right), u^i\left(\varphi_{t_1 - \Delta}^i(\varphi)\right)\right) u^i\left(\varphi_{t_1 - \Delta}^i(\varphi)\right) \right] dt_1. \tag{2.9}$$

Let us now find conditions under which the system (2.4) possesses an invariant torus. Since $\Omega_\tau^\tau(\varphi, i) = E$, the estimate (2.7) implies that the inequalities

$$\left\| y_t^i(\tau, \varphi) \right\| \le e^{(\beta - \varepsilon_1)(\tau - t)}, \quad t \ge \tau;$$

$$\left\| \Omega_\tau^t(\varphi, i) \right\| \le n e^{(\beta - \varepsilon_1)(\tau - t)}, \quad t \ge \tau, \tag{2.10}$$

hold for the solutions $y_t^i(\tau, \varphi)$ and $y_\tau^i(\tau, \varphi) = l_s = (0, 0, \ldots, 1, 0, \ldots, 0)$. Since $u^0(\varphi) = 0$, the next inequality

$$\left\| u^1(\varphi) \right\| \le \int_{-\infty}^{0} \left\| \Omega_{t_1}^0(\varphi, 0) \right\| \left\| c(\varphi, \psi) \right\|_0 dt_1 \le \frac{n \| c(\varphi, \psi) \|_0}{\beta - \varepsilon_1} \tag{2.11}$$

follows form (2.9) for $i = 0$.

Assume that $\| c(\varphi, \psi) \|_0$ is so small that the inequality

$$\frac{n \| c(\varphi, \psi) \|_0}{\beta - \varepsilon_1} \le d_1$$

holds. Then (2.11) yields $\left\| u^1(\varphi) \right\| \le d_1$. We now fix $i = 1$ in (2.9). Then

$$\left\| u^2(\varphi) \right\| \le \int\limits_{-\infty}^{0} \left\| \Omega_{t_1}^0(\varphi, 1) \right\| \left(\left\| b_1(\varphi, \psi, y, z) \right\|_0 \frac{n\|c(\varphi, \psi)\|_0}{\beta - \varepsilon_1} + \|c(\varphi, \psi)\|_0 \right) dt_1$$

$$\le \frac{n\|c(\varphi, \psi)\|_0}{\beta - \varepsilon_1} (1 + q),$$

where

$$q = \frac{n\|b_1(\varphi, \psi, y, z)\|_0}{\beta - \varepsilon_1}.$$

Assuming that

$$q < 1, \quad \frac{n\|c(\varphi, \psi)\|_0}{\beta - \varepsilon_1}(1 + q) \le d_1, \tag{2.12}$$

we find $\| u^2(\varphi) \| \le d_1$. By induction, we can prove that the estimates

$$\left\| u^{i+1}(\varphi) \right\| \le \frac{n\|c(\varphi, \psi)\|_0}{\beta - \varepsilon_1} \sum_{s=0}^{i} q^s \le \frac{n\|c(\varphi, \psi)\|_0}{(\beta - \varepsilon_1)(1 - q)} \le d_1 \tag{2.13}$$

are valid for all $i = 0, 1, 2, \ldots$ Therefore, the functions $u^{i+1}(\varphi)$ exist for all i and satisfy (2.13) provided that the inequalities (2.12) hold.

We now clarify conditions which guarantee the existence of the derivatives

$$\frac{\partial u^{i+1}(\varphi)}{\partial \varphi_j}, \quad j = 1, 2, \ldots; \; m; i = 0, 1, 2, \ldots$$

By differentiating (2.9), we obtain

$$\frac{\partial u^{i+1}(\varphi)}{\partial \varphi_j} = \int\limits_{-\infty}^{0} \frac{\partial \Omega_{t_1}^0(\varphi, i)}{\partial \varphi_j} \Big\{ c\left(\varphi_{t_1}^i, \varphi_{t_1 - \Delta}^i(\varphi)\right)$$

$$- b_1\left[\varphi_{t_1}^i(\varphi), \varphi_{t_1 - \Delta}^i(\varphi), u^i\left(\varphi_{t_1}^i(\varphi)\right), u^i\left(\varphi_{t_1 - \Delta}^i(\varphi)\right)\right] \Big\} u^i\left(\varphi_{t_1 - \Delta}^i(\varphi)\right) \Big\} dt_1$$

$$
+ \int\limits_{-\infty}^{0} \Omega_{t_1}^{0}(\varphi, i) \left\{ \sum_{r=1}^{m} \left[\frac{\partial b_1}{\partial \varphi_{t_1,r}^{i}} \frac{\partial \varphi_{t_1,r}^{i}(\varphi)}{\partial \varphi_j} + \frac{\partial b_1}{\partial \varphi_{t_1-\Delta,r}} \frac{\partial \varphi_{t_1-\Delta,r}^{i}(\varphi)}{\partial \varphi_j} \right. \right.
$$

$$
+ \sum_{k=1}^{m} \left(\frac{\partial b_1}{\partial y_k} \frac{\partial u_k^{i}\left(\varphi_{t_1,r}^{i}(\varphi)\right)}{\partial \varphi_{t_1,r}^{i}} \frac{\partial \varphi_{t_1,r}^{i}(\varphi)}{\partial \varphi_j} \right.
$$

$$
\left. \left. + \frac{\partial b_1}{\partial y_{k\Delta}} \frac{\partial u_k^{i}\left(\varphi_{t_1-\Delta}^{i}(\varphi)\right)}{\partial \varphi_{t_1-\Delta,r}^{i}} \frac{\partial \varphi_{t_1-\Delta,r}^{i}(\varphi)}{\partial \varphi_j} \right) \right] u^{i}\left(\varphi_{t_1-\Delta}^{i}(\varphi)\right)
$$

$$
\left. - b_1 \frac{\partial u^{i}\left(\varphi_{t_1-\Delta}^{i}(\varphi)\right)}{\partial \varphi_{t_1-\Delta}^{i}} \frac{\partial \varphi_{t_1-\Delta}^{i}}{\partial \varphi_j} + \frac{\partial c}{\partial \varphi_{t_1}^{i}} \frac{\partial \varphi_{t_1}^{i}(\varphi)}{\partial \varphi_j} + \frac{\partial c}{\partial \varphi_{t_1-\Delta}^{i}} \frac{\partial \varphi_{t_1-\Delta}^{i}(\varphi)}{\partial \varphi_j} \right\} dt_1
$$

$$
= \int\limits_{-\infty}^{\infty} \frac{\partial \Omega_{t_1}^{0}(\varphi)}{\partial \varphi_j} H_1^{i}(t_1, \varphi) dt_1 + \int\limits_{-\infty}^{0} \Omega_{t_1}^{0}(\varphi, i) H_2^{i}(t_1, \varphi) dt_1. \tag{2.14}
$$

Let us differentiate the solution of the second equation in (2.4)

$$
\frac{d}{dt_1} \left(\frac{\partial \varphi_{t_1}^{i}(\varphi)}{\partial \varphi_j} \right) = \left(\frac{\partial a\left(\varphi_{t_1}^{i}(\varphi), u^{i}\left(\varphi_{t_1-\Delta}^{i}(\varphi)\right) \right)}{\partial \varphi_{t_1}^{i}} \right.
$$

$$
\left. + \frac{\partial a\left(\varphi_{t_1}^{i}(\varphi), u^{i}\left(\varphi_{t_1}^{i}(\varphi)\right) \right)}{\partial y} \frac{\partial u^{i}\left(\varphi_{t_1}^{i}(\varphi)\right)}{\partial \varphi_{t_1}^{i}} \right) \frac{\partial \varphi_{t_1}^{i}(\varphi)}{\partial \varphi_j}. \tag{2.15}
$$

Assuming that

$$
\left\| \frac{\partial u^{i}(\varphi)}{\partial \varphi_j} \right\| \leq R_i < \rho, \tag{2.16}
$$

we obtain from (2.15)

$$
\frac{d}{dt} \left\langle \frac{\partial \varphi_{t_1}^{i}(\varphi)}{\partial \varphi_j}, \frac{\partial \varphi_{t_1}^{i}(\varphi)}{\partial \varphi_j} \right\rangle \geq 2\gamma \left\langle \frac{\partial \varphi_{t_1}^{i}(\varphi)}{\partial \varphi_j}, \frac{\partial \varphi_{t_1}^{i}(\varphi)}{\partial \varphi_j} \right\rangle,
$$

$$\gamma = \alpha - \varepsilon_1 - m A \rho,$$

where

$$A = \left\| \frac{\partial a(\varphi, y)}{\partial y} \right\|_0, \quad \left\| \frac{\partial a\left(\varphi_{t_1}^i(\varphi), u^i\left(\varphi_{t_1}^i(\varphi) \right) \right)}{\partial \varphi_{t_1}^i} - \frac{\partial a\left(\varphi_{t_1}^i(\varphi), 0 \right)}{\partial \varphi_{t_1}^i} \right\| \leq \varepsilon_1$$

for all $\| u^i(\varphi) \| \leq d_2 \leq d$, or

$$\ln \left(\frac{1}{\left\langle \dfrac{\partial \varphi_{t_1}^i(\varphi)}{\partial \varphi_j}, \dfrac{\partial \varphi_{t_1}^i(\varphi)}{\partial \varphi_j} \right\rangle} \right) \geq 2 \int\limits_{t_1}^{0} \gamma \, dt \text{ for } t_1 \leq 0,$$

$$\left\langle \frac{\partial \varphi_{t_1}^i(\varphi)}{\partial \varphi_j}, \frac{\partial \varphi_{t_1}^i(\varphi)}{\partial \varphi_j} \right\rangle \leq e^{2\gamma_1} \quad \text{for } t_1 \leq 0.$$

This yields

$$\left\| \frac{\partial \varphi_{t_1}^i(\varphi)}{\partial \varphi_j} \right\| \leq e^{\gamma_1} \quad \text{for } t_1 \leq 0. \tag{2.17}$$

Let us differentiate the solution $y_t^i(\tau, \varphi)$ $\left(y_t^i(\tau, \varphi) = e_s \right)$ of eqn. (2.5)

$$\frac{d}{dt_1} \left(\frac{\partial y_{t_1}^i(\tau, \varphi)}{\partial \varphi_j} \right) = - b\left(\varphi_{t_1}^i(\varphi), \varphi_{t_1-\Delta}^i(\varphi), u^i\left(\varphi_{t_1}^i(\varphi) \right), u^i\left(\varphi_{t_1-\Delta}^i(\varphi) \right) \right) \frac{\partial y_{t_1}^i(\tau, \varphi)}{\partial \varphi_j}$$

$$- \sum_{r=1}^{m} \left| \frac{\partial b}{\partial \varphi_{t_1, r}} \frac{\partial \varphi_{t_1, r}^i(\varphi)}{\partial \varphi_j} + \frac{\partial b}{\partial \varphi_{t_1-\Delta, r}^i} \frac{\partial \varphi_{t_1-\Delta, r}^i(\varphi)}{\partial \varphi_j} \right.$$

$$+ \sum_{k=1}^{n} \left(\frac{\partial b}{\partial y_k} \frac{\partial u_k^i\left(\varphi_{t_1}^i(\varphi) \right)}{\partial \varphi_{t_1, r}^i} \frac{\partial \varphi_{t_1, r}^i(\varphi)}{\partial \varphi_j} \right.$$

$$+ \frac{\partial b}{\partial y_{\Delta k}} \frac{\partial u_k^i\left(\varphi_{t_1-\Delta}^i(\varphi)\right)}{\partial \varphi_{t_1-\Delta,r}^i} \frac{\partial \varphi_{t_1-\Delta,k}^i(\varphi)}{\partial \varphi_j}\Bigg)\Bigg] y_{t_1}^i(\tau,\varphi)$$

$$\equiv b\left(\varphi_{t_1}^i(\varphi), \varphi_{t_1-\Delta}^i(\varphi),\, u^i\left(\varphi_{t_1}^i(\varphi)\right), u^i\left(\varphi_{t_1-\Delta}^i(\varphi)\right)\right) \frac{\partial y_{t_1}^i(\tau,\varphi)}{\partial \varphi_j} - F^i(t_1,\tau,\varphi). \quad (2.18)$$

The solution of the equation (2.18) can be written as follows

$$\frac{\partial y_{t_1}^i(\tau,\varphi)}{\partial \varphi_j} = -\int_\tau^{t_1} \Omega_{t_2}^{t_1}(\varphi, i) F^i(t_2,\tau,\varphi) dt_2, \; t_1 > \tau. \quad (2.19)$$

Equation (2.19) can be estimated by use of (2.10), (2.11), (2.16), and (2.17); as a result, for all

$$\| u^i(\varphi) \| \leq d_0, \quad d_0 = \min(d_1, d_2), \quad (2.20)$$

we get

$$\left\| \frac{\partial y_{t_1}^i(\tau,\varphi)}{\partial \varphi_j} \right\| \leq \int_\tau^{t_1} \left\| \Omega_{t_0}^{t_1}(\varphi, i) \right\| \left\| F^i(t_2,\tau,\varphi) \right\| dt_2 \leq \frac{n}{\gamma}\left(L_2 + L_2^\Delta e^{-\gamma\Delta}\right)$$

$$+ \left(L_3 + L_3^\Delta e^{-\gamma\Delta}\right)\rho\right) e^{(\beta-\varepsilon_1)(\tau-t_1)}\left(e^{\gamma t_1} - e^{\gamma\tau}\right), \quad (2.21)$$

where

$$L_2 = \sum_{r=1}^m \left\| \frac{\partial b(\varphi,\psi,y,z)}{\partial \varphi_r} \right\|_0, \quad L_2^\Delta = \sum_{r=1}^m \left\| \frac{\partial b(\varphi,\psi,y,z)}{\partial \varphi_{\Delta r}} \right\|_0,$$

$$L_3 = \sum_{k=1}^n m \left\| \frac{\partial b(\varphi,\psi,y,z)}{\partial y_k} \right\|_0, \quad L_3^\Delta = \sum_{k=1}^n m \left\| \frac{\partial b(\varphi,\psi,y,z)}{\partial y_{\Delta k}} \right\|_0.$$

Taking into account (2.21), we find

$$\left\| \frac{d\Omega_r^0(\varphi, i)}{\partial \varphi_j} \right\| \leq \frac{n^2}{\gamma}\left[L_2 + L_2^\Delta e^{-\gamma\Delta} + \left(L_3 + L_3^\Delta e^{-\gamma\Delta}\right)\rho\right] e^{(\beta-\varepsilon_1)\tau}\left(1 - e^{-\gamma\tau}\right) \quad (2.22)$$

for $\tau_1 \leq 0$. Estimating the relation (2.14) for $i = 0$ with the help of (2.10), (2.17), and (2.22), we obtain

$$\left\| \frac{\partial u^1(\varphi)}{\partial \varphi_j} \right\| \leq \int_{-\infty}^{0} \left\| \frac{d\Omega_{t_1}^0(\varphi, 0)}{\partial \varphi_j} \right\| \left\| c(\varphi, \psi) \right\| dt_1 + \int_{-\infty}^{0} \left\| \Omega_{t_1}^0(\varphi, 0) \right\| \left\| \left(\frac{\partial c}{\partial \varphi_{t_1}^i} \frac{\partial \varphi_{t_1}^i(\varphi)}{\partial \varphi_j} \right. \right.$$

$$\left. \left. + \frac{\partial c}{\partial \varphi_{t_1 - \Delta}^i} \frac{\partial \varphi_{t_1 - \Delta}^i(\varphi)}{\partial \varphi_j} \right) \right\| dt_1 \leq \frac{n^2}{\beta(\beta + \gamma)} \left(L_2 + L_2^\Delta e^{-\gamma\Delta} \right) \left\| c(\varphi, \psi) \right\|_0$$

$$+ \frac{n}{\beta + \gamma} \left(L_4 + L_4^\Delta e^{-\gamma\Delta} \right) < R_1, \tag{2.23}$$

where

$$R_1 = \frac{n^2}{(\beta - \varepsilon_1)(\beta + \gamma - \varepsilon_1)} \left(L_2 + L_2^\Delta e^{-\gamma\Delta} \right) \left(\left\| c(\varphi, \psi) \right\|_0 + \left\| b_1(\varphi, \psi, y, z) \right\|_0 d_0 \right)$$

$$+ \frac{n}{\beta + \gamma - \varepsilon_1} \left(L_4 + L_4^\Delta e^{-\gamma\Delta} + \left(L_5 + L_5^\Delta e^{-\gamma\Delta} \right) d_0 \right);$$

$$L_4 = \left\| \frac{\partial c(\varphi, \psi)}{\partial \varphi} \right\|_0 ; L_4^\Delta = \left\| \frac{\partial c(\varphi, \psi)}{\partial \psi} \right\|_0 ;$$

$$L_5 = \sum_{r=1}^{m} \left\| \frac{\partial b_1(\varphi, \psi, y, z)}{\partial \varphi_r} \right\|_0 ; L_5^\Delta = \sum_{r=1}^{m} \left\| \frac{\partial b_1(\varphi, \psi, y, z)}{\partial \psi_r} \right\|_0 .$$

Here, β and ε_1 are so small that the inequality

$$\beta + \gamma - \varepsilon_1 > 0 \tag{2.24}$$

holds. By choosing sufficiently small $\| c \|_0$, $\| b_1 \|_0$, $\left\| \frac{\partial b_1}{\partial \varphi_r} \right\|_0$, $\left\| \frac{\partial b_1}{\partial \psi_r} \right\|_0$, and $\left\| \frac{\partial c}{\partial \psi} \right\|$, one can always guarantee that $i = 1$, and therefore,

$$\left\| \frac{\partial u^1(\varphi)}{\partial \varphi_j} \right\| < \rho. \tag{2.25}$$

From the relation (2.14) with $i = 1$, we get

$$\left\|\frac{\partial u^2(\phi)}{\partial \phi_j}\right\| \le \int_{-\infty}^{0}\left\|\frac{d\Omega_{t_1}^0(\phi,1)}{\partial \phi_j}\right\|\left\|H_1^1(t_1,\phi)\right\|dt_1$$

$$+\int_{-\infty}^{0}\left\|\Omega_{t_1}^0(\phi,1)\right\|\left\|H_2^1(t_1,\phi)\right\|dt_1 \le R_1(1+q_1)=R_2$$

where

$$q_1 = \frac{n^2}{(\beta-\epsilon_1)(\beta+\gamma-\epsilon_1)}\left(L_3+L_3^\Delta e^{-\gamma\Delta}\right)\left(\|c(\phi,\psi)\|_0+\|b(\phi,\psi,y,z)\|_0 d_2\right)$$

$$+\frac{n}{\beta+\gamma-\epsilon_1}\left(L_6 d_0+\left(m\|b_1\|_0+L_6^\Delta d_0\right)e^{-\gamma\Delta}\right);$$

$$L_6 = \sum_{k=1}^{n} m\left\|\frac{\partial b_1(\phi,\psi,y,z)}{\partial y_k}\right\|_0, \quad L_6^\Delta = \sum_{k=1}^{n} m\left\|\frac{\partial b_1(\phi,\psi,y,z)}{\partial z_k}\right\|_0.$$

If the inequalities

$$q_1 < 1 \tag{2.26}$$

and

$$R_1 < \rho(1-q_1), \tag{2.27}$$

hold, then one can easily prove by induction that

$$\left\|\frac{\partial u^i(\phi)}{\partial \phi_j}\right\| \le R_j = \frac{R_1}{1-q_1} \le \rho \tag{2.28}$$

for all $i = 1, 2, \ldots$.

Let us now prove that the functions $\frac{\partial u^{i+1}(\phi)}{\partial \phi_j}$ can always be characterized by same Lipschitz constant. In fact, the relation (2.14) implies that

$$\left\|\frac{\partial u^{i+1}(\overline{\phi})}{\partial \phi_j}-\frac{\partial u^{i+1}(\overline{\overline{\phi}})}{\partial \phi_j}\right\| \le \int_{-\infty}^{0}\left(\left\|\frac{\Omega_{t_1}^0(\overline{\phi},i)}{\partial \phi_j}-\frac{\Omega_{t_1}^0(\overline{\overline{\phi}},i)}{\partial \phi_j}\right\|\left\|H_2^i(t_1,\overline{\phi})\right\|\right.$$

$$+\left\|\frac{\Omega_{t_1}^0(\overline{\phi},i)}{\partial \phi_j}\right\|\left\|H_1^i(t,\overline{\phi})-H_1^i(t,\overline{\overline{\phi}})\right\|\right)dt_1$$

$$+ \int_{-\infty}^{0} \left(\left\| \Omega_{t_1}^0 (\overline{\overline{\varphi}}, i) - \Omega_{t_1}^0 (\overline{\varphi}, i) \right\| \left\| H_2^i (t_1, \overline{\overline{\varphi}}) \right\| \right.$$

$$+ \left\| \Omega_{t_1}^0 (\varphi, i) \right\| \left\| H_2^i (t_1, \overline{\overline{\varphi}}) - H_2^i (t_1, \overline{\varphi}) \right\| \left. \right) dt_1. \tag{2.29}$$

In order to find the bounds for the right-hand side of (2.29), it is necessary to establish some inequalities. We rewrite (2.15) as follows

$$\frac{d}{dt_1} \left(\frac{\partial \varphi_{t_1}^i (\overline{\overline{\varphi}})}{\partial \varphi_j} - \frac{\partial \varphi_{t_1}^i (\overline{\varphi})}{\partial \varphi_j} \right) = \left(\frac{\partial a \left(\varphi_{t_1}^i (\overline{\overline{\varphi}}), u^i \left(\varphi_{t_1}^i (\overline{\overline{\varphi}}) \right) \right)}{\partial \varphi_{t_1}^i} \right.$$

$$+ \frac{\partial a \left(\varphi_{t_1}^i (\overline{\overline{\varphi}}), u^i \left(\varphi_{t_1}^i (\overline{\overline{\varphi}}) \right) \right)}{\partial y} \frac{\partial u^i \left(\varphi_{t_1}^i (\overline{\overline{\varphi}}) \right)}{\partial \varphi_{t_1}^i} \left) \left(\frac{\partial \varphi_{t_1}^i (\overline{\overline{\varphi}})}{\partial \varphi_j} - \frac{\partial \varphi_{t_1}^i (\overline{\varphi})}{\partial \varphi_j} \right) \right.$$

$$+ \left[\frac{\partial a \left(\varphi_{t_1}^i (\overline{\overline{\varphi}}), u^i \left(\varphi_{t_1}^i (\overline{\overline{\varphi}}) \right) \right)}{\partial \varphi_{t_1}^i} - \frac{\partial a \left(\varphi_{t_1}^i (\overline{\varphi}), u^i \left(\varphi_{t_1}^i (\overline{\varphi}) \right) \right)}{\partial \varphi_{t_1}^i} \right.$$

$$+ \frac{\partial a \left(\varphi_{t_1}^i (\overline{\overline{\varphi}}), u^i \left(\varphi_{t_1}^i (\overline{\overline{\varphi}}) \right) \right)}{\partial y} \left(\frac{\partial u^i \left(\varphi_{t_1}^i (\overline{\overline{\varphi}}) \right)}{\partial \varphi_{t_1}^i} - \frac{\partial u^i \left(\varphi_{t_1}^i (\overline{\varphi}) \right)}{\partial \varphi_{t_1}^i} \right)$$

$$+ \left(\frac{\partial a \left(\varphi_{t_1}^i (\overline{\overline{\varphi}}), u^i \left(\varphi_{t_1}^i (\overline{\overline{\varphi}}) \right) \right)}{\partial y} - \frac{\partial a \left(\varphi_{t_1}^i (\overline{\varphi}), u^i \left(\varphi_{t_1}^i (\overline{\varphi}) \right) \right)}{\partial y} \right) \frac{\partial u^i \left(\varphi_{t_1}^i (\overline{\varphi}) \right)}{\partial \varphi_{t_1}^i} \left] \frac{\partial \varphi_{t_1}^i (\overline{\varphi})}{\partial \varphi_{t_1}^i} . \right.$$

This yields

$$\frac{\partial \varphi_{t_1}^i (\overline{\overline{\varphi}})}{\partial \varphi_j} - \frac{\partial \varphi_{t_1}^i (\overline{\varphi})}{\partial \varphi_j} = - \int_{t_1}^{0} \overline{\Omega}_{t_2}^{t_1} (\overline{\overline{\varphi}}, i) \, F_1 (\overline{\overline{\varphi}}, \overline{\varphi}, i) \, dt_2, \tag{2.30}$$

where $\overline{\Omega}_{t_2}^{t_1} (\overline{\varphi}, i)$ is the fundamental matrix of solutions

$$W_{t_2}^{t_1} (\overline{\varphi}, i), \; W_{t_2}^{t_2} (\overline{\varphi}, i) = \overline{e}_p = \left(\underbrace{0, 0, \ldots, 1,}_{p-1} \; \underbrace{0, \ldots, 0}_{m-p} \right)$$

of the system

$$\frac{d}{dt_1} W^{t_1}_{t_2}\left(\overline{\varphi},i\right) = \left(\frac{\partial a\left(\varphi^i_{t_2}\left(\overline{\varphi}\right), u^i\left(\varphi^i_{t_2}\left(\overline{\varphi}\right)\right)\right)}{\partial \varphi^i_{t_2}} \right.$$

$$\left. + \frac{\partial a\left(\varphi^i_{t_2}\left(\overline{\varphi}\right), u^i\left(\varphi^i_{t_2}\left(\overline{\varphi}\right)\right)\right)}{\partial y} \frac{\partial u^i\left(\varphi^i_{t_2}\left(\overline{\varphi}\right)\right)}{\partial \varphi^i_{t_2}} \right) W^{t_1}_{t_2}\left(\overline{\varphi},i\right); \tag{2.31}$$

$$\Omega^{t_1}_{t_2}\left(\overline{\varphi},i\right) = E.$$

From the equation (2.31), we can obtain estimates similar to (2.17), i. e.,

$$\left\| W^{t_1}_{t_2}\left(\overline{\varphi},i\right) \right\| \le e^{\gamma(t_1-t_2)} \quad \text{for } t_1 \le t_2; \tag{2.32}$$

$$\left\| \overline{\Omega}^{t_1}_{t_2}\left(\overline{\varphi},i\right) \right\| \le m e^{\gamma(t_1-t_2)} \quad \text{for } t_1 \le t_2. \tag{2.33}$$

Taking the inequalities (2.17) and (2.28) into account, one can easily find that

$$\left\| u^i\left(\varphi^i_{t_2}\left(\overline{\overline{\varphi}}\right)\right) - u^i\left(\varphi^i_{t_2}\left(\overline{\varphi}\right)\right) \right\| \le m\rho \left\| \varphi^i_{t_2}\left(\overline{\overline{\varphi}}\right) - \varphi^i_{t_2}\left(\overline{\varphi}\right) \right\|;$$

and $\hspace{10cm}$ (2.34)

$$\left\| \varphi^i_{t_2}\left(\overline{\overline{\varphi}}\right) - \varphi^i_{t_2}\left(\overline{\varphi}\right) \right\| \le \left\| \int_0^1 \frac{\partial \varphi^i_{t_2}\left(\overline{\varphi} + \theta\left(\overline{\overline{\varphi}} - \overline{\varphi}\right)\right)}{\partial \varphi} \left(\overline{\overline{\varphi}} - \overline{\varphi}\right) d\theta \right\|$$

$$\le \left\| \overline{\overline{\varphi}} - \overline{\varphi} \right\| m \int_0^1 e^{\gamma t_2} d\theta = \left\| \overline{\overline{\varphi}} - \overline{\varphi} \right\| m e^{\gamma t_2}.$$

Estimating now the right-hand side of (2.30) with the help of (2.17), (2.28), (2.32), (2.33), and (2.34), we get

$$\left\| \frac{\partial \varphi^i_{t_1}\left(\overline{\overline{\varphi}}\right)}{\partial \varphi_j} - \frac{\partial \varphi^i_{t_1}\left(\overline{\varphi}\right)}{\partial \varphi_j} \right\| \le \left\| \overline{\overline{\varphi}} - \overline{\varphi} \right\| \frac{m^2}{\gamma} e^{\gamma t_2} \left(1 - e^{\gamma t_1}\right)\left(K_1 + m A K^i\right)$$

$$+ \left(K_2 + K_3 + m\rho K_4\right) m\rho). \tag{2.35}$$

We have assumed here that

$$\left\| \frac{\partial u^i(\overline{\varphi})}{\partial \varphi_j} - \frac{\partial u^i(\overline{\overline{\varphi}})}{\partial \varphi_j} \right\| \leq K^i \left\| \overline{\varphi} - \overline{\overline{\varphi}} \right\|. \tag{2.36}$$

From (2.18), we obtain the equation

$$\frac{d}{dt_1}\left(\frac{\partial y_{t_1}^i(\tau, \overline{\varphi})}{\partial \varphi_j} - \frac{\partial y_{t_1}^i(\tau, \overline{\overline{\varphi}})}{\partial \varphi_j} \right)$$

$$= - b\left(\varphi_{t_1}^i(\overline{\varphi}), \varphi_{t_1-\Delta}^i(\overline{\varphi}), u^i\left(\varphi_{t_1}^i(\overline{\varphi})\right), u^i\left(\varphi_{t_1-\Delta}^i(\overline{\varphi})\right) \right)\left(\frac{\partial y_{t_1}^i(\tau, \overline{\varphi})}{\partial \varphi_j} \right.$$

$$\left. - \frac{\partial y_{t_1}^i(\tau, \overline{\overline{\varphi}})}{\partial \varphi_j} \right) + \left\{ b\left(\varphi_{t_1}^i(\overline{\varphi}), \varphi_{t_1-\Delta}^i(\overline{\varphi}), u^i\left(\varphi_{t_1}^i(\overline{\varphi})\right), u^i\left(\varphi_{t_1-\Delta}^i(\overline{\varphi})\right) \right) \right.$$

$$\left. - b\left(\varphi_{t_1}^i(\overline{\overline{\varphi}}), \varphi_{t_1-\Delta}^i(\overline{\overline{\varphi}}), u^i\left(\varphi_{t_1}^i(\overline{\overline{\varphi}})\right), u^i\left(\varphi_{t_1-\Delta}^i(\overline{\overline{\varphi}})\right) \right) \right\} \frac{\partial y_{t_1}^i(\tau, \overline{\overline{\varphi}})}{\partial \varphi_j}$$

$$+ F^i\left(t_1, \tau, \overline{\varphi}\right) - F^i\left(t_1, \tau, \overline{\overline{\varphi}}\right). \tag{2.37}$$

Its solution has the form

$$\frac{\partial y_0^i(\tau, \overline{\varphi})}{\partial \varphi_j} - \frac{\partial y_0^i(\tau, \overline{\overline{\varphi}})}{\partial \varphi_j} = \int_\tau^0 \Omega_{t_1}^0(\overline{\varphi}, i)\left\{ \left(b\left(\varphi_{t_1}^i(\overline{\varphi}), \varphi_{t_1-\Delta}^i(\overline{\varphi}), u^i\left(\varphi_{t_1}^i(\overline{\varphi})\right), u^i\left(\varphi_{t_1-\Delta}^i(\overline{\varphi})\right) \right) \right.\right.$$

$$\left.\left. - b\left(\varphi_{t_1}^i(\overline{\overline{\varphi}}), \varphi_{t_1-\Delta}^i(\overline{\overline{\varphi}}), u^i\left(\varphi_{t_1}^i(\overline{\overline{\varphi}})\right), u^i\left(\varphi_{t_1-\Delta}^i(\overline{\overline{\varphi}})\right) \right) \right) \right) \frac{\partial y_{t_1}^i(\tau, \overline{\overline{\varphi}})}{\partial \varphi_j}$$

$$+ F^i\left(t_1, \tau, \overline{\varphi}\right) - F^i\left(t_1, \tau, \overline{\overline{\varphi}}\right)\right\} dt_1 \tag{2.38}$$

Using this relation, one can estimate the difference $\dfrac{\partial y_0^i(\tau, \overline{\varphi})}{\partial \varphi_j} - \dfrac{\partial y_0^i(\tau, \overline{\overline{\varphi}})}{\partial \varphi_j}$ just as above.

Consider now the inequality (2.29). Taking into account the estimates obtained above, one can easily prove by induction that the following inequality holds

$$\left\|\frac{\partial u^{i+1}(\tau,\overline{\overline{\varphi}})}{\partial\varphi_j} - \frac{\partial u^{i+1}(\tau,\overline{\varphi})}{\partial\varphi_j}\right\| \le K\left\|\overline{\overline{\varphi}} - \overline{\varphi}\right\|, \tag{2.39}$$

$$K = \overline{K}/(1-q_2), \tag{2.40}$$

for all $i = 0, 1, \ldots$, where $\overline{K}$ and q_2 are fixed constants; moreover,

$$q_2 < 1. \tag{2.41}$$

This means that under the condition (2.41) all derivatives $\dfrac{\partial u^{i+1}(\tau,\varphi)}{\partial\varphi_j}$, $i = 0, 1, 2, \ldots$

satisfy the Lipschitz condition with the constant K.

Let us now prove the convergence of the sequence of invariant tori (2.3) in norm

$$\|u(\varphi)\|_1 = \|u(\varphi)\|_0 + \max_{1\le j\le m}\left\|\frac{\partial u(\varphi)}{\partial\varphi_j}\right\|_0. \tag{2.42}$$

Using the relation (2.9), we obtain

$$\left\| u^{i+1}(\varphi) - u^i(\varphi) \right\|_1 = \left\| u^{i+1}(\varphi) - u^i(\varphi) \right\|_0 + \max_{1\le j\le m}\left\|\frac{\partial u^{i+1}(\tau,\varphi)}{\partial\varphi_j} - \frac{\partial u^i(\tau,\varphi)}{\partial\varphi_j}\right\|_0$$

$$\le \int\limits_0^\infty \left\{ \left\| \Omega_{t_1}^0(\varphi,i) - \Omega_{t_1}^0(\varphi,i-1) \right\| \left(\left\| b_1 \right\| d_2 + \left\| c \right\| \right) + \left\| \Omega_{t_1}^0(\varphi,i-1) \right\| \right.$$

$$\times \left(\left\| b_1\left(\varphi_{t_1}^i(\varphi), \varphi_{t_1-\Delta}^i(\varphi), u^i\left(\varphi_{t_1}^i(\varphi)\right), u^i\left(\varphi_{t_1-\Delta}^i(\varphi)\right) \right) u^i\left(\varphi_{t_1-\Delta}^i(\varphi)\right) \right.$$

$$- b_1\left(\varphi_{t_1}^{i-1}(\varphi), \varphi_{t_1-\Delta}^{i-1}(\varphi), u^{i-1}\left(\varphi_{t_1}^{i-1}(\varphi)\right), u^{i-1}\left(\varphi_{t_1-\Delta}^{i-1}(\varphi)\right) \right) u^{i-1}\left(\varphi_{t_1-\Delta}^{i-1}(\varphi)\right) \right\|$$

$$+ \left\| c\left(\varphi_{t_1}^i(\varphi), \varphi_{t_1-\Delta}^i(\varphi) \right) - c\left(\varphi_{t_1}^{i-1}(\varphi), \varphi_{t_1-\Delta}^{i-1}(\varphi) \right) \right\| \right) + \left\|\frac{\partial\Omega_{t_1}^0(\varphi,i)}{\partial\varphi_i}\right.$$

$$\left. - \frac{\partial\Omega_{t_1}^0(\varphi,i-1)}{\partial\varphi_i}\right\| \left\| H_1^i(t,\varphi) \right\| + \left\|\frac{\partial\Omega_{t_1}^0(\varphi,i-1)}{\partial\varphi_i}\right\| \left\| H_1^i(t,\varphi) - H_1^{i-1}(t,\varphi) \right\|$$

$$
+ \ \left\| \Omega_{t_1}^0 (\varphi, i) - \Omega_{t_1}^0 (\varphi, i-1) \right\| \ \left\| H_2^i (t_1, \varphi) \right\|
$$

$$
+ \ \left\| \Omega_{t_1}^0 (\varphi, i-1) \right\| \ \left\| H_2^i (t, \varphi) - H_2^{i-1} (t, \varphi) \right\| \Big\}. \tag{2.43}
$$

To estimate the right-hand side of this inequality, we must establish some relations. For the solutions $\varphi_t^i (\varphi)$ of the first equation of the system (2.4), we have

$$
\frac{d \Big(\varphi_{t_1}^i (\varphi) - \varphi_{t_1}^{i-1} (\varphi) \Big)}{dt} = \Bigg(\int_0^1 \frac{\partial a \Big(\varphi_{t_1}^{i-1} (\varphi) + \theta \Big(\varphi_{t_2}^i (\varphi) - \varphi_{t_2}^{i-1} (\varphi) \Big), u^i \big(\varphi_t (\varphi) \big) \Big)}{\partial \varphi_{t_1}^i (\varphi)} d\theta
$$

$$
+ \int_0^1 \frac{\partial a \Big(\varphi_{t_1}^i (\varphi), u^i \Big(\varphi_{t_1}^{i-1} (\varphi) \Big) + \theta \Big(u^i \big(\varphi_{t_1}^i (\varphi) \big) - u^{i-1} \big(\varphi_{t_1}^{i-1} (\varphi) \big) \Big) \Big)}{\partial y} d\theta
$$

$$
\times \int_0^1 \frac{\partial u^{i-1} \Big(\varphi_{t_1}^{i-1} (\varphi) + \theta_1 \Big(\varphi_{t_1}^i (\varphi) - \varphi_{t_1}^{i-1} (\varphi) \Big) \Big)}{\partial \varphi_{t_1}^i (\varphi)} d\theta_1 \Bigg\} \Big(\varphi_{t_1}^i (\varphi) - \varphi_{t_1}^{i-1} (\varphi) \Big)
$$

$$
+ \int_0^1 \frac{\partial a \Big(\varphi_{t_1}^{i-1} (\varphi), u^{i-1} \Big(\varphi_{t_1}^{i-1} (\varphi) \Big) + \theta \Big(u^i \big(\varphi_{t_1}^i (\varphi) \big) - u^{i-1} \big(\varphi_{t_1}^{i-1} (\varphi) \big) \Big) \Big)}{\partial y}
$$

$$
\times \Big(u^i \big(\varphi_{t_1}^i (\varphi) \big) - u^{i-1} \big(\varphi_{t_1}^i (\varphi) \big) \Big) \, d\theta.
$$

This yields

$$
\varphi_{t_1}^i (\varphi) - \varphi_{t_1}^{i-1} (\varphi)
$$

$$
= \int_{t_1}^0 \overline{\Omega}_{t_2}^{t_1} \int_0^1 \frac{\partial a \Big(\varphi_{t_2}^{i-1} (\varphi), u^{i-1} \Big(\varphi_{t_2}^{i-1} (\varphi) \Big) + \theta \Big(u^i \big(\varphi_{t_2}^i (\varphi) \big) - u^{i-1} \big(\varphi_{t_2}^{i-1} (\varphi) \big) \Big) \Big)}{\partial y}
$$

$$
\times \Big(u^i \big(\varphi_{t_2}^i (\varphi) \big) - u^{i-1} \big(\varphi_{t_2}^i (\varphi) \big) \Big) \, d\theta \, dt_2.
$$

By virtue of (2.33), we finally obtain

$$\left\| \varphi_{t_1}^i(\varphi) - \varphi_{t_1}^{i-1}(\varphi) \right\| \leq \left\| u^i(\varphi) - u^{i-1}(\varphi) \right\|_0 A\frac{m}{\gamma}\left(1 - e^{\gamma t_1}\right),\ t_1 \leq 0 \qquad (2.44)$$

Treating the relation (2.15) in exactly the same way, in which the estimate (4.30) has been obtained, we find

$$\left\| \frac{\partial \varphi_{t_1}^i(\varphi)}{\partial \varphi_j} - \frac{\partial \varphi_{t_1}^{i-1}(\varphi)}{\partial \varphi_j} \right\| \leq |t_1| e^{\gamma t_1} m\left\{ \left[K_2 + mpK_4 + \frac{m}{\gamma}A\left(K_1 + mpK_2 + mAK\right.\right.\right.$$

$$\left.\left. + mpK_3 + m^2p^2K_4\right)\right] \left\| u^i(\varphi) - u^{i-1}(\varphi) \right\|_0 + mA\left\| \frac{\partial u^i(\varphi)}{\partial \varphi_j} - \frac{\partial u^{i-1}(\varphi)}{\partial \varphi_j} \right\|_0 \right\}$$

$$+ \left(e^{2\gamma t_1} - e^{\gamma t_1}\right)\frac{m}{\gamma^2}A\left[K_1 + mAK + mp(K_2 + K_3 + mp(K_2 + K_3\right.$$

$$\left. + mpK_4)\right] \| u^i(\varphi) - u^{i-1}(\varphi)\|_0,\quad t_1 \leq 0. \qquad (2.45)$$

It follows from the system (2.5) that

$$\frac{d}{dt}\left(y_{t_1}^i(\tau,\varphi) - y_{t_1}^{i-1}(\tau,\varphi) \right) = -b\left(\varphi_{t_1}^i(\varphi), \varphi_{t_1-\Delta}^i(\varphi), u^i\left(\varphi_{t_1}^i(\varphi)\right), u^i\left(\varphi_{t_1-\Delta}^i(\varphi)\right) \right)\left(y_{t_1}^i(\tau,\varphi) \right.$$

$$- y_{t_1}^{i-1}(\tau,\varphi) \bigg) + \left[b\left(\varphi_{t_1}^{i-1}(\varphi), \varphi_{t_1-\Delta}^{i-1}(\varphi), u^{i-1}\left(\varphi_{t_1}^{i-1}(\varphi)\right), u^{i-1}\left(\varphi_{t_1-\Delta}^{i-1}(\varphi)\right) \right) \right.$$

$$- b\left(\varphi_{t_1}^i(\varphi), \varphi_{t_1-\Delta}^i(\varphi), u^i\left(\varphi_{t_1}^i(\varphi)\right), u^i\left(\varphi_{t_1-\Delta}^i(\varphi)\right) \right)\bigg] y_{t_1}^{i-1}(\tau,\varphi).$$

Using this relation, we obtain (by analogy with (2.19)) the inequality

$$\left\| y_{t_1}^i(\tau,\varphi) - y_{t_1}^{i-1}(\tau,\varphi) \right\| \leq n\left\| u^i(\varphi) - u^{i-1}(\varphi) \right\|_0 e^{(\beta - \varepsilon_1)\chi_{\tau - t_1})}\left\{ (t_1 - \tau)\left[\frac{m}{\gamma}A\left(\left\| \frac{\partial b}{\partial \varphi} \right\|_0 \right.\right.\right.$$

$$\left.\left. + \left\| \frac{\partial b}{\partial \psi} \right\|_0 \right) + \left(1 + \frac{m^2p}{\gamma}A\right)\left(\left\| \frac{\partial b}{\partial y} \right\|_0 + \left\| \frac{\partial b}{\partial z} \right\|_0 \right) \right]$$

$$+ \left(e^{\gamma \tau} - e^{\gamma t_1}\right)\frac{m}{\gamma^2}A\left[\left\| \frac{\partial b}{\partial \psi} \right\|_0 + mp\left\| \frac{\partial b}{\partial y} \right\|_0 \right.$$

$$+e^{-\lambda\Delta}\left(\left\|\frac{\partial b}{\partial\psi}\right\|_0 + m\rho\left\|\frac{\partial b}{\partial z}\right\|_0\right)\right]\right\} \tag{2.46}$$

for all $\tau \le t_1 \le 0$. Taking into account (2.28) and (2.44), we find

$$\left\| b_1\left(\varphi_{t_1}^i(\varphi),\varphi_{t_1-\Delta}^i(\varphi),u^i\left(\varphi_{t_1}^i(\varphi)\right),u^i\left(\varphi_{t_1-\Delta}^i(\varphi)\right)\right) u^i\left(\varphi_{t_1-\Delta}^i(\varphi)\right) \right.$$

$$\left. - b_1\left(\varphi_{t_1}^{i-1}(\varphi),\varphi_{t_1-\Delta}^{i-1}(\varphi),u^{i-1}\left(\varphi_{t_1}^{i-1}(\varphi)\right),u^{i-1}\left(\varphi_{t_1-\Delta}^{i-1}(\varphi)\right)\right) u^{i-1}\left(\varphi_{t_1-\Delta}^{i-1}(\varphi)\right)\right\|$$

$$+ \left\| c\left(\varphi_{t_1}^i(\varphi),\varphi_{t_1-\Delta}^i(\varphi)\right) - c\left(\varphi_{t_1}^{i-1}(\varphi),\varphi_{t_1-\Delta}^{i-1}(\varphi)\right)\right\| \le \left\| u^i(\varphi)-u^{i-1}(\varphi)\right\|_0 \left\{ \|b_1\|_0 \right.$$

$$+ \left(\left\|\frac{\partial b_1}{\partial y}\right\|_0 + \left\|\frac{\partial b_1}{\partial z}\right\|_0\right) d_0 + \frac{m}{\gamma}A\left[L_4+\left(\left\|\frac{\partial b_1}{\partial\varphi}\right\|_0 + m\rho\left\|\frac{\partial b_1}{\partial y}\right\|_0\right) d_2\right](1-e^{\gamma t_1})$$

$$+ \frac{m}{\gamma}A\left[L_4^\Delta + m\rho\|b_1\|_0 + \left(\left\|\frac{\partial b_1}{\partial\psi}\right\|_0 + m\rho\left\|\frac{\partial b_1}{\partial z}\right\|_0\right) d_0\right](1-e^{\gamma(t_1-\Delta)})\right\}. \tag{2.47}$$

We can now estimate the right-hand side of (2.43). We have

$$\left\| u^{i+1}(\varphi)-u^i(\varphi)\right\|_1 \le N_1\left\| u^i(\varphi)-u^{i-1}(\varphi)\right\|_0 + N_2\max_{1\le j\le m}\left\|\frac{\partial u^i(\varphi)}{\partial\varphi_j}-\frac{\partial u^{i-1}(\varphi)}{\partial\varphi_j}\right\|_0, \tag{2.48}$$

where N_1 and N_2 are completely determined variables. The inequality (2.48) can be rewritten as follows

$$\left\| u^{i+1}(\varphi)-u^i(\varphi)\right\|_1 \le N\left\| u^i(\varphi)-u^{i-1}(\varphi)\right\|_1, \tag{2.49}$$

where $N = \max(N_1, N_2)$. If

$$N < 1 \tag{2.50}$$

then (2.49) implies the convergence of the sequence of tori (2.3) in the norm $|\cdot|_1$. It is clear that one can always satisfy the inequalities (2.12), (2.26), (2.27), (2.41), and (2.50) by choosing the functions $b_1(\varphi, \psi, y, z)$ and $c(\varphi, \psi)$ such that they, their first derivatives with respect to all the arguments, and the Lipschitz constants for these derivatives, are sufficiently small in the region (2.2).

Theorem 4.3. (Tsydilo, 1973). *Suppose that the right-hand side of the system of differential equations*

$$\frac{dy(t)}{dt} = -b(\varphi(t), \varphi(t-\Delta), y(t), y(t-\Delta)) \, y(t)$$

$$- b_1(\varphi(t), \varphi(t-\Delta), y(t), y(t-\Delta)) \, y(t-\Delta) + c(\varphi(t), \varphi(t-\Delta));$$

$$\frac{d\varphi(t)}{dt} = a(\varphi(t), y(t))$$

$$(2.51)$$

satisfies the following conditions:

(i) $a(\varphi, y), c(\varphi, \psi), b(\varphi, \psi, y, z), b_1(\varphi, \psi, y, z) \in C^1(\mathfrak{M});$

(ii) the derivatives of the vector functions $a(\varphi, y)$ and $c(\varphi, \psi)$ and of the matrices $b(\varphi, \psi, y, z)$ and $b_1(\varphi, \psi, y, z)$ satisfy the Lipschitz conditions;

(iii)

$$\min_{\|\eta\|=1} \langle b(\varphi, \psi, 0, 0) \eta, \eta \rangle \geq \beta,$$

$$\min_{\|\eta\|=1} \left\langle \frac{\partial a(\varphi, 0)}{\partial \varphi} \eta, \eta \right\rangle \geq \alpha, \ \alpha > 0, \ \beta + 2\alpha \geq 0.$$

The functions $b_1(\varphi, \psi, y, z)$ and $c(\varphi, \psi)$, their first derivatives with respect to all the arguments, and the Lipschitz constants of these derivatives are sufficiently small in the region

$$\varphi \in \mathcal{C}_m, \ \psi \in \mathcal{C}_m, \ \|y\| \leq d, \ \|z\| \leq d,$$

$$(2.52)$$

i. e., such that the inequalities (2.12), (2.26), (2.27), (2.41), and (2.50) hold. Then the system of differential equations (2.51) has an invariant toroidal manifold $y = u(\varphi)$ such that

$$\|u(\varphi)\| \leq d_0, \ \left\|\frac{\partial u(\varphi)}{\partial \varphi_j}\right\| \leq \rho,$$

$$\left\|\frac{\partial u(\bar{\bar{\varphi}})}{\partial \varphi_j} - \frac{\partial u(\bar{\varphi})}{\partial \varphi_j}\right\| \leq K \|\bar{\bar{\varphi}} - \bar{\varphi}\| \ (j = 1, \dots, m)$$

§3. Invariant Toroidal Sets for the Systems of Differential Equations with Lag under Pulse Influence

Consider a system of differential equations of the form

$$\frac{dy(t)}{dt} = A(\varphi)\, y(t) + B(\varphi)\, y(t-\Delta) + c(\varphi(t),\, \varphi(t-\Delta), y(t), y(t-\Delta),\, \mu),\quad t \neq t_j;$$

$$\frac{d\varphi(t)}{dt} = a(\varphi(t), y(t),\, \mu); \tag{3.1}$$

$$\Delta y\big|_{t=t_j} = H_j(\varphi(t_j),\, \varphi(t_j-\Delta), y(t_j), y(t_j-\Delta),\, \mu),$$

where the functions $y = (y_1, \ldots, y_n)$, $\varphi = (\varphi_1, \ldots, \varphi_m)$, $A(\varphi)$, $B(\varphi)$, $c(\varphi(t), \varphi(t-\Delta), y(t), y(t-\Delta), \mu)$, $a(\varphi(t), y(t), \mu)$, and $H_j(\varphi(t_j), \varphi(t_j-\Delta), y(t_j), y(t_j-\Delta), \mu)$ are periodic in φ and $\psi = \varphi(t-\Delta)$ with period 2π; they are defined for all y, $z = y(t-\Delta)$, and μ belonging to the region

$$\|y\| = \left(\sum_{i=1}^{n} y_i^2\right)^{\frac{1}{2}} \le d,\ \ \|z\| \le d,\ \ \mu \in \left[0, \mu_0\right] \tag{3.2}$$

(μ is a small parameter). Henceforth, we assume that the functions H_j, times t_j, and lag Δ satisfy the conditions

$$H_{j+p} = H_j,\ \ t_{j+1} - t_j \ge h,\ \ h > \Delta \ge 0.$$

Let us investigate the conditions under which the system (3.1) possesses an invariant toroidal set for $\mu \neq 0$ (we assume that for $\mu = 0$ this system has the invariant torus $y \equiv 0$). We call the set $\mathcal{T}(\mu)$: $y = u(\varphi, \mu)$ invariant if $u(\varphi, \mu)$ is a bounded continuous function periodic in φ with period 2π. This function defines an invariant toroidal set $\mathcal{T}(\mu)$ of the system of differential equations (3.1) provided that the identity

$$\frac{du\big(\varphi_t(\tau,\, \varphi), \mu\big)}{dt} \equiv A(\varphi_t(\tau,\, \varphi))\, u(\varphi_t(\tau,\, \varphi),\, \mu) + B(\varphi_t(\tau,\, \varphi))\, u(\varphi_{t-\Delta}(\tau,\, \varphi),\, \mu)$$

$$+\, c(\varphi_t(\tau,\, \varphi),\, \varphi_{t-\Delta}(\tau,\, \varphi), u(\varphi_t(\tau,\, \varphi),\, \mu), u(\varphi_{t-\Delta}(\tau,\, \varphi),\, \mu),\, \mu),\quad t \neq t_j, \tag{3.3}$$

$$\Delta u(\varphi_t(\tau,\, \varphi),\, \mu)|_{t=t_j} = H_j(\varphi_{t_j}(\tau,\, \varphi),\, \varphi_{t_j-\Delta}(\tau,\, \varphi),\, u(\varphi_{t_j}(\tau,\, \varphi),\, \mu),\, u(\varphi_{t_j-\Delta}(\tau,\, \varphi),\, \mu),\, \mu),$$

holds for all $t \in (-\infty,\, \infty)$. Here, $\varphi_t(\tau,\, \varphi)$, $\varphi_t(\tau,\, \varphi) = \varphi$ is a solution of the system of differential equations

$$\frac{d\varphi(t)}{dt} = a(\varphi,\, u(\varphi,\, \mu),\, \mu),$$

with τ and φ being arbitrary constants.

Consider the system of differential equations

$$\frac{dy(t)}{dt} = A(\varphi)\, y(t) + B(\varphi)\, y(t-\Delta) + c_1(\varphi);$$

$$(3.4)$$

$$\frac{d\varphi(t)}{dt} = a_1(\varphi), \quad t \neq t_j; \quad \Delta y\,|_{t=t_j} = H_j(\varphi).$$

Denote by $G_t(\tau\, \varphi)$ a real matrix function continuous in $\tau \neq t$ and φ which is periodic in φ with period 2π and satisfies the following conditions:

(i) for $t \neq \tau$, the function $G_t(\tau\, \varphi)$ is a solution of the equation

$$\frac{dy(t)}{dt} = A(\varphi_t(\tau,\, \varphi))\, y(t) + B(\varphi_t(\tau,\, \varphi))\, y(t-\Delta);$$

(ii) $$G_t(\tau\, \varphi)|_{t=\tau+0} - G_t(\tau\, \varphi)|_{t=\tau-0} = E, \qquad (3.5)$$

where E is the unit matrix;

(iii) $$\| G_t(\tau\, \varphi) \| \leq K\, e^{-\gamma|t-\tau|} \qquad (3.6)$$

for positive $K > 0$ and $\gamma > 0$ independent of t, τ, φ, and for all $(t,\, \tau) \in (-\infty,\, \infty)$.

The function $G_0(\tau\, \varphi) = G_t(\tau\, \varphi)|_{t=0}$ is called a Green's function for the problem concerning invariant toroidal sets of the system (3.4). The invariant toroidal set $y = u(\varphi,\, \mu)$ for the system (3.4) is defined by

$$u(\varphi,\mu) = \int_{-\infty}^{\infty} G_0(s,\varphi)c_1\big(\varphi_s(\varphi)\big)\, ds + \sum G_0\big(t_j,\varphi\big)\, H_j\Big(\varphi_{t_j}(\varphi)\Big) \qquad (3.7)$$

In order to find an invariant toroidal set of the system (3.1), we shall use (as before) the Green's function for the problem of invariant toroidal sets of the linear system (3.4) and employ the iteration process which enables us to construct $\mathcal{T}(\mu)$ as the limit of a sequence of invariant toroidal sets $\mathcal{T}^0(\mu) \equiv 0, \mathcal{T}^1(\mu), ..., \mathcal{T}^i(\mu), ...,$ each being an invariant toroidal set

$$\mathcal{T}^{i+1}(\mu): y = u^{i+1}(\varphi, \mu), \quad = 0, 1, ... \tag{3.8}$$

of the linear system of differential equations

$$\frac{dy(t)}{dt} = A(\varphi(t)) y(t) + B(\varphi(t)) y(t - \Delta)$$

$$+ c(\varphi(t), \varphi(t - \Delta), u^i(\varphi(t), \mu), u^i(\varphi(t-\Delta), \mu), \mu), \quad t \neq t_j;$$

$$\Delta y\,|_{t=t_j} = H_j(\varphi(t_j), \varphi(t_j - \Delta), u^i(\varphi(t_j - \Delta), \mu), u^i(\varphi_{t_j-\Delta}, \mu), \mu); \tag{3.9}$$

$$\frac{d\varphi(t)}{dt} = a(\varphi(t), u^i(\varphi(t), \mu), \mu).$$

Suppose that $\varphi_t^i(\varphi), \varphi_\tau^i(\varphi) = \varphi$ is a solution of the second equation of (3.9). Then, taking into account (3.7), we obtain the following formula for the invariant toroidal set of the system (3.9)

$$y = u^{i+1}(\varphi,\mu) = \int_{-\infty}^{\infty} G_0^i(s,\varphi)\, c\Big(\varphi_s^i(\varphi), \varphi_{s-\Delta}^i(\varphi), u^i\big(\varphi_s^i(\varphi),\mu\big), u^i\big(\varphi_{s-\Delta}^i(\varphi),\mu\big),\mu\Big)\, ds$$

$$+ \sum_{-\infty < i < \infty} G_0^i(t_j,\varphi)\, H_j\Big(\varphi_{t_j}^i(\varphi), \varphi_{t_j-\Delta}^i(\varphi), u^i\big(\varphi_{t_j}^i(\varphi),\mu\big), u^i\big(\varphi_{t_j-\Delta}^i(\varphi),\mu\big),\mu\Big). \tag{3.10}$$

Henceforth, we assume that the system (3.1) satisfies the conditions

$$\| a(\varphi, y, \mu) \| + \| c(\varphi, \psi, y, z, \mu) \| + \| H_j(\varphi, \psi, y, z, \mu) \| \le M(d, \mu);$$

$$\| c(\varphi', \psi', y', z', \mu) - c(\varphi'', \psi'', y'', z'', \mu) \| + \| H_j(\varphi', \psi', y', z', \mu)$$

$$- H_j(\varphi'', \psi'', y'', z'', \mu) \| \le L(d, \mu)\, [\| \varphi' - \varphi'' \| + \| \psi' - \psi'' \| + \| y' - y'' \| + \| z' - z'' \|];$$

$$\| a(\varphi', y', \mu) - a(\varphi'', y'', \mu) \| \le L(d, \mu)\, [\| \varphi' - \varphi'' \| + \| y' - y'' \|];$$

$$\| A(\varphi') - A(\varphi'') \| \leq A_1 \| \varphi' - \varphi'' \|, \, A_1 > 0;$$

(3.11)

$$\| B(\varphi') - B(\varphi'') \| \leq B_1 \| \varphi' - \varphi'' \|, \, B_1 > 0,$$

where $M(d, \mu)$ and $L(d, \mu)$ are positive monotonically decreasing functions such that $L(d, \mu) \to 0$ and $M(d, \mu) \to 0$ as $d \to 0$ and $\mu \to 0$.

Under conditions (3.11), the method of finding $\mathcal{T}(\mu)$ (as a limit of the sequence (3.8)) is justified by Lemma 4.1, according to which, the limiting function $u = \lim_{i \to \infty} u^i(\varphi, \mu)$ defines the invariant set $\mathcal{T}(\mu)$ provided that the sequence of invariant sets (3.8) converges uniformly.

The existence of the sequence of toroidal sets (3.8) is established by the following lemma.

Lemma 4.3. (Martinyuk and Tsyganovsky, 1979a). *Suppose that the above-mentioned conditions imposed on the right-hand side of* (3.1) *are satisfied. Then one can always find positive* μ_1 $(0 \leq \mu_1 \leq \mu_0)$ *such that for all* $\mu \leq \mu_1$ *the sequence* (3.10) *defines the sequence of invariant sets, each satisfying the inequality*

$$\left\| u^{i+1}(\varphi, \mu) \right\| \leq \frac{\alpha M(d, \mu)}{1 - 2\alpha L(d, \mu)},$$

(3.12)

where

$$\alpha = \frac{2K}{\gamma} \left[1 + \frac{\gamma}{1 - e^{-\gamma h}} \right], \, 2\alpha L(d, \mu) < 1.$$

Proof. Taking into account the fact that $u^0(\varphi, \mu) \equiv 0$, for $i = 0$, we obtain

$$\| u^1(\varphi, \mu) \| \leq K \int_{-\infty}^{\infty} e^{-\gamma |s|} \left\| c\left(\varphi_s^0(\varphi), \varphi_{s-\Delta}^0(\varphi), 0, 0, \mu \right) \right\| ds$$

$$+ K \sum_{-\infty < j < \infty} e^{-\gamma |t_j|} \left\| H_j \left(\varphi_{t_j}^0(\varphi), \varphi_{t_j - \Delta}^0(\varphi), 0, 0, \mu \right) \right\|$$

$$\leq K M(d, \mu) \int_{-\infty}^{\infty} e^{-\gamma |s|} ds + K M(d, \mu) \sum_{-\infty < j < \infty} e^{-\gamma |t_j|}$$

$$\leq K M(d, \mu) \left[\frac{2}{\gamma} + \frac{2}{1 - s^{-\gamma h}} \right] \leq \alpha M(d, \mu) \leq \frac{\alpha M(d, \mu)}{1 - 2\alpha L(d, \mu)},$$

(3.13)

provided that $2\alpha\, L(d,\,\mu) < 1$. Assuming that (3.13) holds for $i = k$, we must prove that the invariant set $\mathcal{C}^{k+1}(\mu)$: $y = u^{k+1}(\varphi,\,\mu)$ exists and satisfies the inequality (3.13). Indeed, employing the inequalities $t_{j+1} - t_j \geq h$, (3.11), and (3.13), we can derive from (3.10) the following estimate

$$
\| u^{i+1}(\varphi,\,\mu)\,\| \leq \int\limits_{-\infty}^{\infty} \left\| G_0^i(s,\varphi)\right\| \left\| c\!\left(\varphi_s^i(\varphi),\varphi_{s-\Delta}^i(\varphi),u^i\!\left(\varphi_s^i(\varphi),\mu\right),u^i\!\left(\varphi_{s-\Delta}^i(\varphi),\mu\right),\mu\right)\right.
$$

$$
-\ c\!\left(\varphi_s^i(\varphi),\varphi_{s-\Delta}^i(\varphi),0,0,\mu\right)\Big\|\,ds
$$

$$
+\ \int\limits_{-\infty}^{\infty} \left\| G_0^i(s,\varphi)\right\| \left\| c\!\left(\varphi_s^i(\varphi),\varphi_{s-\Delta}^i(\varphi),0,0,\mu\right)\right\| ds \ +\ \sum_{-\infty<j<\infty} \left\| G_0^i(t_j,\varphi)\right\|
$$

$$
\times\left\| H_j\!\left(\varphi_{t_j}^i(\varphi),\varphi_{t_j-\Delta}^i(\varphi),u^i\!\left(\varphi_{t_j}^i(\varphi),\mu\right),u^i\!\left(\varphi_{t_j-\Delta}^i(\varphi),\mu\right),\mu\right)\right.
$$

$$
-\ H_j\!\left(\varphi_{t_j}^i(\varphi),\varphi_{t_j-\Delta}^i(\varphi),0,0,\mu\right)\Big\| \ +\ \sum_{-\infty<j<\infty} \left\| G_0^i(t_j,\varphi)\right\|
$$

$$
\times\left\| H_j\!\left(\varphi_{t_j}^i(\varphi),\varphi_{t_j-\Delta}^i(\varphi),0,0,\mu\right)\right\| \ \leq\ K\,L(d,\mu)\int\limits_{-\infty}^{\infty} e^{-\gamma|s|}\Big[\big\| u^i\!\left(\varphi_s^i(\varphi),\mu\right)\big\|
$$

$$
-\ \big\| u^i\!\left(\varphi_{s-\Delta}^i(\varphi),\mu\right)\big\|\Big]\,ds \ +\ K\,L(d,\mu)\sum_{-\infty<j<\infty} e^{-\gamma|t_j|}\Big[\big\| u^i\!\left(\varphi_{t_j}^i(\varphi),\mu\right)\big\|
$$

$$
+\ \big\| u^i\!\left(\varphi_{t_j-\Delta}^i(\varphi),\mu\right)\big\|\Big] \ +\ K\,M(d,\mu)\int\limits_{-\infty}^{\infty} e^{-\gamma|s|}\,ds \ +\ K\,M(d,\mu)\sum_{-\infty<j<\infty} e^{-\gamma|t_j|}
$$

$$
\leq\ 2\,K\,L(d,\mu)\frac{\alpha\,M(d,\mu)}{1-2\alpha\,L(d,\mu)}\left[\frac{2}{\gamma} + \frac{2}{1-e^{-\gamma h}}\right] + \alpha\,M(d,\mu)
$$

$$
=\ \frac{2\alpha^2\,L(d,\mu)\,M(d,\mu)}{1-2\alpha\,L(d,\mu)} + \alpha\,M(d,\mu) \ =\ \frac{\alpha\,M(d,\mu)}{1-2\alpha\,L(d,\mu)}
$$

By induction, we conclude that the estimate (3.13) holds for all $i = 0, 1, 2, \ldots$, if $2\alpha L(d,\,\mu) < 1$.

Since $M(d,\,\mu) \to 0$ and $N(d,\,\mu) \to 0$ as $\mu \to 0$, one can always find μ_1 such that the inequalities

$$\frac{\alpha M(d,\mu)}{1-2\alpha L(d,\mu)} \le d, \quad 2\alpha L(t,\mu) < 1; \tag{3.14}$$

hold. If μ_1 is so chosen, then for all $\mu \in [0, \mu_1]$ and $i = 0, 1, 2, \ldots$, the system (3.9) has the invariant toroidal set $\mathcal{T}^{k+1}(\mu)$: $y = u^{k+1}(\varphi, \mu)$ satisfying the estimate (3.13). Furthermore, (3.13) involves the uniform boundedness of the sequence of invariant toroidal sets (3.8).

In order to prove that the sequence (3.8) is convergent, we need the following auxiliary statement.

Lemma 4.4. (Martinyuk and Tsyganovsky, 1979a). *There exist positive numbers* $0 \le \mu_2 \le \mu_1$ *and* $N(\mu)$ *(* $N(\mu) \to 0$ *as* $\mu \to 0$ *) such that the inequalities*

$$\| u^{i+1}(\varphi', \mu) - u^{i+1}(\varphi'', \mu) \| \le N(\mu) \| \varphi' - \varphi'' \|; \tag{3.15}$$

$$\left\| \varphi_t^i(\tau, \varphi') - \varphi_t^i(\tau, \varphi'') \right\| \le \left\| \varphi' - \varphi'' \right\| e^{L(d,\mu)(1+N(\mu))(t-\tau)} \tag{3.16}$$

hold for all $0 \le \mu \le \mu_2$, $-\infty < t < \infty$, $-\infty < \tau < \infty$, φ', φ'', *and* $i = 0, 1, \ldots$

Proof. From the second equation of (3.9) with $i = 0$, we obtain

$$\left\| \varphi_t^0(\tau, \varphi') - \varphi_t^0(\tau, \varphi'') \right\| \le \left\| \varphi' - \varphi'' \right\| + L(d,\mu) \left| \int_\tau^t \left\| \varphi_s^0(\tau, \varphi') - \varphi_s^0(\tau, \varphi'') \right\| ds \right|,$$

and this yields the estimate

$$\left\| \varphi_t^0(\tau, \varphi') - \varphi_t^0(\tau, \varphi'') \right\| \le \left\| \varphi' - \varphi'' \right\| e^{L(d,\mu)|t-\tau|}. \tag{3.17}$$

Since $G_t(\tau, \varphi)$ is a solution of the equation (3.5), the difference $G_t^0(\tau, \varphi') - G_t^0(\tau, \varphi'')$ is a solution of the equation

$$\frac{d}{dt}\left[G_t^0(\tau, \varphi') - G_t^0(\tau, \varphi'') \right] = A\left(\varphi_t^0(\tau, \varphi')\right)\left[G_t^0(\tau, \varphi') - G_t^0(\tau, \varphi'') \right]$$

$$+ B\left(\varphi_t^0(\tau, \varphi')\right)\left[G_{t-\Delta}^0(\tau, \varphi') - G_{t-\Delta}^0(\tau, \varphi'') \right]$$

$$+ \left[A\left(\varphi_t^0(\tau, \varphi')\right) - A\left(\varphi_t^0(\tau, \varphi'')\right) \right] G_t^0(\tau, \varphi'')$$

$$+ \left[B\left(\varphi_t^0(\tau, \varphi')\right) - B\left(\varphi_t^0(\tau, \varphi'')\right) \right] G_{t-\Delta}^0(\tau, \varphi''). \tag{3.18}$$

Therefore, taking (3.6) and (3.17) into account, we find

$$\left\| G_t^0(\tau,\varphi') - G_t^0(\tau,\varphi'') \right\| \leq \left\| \varphi' - \varphi'' \right\| \Biggl\{ K^2 A_1 \int_{-\infty}^{\infty} e^{-\gamma|t-s|} e^{-(\gamma-L(d,\mu))|s-\tau|} \, ds$$

$$+ K^2 B_1 \int_{-\infty}^{\infty} e^{-\gamma|t-s|} e^{-\gamma|s-\tau-\Delta|} e^{L(d,\mu)|s-\tau|} \, ds \Biggr\}$$

$$\leq \left\| \varphi' - \varphi'' \right\| \Biggl\{ K^2 A_1 \left[\frac{2}{2\gamma - L(d,\mu)} + \frac{1-e^{-L(d,\mu)|t-\tau|}}{L(d,\mu)} \right] e^{-(\gamma-L(d,\mu))|t-\tau|}$$

$$+ K^2 B_1 e^{L(d,\mu)\Delta} \left[\frac{1+e^{L(d,\mu)\Delta}}{2\gamma - L(d,\mu)} + \frac{e^{(2\gamma+L(d,\mu))\Delta} - 1}{2\gamma + L(d,\mu)} \right.$$

$$\left. + \frac{1-e^{-L(d,\mu)\Delta}}{L(d,\mu)} + \frac{1-e^{-L(d,\mu)|t-\tau-\Delta|}}{L(d,\mu)} \right] e^{-(\gamma-L(d,\mu))|t-\tau-\Delta|} \Biggr\}. \qquad (3.19)$$

Let us consider the representation (3.10) for $i = 0$. Using (3.17) and (3.19), we obtain

$$\left\| u^1(\varphi',\mu) - u^1(\varphi'',\mu) \right\| \leq N_1(\mu) \left\| \varphi' - \varphi'' \right\|, \qquad (3.20)$$

where

$$N_1(\mu) = KL(d,\mu) \Biggl\{ \frac{2+e^{L(d,\mu)\Delta} + e^{-\gamma\Delta}}{\gamma - L(d,\mu)} + \frac{2+e^{-L(d,\mu)\Delta} + e^{-\gamma\Delta}}{1-e^{-(\gamma-L(d,\mu))h}} + \frac{e^{L(d,\mu)\Delta} - e^{-\gamma\Delta}}{\gamma + L(d,\mu)}$$

$$+ \frac{e^{L(d,\mu)\Delta} - e^{-\gamma\Delta}}{1-e^{-(\gamma+L(d,\mu))h}} \Biggr\} + K^2 M(d,\mu) \Biggl\{ \frac{2A_1}{2\gamma - L(d,\mu)} + B_1 e^{L(d,\mu)\Delta} \left[\frac{1+e^{L(d,\mu)\Delta}}{2\gamma - L(d,\mu)} \right.$$

$$\left. + \frac{e^{(2\gamma+L(d,\mu))\Delta} - 1}{2\gamma + L(d,\mu)} + \frac{1-e^{-L(d,\mu)\Delta}}{L(d,\mu)} \right] \Biggr\} \left[\frac{2}{\gamma - L(d,\mu)} + \frac{2}{1-e^{-(\gamma-L(d,\mu))}} \right]$$

$$+ K^2 M(d,\mu) \left(A_1 + B_1 e^{L(d,\mu)\Delta} \right) \frac{1}{L(d,\mu)} \left[\frac{2}{\gamma - L(d,\mu)} \right.$$

$$\left. + \frac{2}{\gamma} \frac{2}{1-e^{-(\gamma-L(d,\mu))h}} - \frac{2}{1-e^{-\gamma h}} \right]. \qquad (3.21)$$

Using the second equation in (3.9) with $i = 1$ and (3.20), we obtain the inequality

$$\left\| \varphi_t^1(\tau,\varphi') - \varphi_t^1(\tau,\varphi'') \right\| \le \left\| \varphi' - \varphi'' \right\| + L(d,\mu)\left(1 + N_1(\mu)\right)$$

$$\times \left| \int_\tau^t \left\| \varphi_s^1(\tau,\varphi') - \varphi_s^1(\tau,\varphi'') \right\| ds \right|,$$

which yields

$$\left\| \varphi_t^1(\tau,\varphi') - \varphi_t^1(\tau,\varphi'') \right\| \le \left\| \varphi' - \varphi'' \right\| e^{L(d,\mu)\left(1 + N_1(\mu)\right)|t-\tau|}. \tag{3.22}$$

Estimating the difference $G_t^1(\tau,\varphi') - G_t^1(\tau,\varphi'')$ by use of (3.22), we obtain

$$\left\| G_t^1(\tau,\varphi') - G_t^1(\tau,\varphi'') \right\| \le \left\| \varphi' - \varphi'' \right\| \left\{ K^2 A_1 \left[\frac{2}{2\gamma - L(d,\mu)\left(1 + N_1(\mu)\right)} \right. \right.$$

$$+ \frac{1 - e^{-L(d,\mu)\left(1 + N_1(\mu)\right)|t-\tau|}}{L(d,\mu)\left(1 + N_1(\mu)\right)} \left] e^{-\left(\gamma - L(d,\mu)\left(1 + N_1(\mu)\right)\right)|t-\tau|} \right.$$

$$+ K^2 B_1 \, e^{L(d,\mu)\left(1 + N_1(\mu)\right)\Delta} \left[\frac{1 + e^{L(d,\mu)\left(1 + N_1(\mu)\right)\Delta}}{2\gamma - L(d,\mu)\left(1 + N_1(\mu)\right)} \right.$$

$$+ \frac{e^{\left(2\gamma + L(d,\mu)\left(1 + N_1(\mu)\right)\right)\Delta} - 1}{2\gamma + L(d,\mu)\left(1 + N_1(\mu)\right)} + \frac{1 - e^{-L(d,\mu)\left(1 + N_1(\mu)\right)\Delta}}{L(d,\mu)\left(1 + N_1(\mu)\right)}$$

$$+ \frac{1 - e^{-L(d,\mu)\left(1 + N_1(\mu)\right)|t-\tau-\Delta|}}{L(d,\mu)\left(1 + N_1(\mu)\right)} \right]$$

$$\times e^{-\left(\gamma - L(d,\mu)\left(1 + N_1(\mu)\right)\right)|t-\tau-\Delta|} \right\}, \tag{3.23}$$

where $L(d,\mu)\left(1 + N_1(\mu)\right) < \gamma$. The representation (3.10) with $i = 1$ and the inequalities (3.22), (3.23) imply that

$$\| u^2(\varphi',\mu) - u^2(\varphi'',\mu) \| \le \| \varphi' - \varphi'' \| N_2(\mu), \tag{3.24}$$

where

$$
N_2(\mu) = L(d, \mu) \, (1 + N_1(\mu)) \, K \left\{ \frac{2 + e^{L(d,\mu)(1 + N_1(\mu))\Delta} + e^{-\gamma\Delta}}{\gamma - L(d,\mu)(1 + N_1(\mu))} \right.
$$

$$
+ \frac{2 + e^{L(d,\mu)(1 + N_1(\mu))\Delta} + e^{-\gamma\Delta}}{1 - e^{-(\gamma - L(d,\mu)(1 + N_1(\mu)))h}} + \frac{e^{L(d,\mu)(1 + N_1(\mu))\Delta} - e^{-\gamma\Delta}}{\gamma + L(d,\mu)(1 + N_1(\mu))}
$$

$$
+ \left. \frac{e^{L(d,\mu)(1 + N_1(\mu))\Delta} - e^{-\gamma\Delta}}{1 - e^{-(\gamma + L(d,\mu)(1 + N_1(\mu)))h}} \right\} + M(d, \mu) \, K^2 \left\{ \frac{2 A_1}{2\gamma - L(d,\mu)(1 + N_1(\mu))} \right.
$$

$$
+ B_1 e^{L(d,\mu)(1 + N_1(\mu))\Delta} \left[\frac{1 + e^{L(d,\mu)(1 + N_1(\mu))\Delta}}{2\gamma - L(d,\mu)(1 + N_1(\mu))} \right.
$$

$$
+ \left. \left. \frac{e^{(2\gamma + L(d,\mu)(1 + N_1(\mu)))\Delta} - 1}{2\gamma + L(d,\mu)(1 + N_1(\mu))} + \frac{1 - e^{-L(d,\mu)(1 + N_1(\mu))\Delta}}{L(d,\mu)(1 + N_1(\mu))} \right] \right\}
$$

$$
\times \left[\frac{2}{\gamma - L(d,\mu)(1 + N_1(\mu))} + \frac{2}{1 - e^{-(\gamma - L(d,\mu)(1 + N_1(\mu)))h}} \right]
$$

$$
+ K^2 \left(A_1 + B_1 \, e^{L(d,\mu)(1 + N_1(\mu))\Delta} \right) \frac{1}{L(d,\mu)(1 + N_1(\mu))}
$$

$$
\left[\frac{2}{\gamma - L(d,\mu)(1 + N_1(\mu))} - \frac{2}{\gamma} + \frac{2}{1 - e^{-(\gamma - L(d,\mu)(1 + N_1(\mu)))h}} - \frac{2}{1 - e^{-\gamma h}} \right].
$$

One can easily show by induction that the inequalities

$$
\left\| \varphi_t^i(\tau, \varphi') - \varphi_t^i(\tau, \varphi'') \right\| \le \left\| \varphi' - \varphi'' \right\| e^{L(d,\mu)(1 + N_i(\mu))|t - \tau|}; \tag{3.25}
$$

$$
\left\| G_t^i(\tau, \varphi') - G_t^i(\tau, \varphi'') \right\| \le \left\| \varphi' - \varphi'' \right\| \left\{ K^2 A_1 \left[\frac{2}{2\gamma - L(d,\mu)(1 + N_i(\mu))} \right. \right.
$$

$$+ \left. \frac{1 - e^{-L(d,\mu)(1+N_i(\mu))|t-\tau|}}{L(d,\mu)(1+N_i(\mu))} \right] e^{-(\gamma - L(d,\mu)(1+N_i(\mu)))|t-\tau|}$$

$$+ K^2 B_1 e^{L(d,\mu)(1+N_i(\mu))\Delta} \left[\frac{1 + e^{L(d,\mu)(1+N_i(\mu))\Delta}}{2\gamma - L(d,\mu)(1+N_i(\mu))} \right.$$

$$+ \frac{e^{(2\gamma + L(d,\mu)(1+N_i(\mu)))\Delta} - 1}{2\gamma + L(d,\mu)(1+N_i(\mu))} + \frac{1 - e^{-L(d,\mu)(1+N_i(\mu))\Delta}}{L(d,\mu)(1+N_i(\mu))}$$

$$+ \left. \frac{1 - e^{-L(d,\mu)(1+N_i(\mu))|t-\tau-\Delta|}}{L(d,\mu)(1+N_i(\mu))} \right] e^{-(\gamma - L(d,\mu)(1+N_i(\mu)))|t-\tau-\Delta|} \bigg\} \tag{3.26}$$

$$\left\| u^{i+1}(\varphi',\mu) - u^{i+1}(\varphi'',\mu) \right\| \leq N_{i+1} \left\| \varphi' - \varphi'' \right\| \tag{3.27}$$

hold for all $i = 0, 1, 2, \dots$ provided that $L(d, \mu)(1 + N_i(\mu)) < \gamma$, where $N_{i+1}(\mu)$ can be found from the recursion relation

$$N_{i+1}(\mu) = L(d, \mu)(1 + N_i(\mu)) M_1(\mu, N_i(\mu)) + M(d, \mu) M_2(\mu, N_i(\mu)) \tag{3.28}$$

Here,

$$M_1(\mu, N_i(\mu)) = K \left\{ \frac{2 + e^{L(d,\mu)(1+N_i(\mu))\Delta} + e^{-\gamma\Delta}}{\gamma - L(d,\mu)(1+N_i(\mu))} + \frac{2 + e^{L(d,\mu)(1+N_i(\mu))\Delta} + e^{-\gamma\Delta}}{1 - e^{-(\gamma - L(d,\mu)(1+N_i(\mu)))h}} \right.$$

$$+ \left. \frac{e^{L(d,\mu)(1+N_i(\mu))\Delta} - e^{-\gamma\Delta}}{\gamma + L(d,\mu)(1+N_i(\mu))} + \frac{e^{L(d,\mu)(1+N_i(\mu))\Delta} - e^{-\gamma\Delta}}{1 - e^{-(\gamma + L(d,\mu)(1+N_i(\mu)))h}} \right\};$$

$$\tag{3.29}$$

$$M_2(\mu, N_i(\mu)) = K^2 \left\{ \frac{2 A_1}{2\gamma - L(d,\mu)(1+N_i(\mu))} + B_1 e^{L(d,\mu)(1+N_i(\mu))\Delta} \right.$$

$$\times \left[\frac{1 + e^{L(d,\mu)(1+N_i(\mu))\Delta}}{2\gamma - L(d,\mu)(1+N_i(\mu))} + \frac{e^{(2\gamma + L(d,\mu)(1+N_i(\mu)))\Delta} - 1}{2\gamma + L(d,\mu)(1+N_i(\mu))} \right.$$

$$+ \frac{1 - e^{-L(d,\mu)(1+N_i(\mu))\Delta}}{L(d,\mu)(1+N_i(\mu))} + \frac{2}{\gamma - L(d,\mu)(1+N_i(\mu))}$$

$$+ \frac{2}{1 - e^{-(\gamma - L(d,\mu)(1+N_i(\mu)))h}} \Bigg]\Bigg\} + K^2\left(A_1 + B_1 e^{L(d,\mu)(1+N_i(\mu))\Delta}\right)$$

$$\times \frac{1}{L(d,\mu)(1+N_i(\mu))} \left[\frac{2}{\gamma - L(d,\mu)(1+N_i(\mu))}\right.$$

$$-\frac{2}{\gamma} + \frac{2}{1 - e^{-(\gamma - L(d,\mu)(1+N_i(\mu)))h}} - \frac{2}{1 - e^{-\gamma h}}\left.\right].$$

Let us choose $\mu_2 \in [0, \mu_1]$ so small that the inequalities

$$L(d, \mu), M_1(\mu, 1) < \frac{1}{3}, \quad M(d, \mu), M_2(\mu, 1) < \frac{1}{3}, \quad 2L(d, \mu) < \gamma \qquad (3.30)$$

hold for all $\mu \in [0, \mu_2]$. This is possible, since the relations (3.30) yield

$$M_1(\mu, 1) \rightarrow K\left(\frac{4}{\gamma} + \frac{4}{1 - e^{-\gamma h}}\right);$$

$$M_2(\mu, 1) \rightarrow K^2\Bigg\{A_1\left[\frac{4}{\gamma^2} + \frac{2}{\gamma(1 - e^{-\gamma h})} + \frac{2he^{-\gamma}}{(1 - e^{-\gamma h})^2}\right]$$

$$+ B_1\left[\frac{3 + e^{2\gamma\Delta}}{\gamma^2} + \frac{1 + e^{2\gamma\Delta}}{\gamma(1 - e^{-\gamma h})} + \frac{2he^{-\gamma h}}{(1 - e^{-\gamma h})^2}\right]\Bigg\}.$$

Then

$$N_{i+1}(\mu) \leq 2L(d, \mu) M_1(\mu, 1) + M(d, \mu) M_2(\mu, 1),$$

Therefore, it suffices to set

$$N(\mu) = 2L(d, \mu) M_1(\mu, 1) + M(d, \mu) M_2(\mu, 1) \qquad (3.31)$$

to complete the proof of Lemma 4.4.

Let us now prove that the sequence of invariant toroidal sets (3.8) is convergent. For this purpose, we estimate the difference $u^{i+1}(\varphi, \mu) - u^i(\varphi, \mu)$. By using (3.10), we obtain

$$
\left\| u^{i+1}(\varphi,\mu) - u^i(\varphi,\mu) \right\| \le \int_{-\infty}^{\infty} \Big\{ \left\| G_0^i(s,\varphi) \right\| \left\| c\Big(\varphi_s^i(\varphi), \varphi_{s-\Delta}^i(\varphi), u^i\big(\varphi_s^i(\varphi)\big), u^i\big(\varphi_{s-\Delta}^i(\varphi)\big), \mu\Big) \right.
$$

$$
\left. - c\Big(\varphi_s^{i-1}(\varphi), \varphi_{s-\Delta}^{i-1}(\varphi), u^{i-1}\big(\varphi_s^{i-1}(\varphi)\big), u^{i-1}\big(\varphi_{s-\Delta}^{i-1}(\varphi)\big), \mu\Big) \right\| + \left\| G_0^i(s,\varphi) \right.
$$

$$
\left. - G_0^{i-1}(s,\varphi) \right\| \left\| c\Big(\varphi_s^{i-1}(\varphi), \varphi_{s-\Delta}^{i-1}(\varphi), u^{i-1}\big(\varphi_s^{i-1}(\varphi), \mu\big), u^{i-1}\big(\varphi_{s-\Delta}^{i-1}(\varphi), \mu\big), \mu\Big) \right\| \Big\} ds
$$

$$
+ \sum_{-\infty < j < \infty} \Big\{ \left\| G_0^i(t_j, \varphi) \right\|
$$

$$
\times \left\| H_j\Big(\varphi_{t_j}^{i-1}(\varphi), \varphi_{t_j - \Delta}^i(\varphi), u^i\big(\varphi_{t_j}^i(\varphi), \mu\big), u^i\big(\varphi_{t_j - \Delta}^i(\varphi), \mu\big), \mu\Big) \right.
$$

$$
\left. - H_j\Big(\varphi_t^{i-1}(\varphi), \varphi_{t-\Delta}^{i-1}(\varphi), u^{i-1}\big(\varphi_{t_j}^{i-1}(\varphi), \mu\big), u^{i-1}\big(\varphi_{t_j - \Delta}^{i-1}(\varphi), \mu\big), \mu\Big) \right\|
$$

$$
+ \left\| G_0^i(t_j, \varphi) - G_0^{i-1}(t_j, \varphi) \right\|
$$

$$
\times \left\| H_j\Big(\varphi_{t_j}^{i-1}(\varphi), \varphi_{t_j - \Delta}^{i-1}(\varphi), u^{i-1}\big(\varphi_{t_j}^{i-1}(\varphi), \mu\big), u^{i-1}\big(\varphi_{t_j}^{i-1}(\varphi), \mu\big), \mu\Big) \right\| \Big\}. \quad (3.32)
$$

Employing here (3.6), (3.11), and (3.15), we get

$$
\left\| u^{i+1}(\varphi,\mu) - u^i(\varphi,\mu) \right\| \le KL(d,\mu) \int_{-\infty}^{\infty} e^{-\gamma |s|} \Big\{ (1 + N(\mu)) \Big[\left\| \varphi_s^i(\varphi) - \varphi_s^{i-1}(\varphi) \right\|
$$

$$
+ \left\| \varphi_{s-\Delta}^i(\varphi) - \varphi_{s-\Delta}^{i-1}(\varphi) \right\| \Big] + \left\| u^i(\varphi,\mu) - u^{i-1}(\varphi,\mu) \right\|_0
$$

$$
+ \left\| u^i(\varphi,\mu) - u^{i-1}(\varphi,\mu) \right\|_0 \Big\} ds + M(d,\mu)
$$

$$
\times \int_{-\infty}^{\infty} \left\| G_0^i(s,\varphi) - G_0^{i-1}(s,\varphi) \right\| + KL(d,\mu)
$$

$$\times \sum_{-\infty < j < \infty} e^{-\gamma |t_j|} \left\{ (1+N(\mu)) \left[\left\| \varphi^i_{t_j}(\varphi) - \varphi^{i-1}_{t_j}(\varphi) \right\| \right. \right.$$

$$+ \left\| \varphi^i_{t_j-\Delta}(\varphi) - \varphi^{i-1}_{t_j-\Delta}(\varphi) \right\| \right] + \left\| u^i(\varphi,\mu) - u^{i-1}(\varphi,\mu) \right\|_0$$

$$+ \left\| u^i(\varphi,\mu) - u^{i-1}(\varphi,\mu) \right\|_0 \right\} ds + M(d,\mu)$$

$$\times \sum_{-\infty < j < \infty} \left\| G^i_0(t_j,\varphi) - G^{i-1}_0(t_j,\varphi) \right\| \tag{3.33}$$

where $\| u(\varphi, \mu) \|_0 = \sup \| u(\varphi, \mu) \|$. Let us now estimate the differences $\varphi^i_t(\tau, \varphi) - \varphi^{i-1}_t(\tau,\varphi)$ and $\left\| G^i_0(\tau,\varphi) - G^{i-1}_0(\tau,\varphi) \right\|$. Using the second equation in (3.9) and the relations (3.11) and (3.15), we find

$$\left\| \varphi^i_t(\tau,\varphi) - \varphi^{i-1}_t(\tau,\varphi) \right\| \le L(d,\mu) \left| \int_\tau^t \left\{ \left\| u^i(\varphi,\mu) - u^{i-1}(\varphi,\mu) \right\|_0 \right. \right.$$

$$+ (1+N(\mu)) \left\| \varphi^i_s(\tau,\varphi) - \varphi^{i-1}_s(\tau,\varphi) \right\| \right\} ds \Big| \tag{3.34}$$

Applying Theorem 1.2 (Filatov and Sharova, 1975) to the inequality (3.34), we obtain

$$\left\| \varphi^i_t(\tau,\varphi) - \varphi^{i-1}_t(\tau,\varphi) \right\| \le \frac{1}{1+N(\mu)} \left\| u^i(\varphi,\mu) - u^{i-1}(\varphi,\mu) \right\|_0$$

$$\times \left(e^{L(d,\mu)(1+N(\mu))|t-\tau|} - 1 \right). \tag{3.35}$$

Since $G^i_t(\tau,\varphi)$ satisfies the equation (3.5), the difference $G^i_t(\tau,\varphi) - G^{i-1}_t(\tau,\varphi)$ is a solution of the equation

$$\frac{d}{dt} \left[G^i_t(\tau,\varphi) - G^{i-1}_t(\tau,\varphi) \right] = A\big(\varphi^i_t(\tau,\varphi)\big) \left[G^i_t(\tau,\varphi) - G^{i-1}_t(\tau,\varphi) \right]$$

$$+ B\big(\varphi^i_t(\tau,\varphi)\big) \left[G^i_{t-\Delta}(\tau,\varphi) - G^{i-1}_{t-\Delta}(\tau,\varphi) \right]$$

$$+ \left[A\big(\varphi^i_t(\tau,\varphi)\big) - A\big(\varphi^{i-1}_t(\tau,\varphi)\big) \right] G^{i-1}_t(\tau,\varphi) \tag{3.36}$$

$$+ \left[B\big(\varphi^i_t(\tau,\varphi)\big) - B\big(\varphi^{i-1}_t(\tau,\varphi)\big) \right] G^{i-1}_{t-\Delta}(\tau,\varphi)$$

and, consequently,

$$
\left\| G_t^i(\tau,\varphi) - G_t^{i-1}(\tau,\varphi) \right\| \leq \frac{\left\| u^i(\varphi,\mu) - u^{i-1}(\varphi,\mu) \right\|_0}{1+N(\mu)} \left\{ K^2 A_1 \left[\frac{2}{2\gamma - L(d,\mu)(1+N(\mu))} \right. \right.
$$

$$
+ \left. \frac{1 - e^{-L(d,\mu)(1+N(\mu))\Delta}}{2\gamma - L(d,\mu)(1+N(\mu))} \right] e^{-(\gamma - L(d,\mu)(1+N(\mu)))|t-\tau|}
$$

$$
+ K^2 B_1 e^{L(d,\mu)(1+N(\mu))\Delta} \left[\frac{1 + e^{L(d,\mu)(1+N(\mu))\Delta}}{2\gamma - L(d,\mu)(1+N(\mu))} \right.
$$

$$
+ \frac{e^{(2\gamma + L(d,\mu)(1+N(\mu)))\Delta} - 1}{2\gamma - L(d,\mu)(1+N(\mu))} + \frac{1 - e^{-L(d,\mu)(1+N(\mu))\Delta}}{L(d,\mu)(1+N(\mu))}
$$

$$
+ \left. \left. \frac{1 - e^{-L(d,\mu)(1+N(\mu))|t-\tau-\Delta|}}{L(d,\mu)(1+N(\mu))} \right] e^{-(\gamma - L(d,\mu)(1+N(\mu)))|t-\tau-\Delta|} \right\}. \tag{3.37}
$$

Taking the estimates (3.35) and (3.37) into account, we rewrite (3.33) as follows

$$
\| u^{i+1}(\varphi,\mu) - u^i(\varphi,\mu) \|_0 \leq S(\mu) \| u^i(\varphi,\mu) - u^{i-1}(\varphi,\mu) \|_0, \tag{3.38}
$$

where

$$
S(\mu) = L(d,\mu) K \left[\frac{2 + e^{L(d,\mu)(1+N(\mu))\Delta}}{\gamma - L(d,\mu)(1+N(\mu))} + \frac{e^{L(d,\mu)(1+N(\mu))\Delta} - e^{-\gamma\Delta}}{\gamma + L(d,\mu)(1+N(\mu))} \right.
$$

$$
+ \left. \frac{2 + e^{L(d,\mu)(1+N(\mu))\Delta} + e^{-\gamma\Delta}}{1 - e^{-(\gamma - L(d,\mu)(1+N(\mu)))h}} + \frac{e^{L(d,\mu)(1+N(\mu))\Delta} - e^{-\gamma\Delta}}{1 - e^{-(\gamma + L(d,\mu)(1+N(\mu)))h}} \right]
$$

$$
+ \frac{M(d,\mu)K^2}{1+N(\mu)} \left\{ \left[\frac{2A_1}{2\gamma - L(d,\mu)(1+N(\mu))} + B_1 e^{L(d,\mu)(1+N(\mu))\Delta} \right] \right.
$$

$$
\times \frac{1 + e^{L(d,\mu)(1+N(\mu))\Delta}}{2\gamma - L(d,\mu)(1+N(\mu))} + \frac{e^{2\gamma + L(d,\mu)(1+N(\mu))\Delta} - 1}{2\gamma + L(d,\mu)(1+N(\mu))}
$$

$$+ \frac{1 - e^{-L(d,\mu)(1+N(\mu))\Delta}}{L(d,\mu)(1+N(\mu))} \Bigg] \Bigg(\frac{2}{\gamma - L(d,\mu)(1+N(\mu))}$$

$$+ \frac{2}{1 - e^{-(\gamma - L(d,\mu)(1+N(\mu)))h}} \Bigg) + \frac{A_1 + B_1 e^{L(d,\mu)(1+N(\mu))\Delta}}{L(d,\mu)(1+N(\mu))}$$

$$\times \Bigg[\frac{2}{\gamma - L(d,\mu)(1+N(\mu))} - \frac{2}{\gamma} + \frac{2}{1 - e^{-(\gamma - L(d,\mu)(1+N(\mu)))h}} - \frac{2}{1 - e^{-\gamma h}} \Bigg] \Bigg\}.$$

Iterating the inequality (3.38), we find that

$$\| u^{i+1}(\varphi, \mu) - u^i(\varphi, \mu) \|_0 \leq [S(\mu)]^i \| u^1(\varphi, \mu) \|_0 \qquad (3.39)$$

for all $i = 0, 1, 2, \ldots$

We choose $\mu^0 \in [0, \mu_2]$ so small that the inequalities

$$L(d, \mu)(1+N(\mu)) < \gamma, \quad S(\mu) < 1,$$

hold for all $\mu^0 \in [0, \mu_2]$. Then, taking the estimate (3.13) into account, we conclude that the sequence of functions $u^i(\varphi, \mu)$ converges uniformly as $i \to \infty$.

The existence of the convergent series of invariant toroidal sets (3.8) yields the theorem on the existence of an invariant toroidal set of the system (3.1).

Theorem 4.4. (Martinyuk, Tsyganovsky, 1979a, b). *Suppose that the right-hand side of* (3.1) *satisfies the following conditions:*

(i) for all $\mu^0 \in [0, \mu_2]$, *the functions* $A(\varphi), B(\varphi), a(\varphi, y, \mu), c(\varphi, \psi, y, z, \mu)$, *and* $H_j(\varphi, \psi, y, z, \mu)$ *are periodic in* φ *and* $\psi = \varphi(t - \Delta)$ *with period* 2π; *they are defined for all* y *and* $z = y(t - \Delta)$ *belonging to the region*

$$\| y \| = \left(\sum_{i=1}^n y_i^2 \right)^{\frac{1}{2}} \leq d, \| z \| \leq d, \mu \in [0, \mu_0];$$

(ii) the functions $A(\varphi), B(\varphi), a(\varphi, y, \mu), c(\varphi, \psi, y, z, \mu)$, *and* $H_j(\varphi, \psi, y, z, \mu)$ *satisfy the conditions*

$$\| a(\varphi, y, \mu) \| + \| c(\varphi, \psi, y, z, \mu) \| + \| H_j(\varphi, \psi, y, z, \mu) \| \leq M(d, \mu);$$

$$\| c(\varphi', \psi', y', z', \mu) - c(\varphi'', \psi'', y'', z'', \mu) \| + \| H_j (\varphi', \psi', y', z', \mu)$$

$$- H_j (\varphi', \psi', y', z', \mu) \| \le L (d, \mu) \{\| \varphi' - \varphi'' \| + \| \psi' - \psi'' \| + \| y' - y'' \| + \| z' - z'' \|\};$$

$$\| a(\varphi', y', \mu) - a(\varphi'', y'', \mu) \| \le L(d, \mu) (\| \varphi' - \varphi'' \| + \| y' - y'' \|);$$

$$\| A (\varphi') - A (\varphi'') \| \le A_1 \| \varphi' - \varphi'' \|; \quad A_1 > 0,$$

$$\| B (\varphi') - B (\varphi'') \| \le B_1 \| \varphi' - \varphi'' \|; \quad B_1 > 0,$$

where $M (d, \mu)$ and $L (d, \mu)$ are positive monotonically decreasing functions such that $L(d, \mu) \to 0$ and $M (d, \mu) \to 0$ as $d \to 0$ and $\mu \to 0$;

(iii) for arbitrary functions $c_1(\varphi)$ and $a_1(t, \varphi)$ sufficiently small in norm, the system of differential equations

$$\frac{dy(t)}{dt} = A(\varphi)y(t) + B(\varphi)y(t-\Delta) + c_1(\varphi)$$

$$\frac{d\varphi}{dt} = a_1(\varphi)$$

possesses the Green's function $G_t(\tau, \varphi)$ for the problem on invariant tori which satisfies the inequality

$$\| G_t(\tau, \varphi) \| \le K e^{-\gamma | t - \tau |}, \quad K > 0, \quad \gamma > 0$$

for all $t, \tau \in (-\infty, \infty)$;

(iv) the times t_j, the functions H_j, and a lag Δ satisfy the inequalities

$$H_{j+1} = H_j, \quad t_{j+1} - t_j \ge h, \quad h > \Delta \ge 0.$$

Then there exists $\mu^0 \in [0, \mu_0]$ such that for all $\mu \in [0, \mu^0]$ the system of equations (3.1) possesses the invariant toroidal set $\mathcal{T}(\mu)$: $y = u(\varphi, \mu)$ satisfying the Lipschitz condition with respect to φ and such that

$$\lim_{\mu \to 0} \| u(\varphi, \mu) \| = 0.$$

§4. Behavior of Solutions of Nonlinear Systems with Lag in the Vicinity of Exponentially Stable Toroidal Manifolds

Consider a system of differential equations

$$\frac{dy(t)}{dt} = b(\varphi(t), \varphi(t - \Delta), y(t), y(t - \Delta)) \, y(t)$$

$$+ \, b_1(\varphi(t), \varphi(t - \Delta), y(t), y(t - \Delta)) \, y(t - \Delta) + c(\varphi(t), \varphi(t - \Delta));$$

$$\frac{d\varphi(t)}{dt} = a(\varphi(t), y(t)); \tag{4.1}$$

where $y = (y_1, y_2, \ldots, y_n)$; the functions a, b, b_1, and c are periodic in $\varphi(t)$ and $\varphi(t - \Delta)$ with period 2π, they are defined for all y and $z = y(t - \Delta)$ belonging to the region

$$D: \|y\| \le d, \quad \|z\| \le d; \tag{4.2}$$

Δ is a constant which characterizes the lag in the system. Suppose that (4.1) possesses a sufficiently smooth invariant torus

$$\mathcal{T}_m : y = u(\varphi), \tag{4.3}$$

where $u(\varphi)$ is a function periodic in φ with period 2π. Suppose also that a function $\psi(t) = \psi(t, \psi_0)$ defines a flow of trajectories on the torus $\mathcal{T}_m$ and, consequently, that it is a solution of the following system of differential equations on the torus

$$\frac{d\psi}{dt} = a(\psi, u(\psi)), \tag{4.4}$$

which is equal to ψ_0 at $t = t_0 = 0$.

Let us transform the system (4.1) assuming that the functions a, b, b_1, and c are sufficiently smooth in the vicinity of the torus $\mathcal{T}_m$. We introduce new functions x and θ by

$$y = u(\varphi) + \mu^2 x, \quad \varphi = \psi + \mu\theta. \tag{4.5}$$

Here, μ is a small positive parameter. After the change of variables the system (4.1) takes the form

$$\frac{dx(t)}{dt} = B(\psi(t), \psi(t-\Delta), \theta(t), \theta(t-\Delta), x(t), x(t-\Delta), \mu) \, x(t)$$

$$+ B_1(\psi(t), \psi(t-\Delta), \theta(t), \theta(t-\Delta), x(t), x(t-\Delta), \mu) \, x(t-\Delta);$$

$$\frac{d\theta(t)}{dt} = A_1(\psi(t), \theta(t), \mu) \, \theta(t) + \mu A_2(\psi(t), \theta(t), x(t), \mu) \, x(t),$$

(4.6)

where

$$A_1(\psi, \theta, \mu) = \int_0^1 \frac{\partial a\big(\psi + \tau\mu\theta, u(\psi + \mu\theta)\big)}{\partial \varphi} \, d\tau + \frac{\partial a\big(\psi, u(\psi)\big)}{\partial x} \int_0^1 \frac{\partial u(\psi + \tau\mu\theta)}{\partial \varphi} \, d\tau,$$

$$A_2(\psi, \theta, x, \mu) = \int_0^1 \frac{\partial a\big(\psi + \mu\theta, u(\psi + \mu\theta) + \tau\mu^2 x\big)}{\partial y} \, d\tau,$$

(4.7)

$$B\,(\psi(t), \psi(t-\Delta), \theta(t), \theta(t-\Delta), x(t), x(t-\Delta), \mu) \equiv b(\psi + \mu\theta, \psi_\Delta + \mu\theta_\Delta, \mu^2 x, \mu^2 x_\Delta)$$

$$= b(\psi + \mu\theta, \psi_\Delta + \mu\theta_\Delta, u(\varphi) + \mu^2 x, u(\varphi_\Delta) + \mu^2 x_\Delta)$$

$$+ \int_0^1 \frac{\partial b\big(\psi + \mu\theta, \psi_\Delta + \mu\theta_\Delta, u(\varphi) + \tau\mu^2 x, u(\varphi_\Delta) + \tau\mu^2 x_\Delta\big)}{\partial y} \, d\tau \, u(\varphi)$$

$$+ \int_0^1 \frac{\partial b_1\big(\psi + \mu\theta, \psi_\Delta + \mu\theta_\Delta, u(\varphi) + \tau\mu^2 x, u(\varphi_\Delta) + \tau\mu^2 x_\Delta\big)}{\partial y} \, d\tau \, u(\varphi_\Delta)$$

$$- \frac{\partial u}{\partial \varphi} \int_0^1 \frac{\partial a\big(\varphi, u(\varphi) + \tau\mu^2 x\big)}{\partial y} \, d\tau;$$

(4.8)

$$B_1(\psi(t), \psi(t-\Delta), \theta(t), \theta(t-\Delta), x(t), x(t-\Delta), \mu) \equiv b_1(\psi + \mu\theta, \psi_\Delta + \mu\theta_\Delta, \mu^2 x, \mu^2 x_\Delta)$$

$$= b_1(\psi + \mu\theta, \psi_\Delta + \mu\theta_\Delta, u(\varphi) + \mu^2 x, u(\varphi_\Delta) + \mu^2 x_\Delta)$$

$$+ \int_0^1 \frac{\partial b\big(\psi + \mu\theta, \psi_\Delta + \mu\theta_\Delta, u(\varphi) + \tau\mu^2 x, u(\varphi_\Delta) + \tau\mu^2 x_\Delta\big)}{\partial y_\Delta} \, d\tau \, u(\varphi)$$

$$+ \int_0^1 \frac{\partial b_1\left(\psi + \mu\theta, \psi_\Delta + \mu\theta_\Delta, u(\varphi) + \tau\mu^2 x, u(\varphi_\Delta) + \tau\mu^2 x_\Delta\right)}{\partial y_\Delta} \, d\tau \, u(\varphi_\Delta).$$

Henceforth, we assume that the right-hand side of (4.1) satisfies in the region D (see (4.2)) the conditions which ensure that in the region

$$\| x \| \leq 1, \quad \| x_\Delta \| \leq 1, \quad \mu \leq \mu_0, \tag{4.9}$$

for some sufficiently small μ_0, the following inequalities hold for the right-hand side of the system (4.6)

$$\| A_1(\psi, \theta, \mu) \| \leq M_1; \quad \| A_2(\psi, \theta, x, \mu) \| \leq M_1;$$

$$\| B(\psi, \psi_\Delta, \theta, \theta_\Delta, x, x_\Delta, \mu) \| \leq M_2; \quad \| B_1(\psi, \psi_\Delta, \theta, \theta_\Delta, x, x_\Delta, \mu) \| \leq M_2;$$

$$\| A_1(\psi, \theta, \mu) - A_1(\psi, \overline{\theta}, \mu) \| \leq \mu K_1 \| \theta - \overline{\theta} \|;$$

$$\| A_2(\psi, \theta, x, \mu) - A_2(\psi, \overline{\theta}, \overline{x}, \mu) \| \leq \mu K_1(\| \theta - \overline{\theta} \| + \| x - \overline{x} \|);$$

$$\| B(\psi, \psi_\Delta, \theta, \theta_\Delta, x, x_\Delta, \mu) - B(\psi, \psi_\Delta, \overline{\theta}, \overline{\theta}_\Delta, \overline{x}, \overline{x}_\Delta, \mu) \| \tag{4.10}$$

$$\leq \mu K_2(\| \theta - \overline{\theta} \| + \| \theta_\Delta - \overline{\theta}_\Delta \| + \| x - \overline{x} \| + \| x_\Delta - \overline{x}_\Delta \|);$$

$$\| B_1(\psi, \psi_\Delta, \theta, \theta_\Delta, x, x_\Delta, \mu) - B_1(\psi, \psi_\Delta, \overline{\theta}, \overline{\theta}_\Delta, \overline{x}, \overline{x}_\Delta, \mu) \|$$

$$\leq \mu K_3(\| \theta - \overline{\theta} \| + \| \theta_\Delta - \overline{\theta}_\Delta \| + \| x - \overline{x} \| + \| x_\Delta - \overline{x}_\Delta \|).$$

The relations (4.7) and (4.8) yield the following identities

$$A_1(\varphi, \theta, 0) \equiv A_1(\varphi, 0, \mu) \equiv \frac{\partial a\left(\psi, u(\psi)\right)}{\partial \varphi} + \frac{\partial a\left(\psi, u(\psi)\right)}{\partial y} \frac{\partial u(\psi)}{\partial \varphi};$$

$$A_2(\varphi, \theta, x, 0) \equiv A_2(\varphi, 0, 0, \mu) \equiv \frac{\partial a\left(\psi, u(\psi)\right)}{\partial y};$$

$$B(\psi, \psi_\Delta, \theta, \theta_\Delta, x, x_\Delta, 0) \equiv B(\psi, \psi_\Delta, 0, 0, 0, 0, \mu) \equiv b(\psi, \psi_\Delta, 0, 0)$$

$$\equiv b(\psi, \psi_\Delta, u(\varphi), u(\psi_\Delta)) + \sum_{\nu=1}^{n} \frac{\partial b\big(\psi, \psi_\Delta, u(\psi), u(\psi_\Delta)\big)}{\partial u_\nu(\psi)} u_\nu(\psi)$$

$$+ \sum_{\nu=1}^{n} \frac{\partial b_1\big(\psi, \psi_\Delta, u(\psi), u(\psi_\Delta)\big)}{\partial u_\nu(\psi)} u_\nu(\psi_\Delta) - \frac{\partial u}{\partial \varphi} \frac{\partial a\big(\psi, u(\psi)\big)}{\partial y}$$

$$\tag{4.11}$$

$$B_1(\psi, \psi_\Delta, \theta, \theta_\Delta, x, x_\Delta, 0) \equiv B_1(\psi, \psi_\Delta, 0, 0, 0, 0, \mu) \equiv b_1(\psi, \psi_\Delta, 0, 0)$$

$$\equiv b_1(\psi, \psi_\Delta, u(\psi), u(\psi_\Delta)) + \sum_{\nu=1}^{n} \frac{\partial b\big(\psi, \psi_\Delta, u(\psi), u(\psi_\Delta)\big)}{\partial u_\nu(\psi_\Delta)} u_\nu(\psi)$$

$$+ \sum_{\nu=1}^{n} \frac{\partial b_1\big(\psi, \psi_\Delta, u(\psi), u(\psi_\Delta)\big)}{\partial u_\nu(\psi_\Delta)} u_\nu(\psi_\Delta).$$

For the sake of brevity, we denote the right-hand sides of (4.11) as follows:

$$A_1(\psi, \theta, 0) = A_1^0(\psi); \quad A_2(\psi, \theta, x, 0) = A_2^0(\psi);$$

$$B(\psi, \psi_\Delta, \theta, \theta_\Delta, x, x_\Delta, 0) = b^0(\psi, \psi_\Delta);$$

$$B(\psi, \psi_\Delta, \theta, \theta_\Delta, x, x_\Delta, 0) = b_1^0(\psi, \psi_\Delta).$$

Theorem 4.5. (Ordynskaya, 1976) *Suppose that the right-hand side of* (4.1) *satisfies the following conditions:*

(i) *the functions* $a(\varphi, y)$, $b(\varphi, \varphi_\Delta, y, y_\Delta)$, $b_1(\varphi, \varphi_\Delta, y, y_\Delta)$, *and* $c(\varphi, \varphi_\Delta)$ *are defined in the region* D; *they are periodic in* φ_ν *and* $\varphi_{\nu\Delta}$ ($\nu = 1, 2, \dots, m$) *with period* 2π *and belong to the space* $C_{\text{Lip}}^1(D)$;

(ii) *the system* (4.1) *possesses the invariant torus* $\mathcal{T}_m$: $y = u(\varphi)$, *in* $C_{\text{Lip}}^1(D)$; *this torus lies in the region* D *together with some its vicinity;*

(iii) *for the linear system of equations*

$$\frac{d\theta(t)}{dt} = A_1^0(\psi)\theta(t);$$

$$\frac{dx(t)}{dt} = b^0\big(\psi(t), \psi(t-\Delta)\big)\, x(t) + b_1^0\big(\psi(t), \psi(t-\Delta)\big)\, x(t-\Delta), \tag{4.12}$$

where $\psi(t) = \psi(t, \psi_0)$ *is a flow of trajectories on the torus defined by the system* (4.4), *we have:*

(a) $$\min_{|\xi|=1}\Big\langle A_1^0(\psi)\xi, \xi\Big\rangle = \alpha(\varphi) \ge \min_{\psi} \alpha(\varphi) = \alpha \tag{4.13}$$

(b) *the second equation of* (4.12) *possesses the Green's function* $G_0(t, s)$ *for the problem of bounded solutions, such that* $G_0(t, s) \equiv 0$ *for* $t < s$; $G_0(s, s) = E$ *and*

$$\| G_0(t, s) \| \le ce^{-\gamma(t-s)} \text{ for } t > s, \tag{4.14}$$

where c *and* γ *are positive constants; moreover*

$$\alpha + \gamma > 0. \tag{4.15}$$

Then one can find μ^0 *such that for all* $0 < \mu \le \mu^0$ *any solution*

$$\varphi(t) = \varphi_t(y_0(t), \varphi_0(t)), \quad y(t) = y_t(y_0(t), \varphi_0(t)) \tag{4.16}$$

of the system of differential equations (4.1) *with initial functions* $\varphi(t) = \varphi_0(t)$, $y(t) = y_0(t)$, *which are given on the initial set* $-\Delta \le t \le 0$ *and belong to the region*

$$\| y_0(t) - u(\varphi_0(t)) \| \le \mu^2, \quad \| \varphi_0(t) - \psi(t) \| \le \mu,$$

is exponentially attracted to some solution $\psi(t) = \psi(t, \psi_0)$, $y(t) = u(\psi)$ *on the torus* $\mathfrak{T}_m$ *according to the law*

$$\| \varphi_t(y_0(t), \varphi_0(t)) - \psi(t, \psi_0) \| \le \mu e^{-(\gamma-e)t},$$

$$\| x_t(x_0(t), \varphi_0(t)) - u(\psi(t, \psi_0)) \| \le \overline{K}\mu e^{-(\gamma-e)t}, \; t \ge 0 \tag{4.17}$$

where $\varepsilon = \varepsilon(\mu^0)$ *is a sufficiently small positive variable such that* $\varepsilon(\mu^0) \to 0$ *as* $\mu^0 \to 0$, *and* $\overline{K}$ *is a positive variable which does not depend on* μ.

Proof. We use the standard iterative process. Choosing $\theta^{(0)}(t) \equiv 0$ and $y^{(0)}(t) \equiv 0$ as the zero approximation, we put $y^{(n)}(t) = w(t)$ and $\theta^{(n)}(t) = v(t)$ for $-\Delta \le t \le 0$ in all successive approximations (here, $w(t)$ and $v(t)$ are some functions from the class $C(-\Delta, 0)$). Successive approximations to solutions of the system (4.6) can be determined by the formulas

$$\frac{dx^{(n)}(t)}{dt} = B(\psi(t), \psi(t-\Delta), \theta^{(n-1)}(t), \theta^{(n-1)}(t-\Delta), x^{(n-1)}(t), x^{(n-1)}(t-\Delta), \mu) x^{(n)}(t)$$

$$+ B_1(\psi(t), \psi(t-\Delta), \theta^{(n-1)}(t), \theta^{(n-1)}(t-\Delta), x^{(n-1)}(t), x^{(n-1)}(t-\Delta), \mu) x^{(n)}(t-\Delta);$$

$$\text{(4.18)}$$

$$\frac{d\theta^{(n)}(t)}{dt} = A_1(\psi(t), \theta^{(n-1)}(t), \mu) \theta^{(n)}(t) + \mu A_2(\psi(t), \theta^{(n-1)}(t), x^{(n)}(t), \mu) x^{(n)}(t).$$

If we set $n = 1$ in (4.18), then we obtain the system of equations for $x^{(1)}(t)$ and $\theta^{(1)}(t)$

$$\frac{dx^{(1)}(t)}{dt} = b^0(\psi(t), \psi(t-\Delta)) x^{(1)}(t) + b_1^0(\psi(t), \psi(t-\Delta)) x^{(1)}(t-\Delta);$$

$$\frac{d\theta^{(1)}(t)}{dt} = A_1^0(\psi(t), \theta^{(1)}(t)) + \mu A_2(\psi(t), 0, x^{(1)}(t), \mu) x^{(1)}(t),$$

Its bounded solution can be written as follows

$$x^{(1)}(t) = G_0(t, 0)\, w(0) + \int_{-\Delta}^{0} G_0(t, s+\Delta)\, b_1^0\big(\psi(s+\Delta), \psi(s)\big) w(s)\, ds;$$

$$\text{(4.19)}$$

$$\theta^{(1)}(t) = -\mu \int_{0}^{\infty} \Omega_s^t\big(A_1^0\big)\, A_2\big(\psi(s), 0, x^{(1)}(s), \mu\big) x^{(1)}(s)\, ds,$$

where $\Omega_s^t\big(A_1^0\big)$ is the fundamental matrix of solutions of the first equation of (4.12) and $G_0(t, s)$ is the Green's function for the problem on bounded solutions.

Using the first relation of (4.19) and the inequalities (4.10) and (4.14), we get

$$\left\| x^{(1)}(t) \right\| \le ce^{-\gamma t}\delta + \int_{-\Delta}^{0} Ke^{-\gamma(t-s-\Delta)} M_3 \delta\, ds \le c^1 e^{-\gamma t}\delta$$

$$\text{(4.20)}$$

for all $t \ge 0$; here we have denoted

$$c^1 = c\left[1 + \frac{M_3}{\gamma}\left(e^{\gamma\Delta} - 1\right) \right], \quad \delta = \max_{-\Delta \le t \le 0} \| w(t) \|.$$

Let $c_1 = \max(c^1, 1)$. Then

$$\left\| x^{(1)}(t) \right\| \leq c_1 e^{-\gamma t} \delta \tag{4.21}$$

for all $t \geq -\Delta$. Employing now the second relation of (4.19) and the inequalities (4.10) and (4.21), we obtain the estimate

$$\left\| \theta^{(1)}(t) \right\| \leq \mu \int_t^\infty e^{\alpha(t-s)} M_1 c_1 e^{-\gamma s} \delta \, ds \leq \mu M_1 c_1 \delta e^{\alpha t} \int_t^\infty e^{-(\alpha-\gamma)s} \, ds \leq \frac{\mu c_1 M_1}{\alpha+\gamma} e^{-\gamma t} \delta, \tag{4.22}$$

for all $t \geq 0$, since the equation (4.13) implies the inequality

$$\left\| \Omega_s^t \left(A_1^0 \right) \theta_0 \right\| \leq e^{\alpha(t-s)} \left\| \theta_0 \right\|, \ 0 \leq t \leq s.$$

for all θ_0.

If we choose the initial function $v(t)$, $-\Delta \leq t \leq 0$, such that $\max_{-\Delta \leq t \leq 0} \left\| v(t) \right\| \leq e^{-\gamma \Delta}$ and if μ_0 is so small that the inequalities

$$\delta c_1 < 1, \quad \mu_0 \mu_1 / (\alpha + \gamma) < 1$$

hold, then the functions $\theta^{(1)}(t)$ and $x^{(1)}(t)$ belong to the region (4.9) and satisfy the inequalities

$$\| x^{(1)}(t) \| \leq e^{-\gamma t}, \quad \| \theta^{(1)}(t) \| \leq e^{-\gamma t}. \tag{4.23}$$

for all $t \geq -\Delta$.

We now fix a sufficiently small positive number ($2\varepsilon < \alpha + \gamma$ and $\varepsilon < \gamma$) and take μ_0 so small that

$$\frac{\mu_0 \mu_1}{\alpha+\gamma-2\varepsilon} < 1, \ 2\mu_0 K_1 < \varepsilon, \ 4\mu_0 K_2 < \varepsilon, \ 4\mu_0 e^{\gamma\Delta} K_3 < \varepsilon. \tag{4.24}$$

Then, for all $t \geq 0$ and $\mu \leq \mu_0$, we have the following estimates

$$\left\| B\left(\psi, \psi_\Delta, \theta^{(1)}, \theta_\Delta^{(1)}, x^{(1)}, x_\Delta^{(1)}, \mu \right) - b^0 \left(\psi, \psi_\Delta \right) \right\| \leq \mu K_2 \left(\left\| \theta^{(1)} \right\| + \left\| \theta_\Delta^{(1)} \right\| \right)$$

$$+ \left\| x^{(1)} \right\| + \left\| x_\Delta^{(1)} \right\| \right) \leq 4\mu K_2 < \varepsilon;$$

$$\left\| B_1\left(\psi,\psi_\Delta,\theta^{(1)},\theta^{(1)}_\Delta,x^{(1)},x^{(1)}_\Delta,\mu\right) - b_1^0\left(\psi,\psi_\Delta\right) \right\| \leq \mu K_3\left(\left\|\theta^{(1)}\right\| + \left\|\theta^{(1)}_\Delta\right\| \right.$$

$$\left. + \left\|x^{(1)}\right\| + \left\|x^{(1)}_\Delta\right\| \right) \leq 4\mu K_3 < \varepsilon; \qquad (4.25)$$

$$\| A_1(\psi, \theta^{(1)}, \mu) - A_1^{(0)}(\psi) \| \leq \mu K_1 < \varepsilon.$$

If we set $n = 2$ in (4.18), then we obtain the system of equations for the functions $x^{(2)}(t)$ and $\theta^{(2)}(t)$

$$\frac{dx^{(2)}(t)}{dt} = B\left(\psi(t),\psi(t-\Delta),\theta^{(1)}(t),\theta^{(1)}(t-\Delta),x^{(1)}(t),x^{(1)}(t-\Delta),\mu\right) x^{(2)}(t)$$

$$+ B_1\left(\psi(t),\psi(t-\Delta),\theta^{(1)}(t),\theta^{(1)}(t-\Delta),x^{(1)}(t),x^{(1)}(t-\Delta),\mu\right) x^{(2)}(t-\Delta);$$
$$(4.26)$$

$$\frac{d\theta^{(2)}(t)}{dt} = A_1(\psi(t), \theta^{(1)}(t), \mu)\, \theta^{(2)}(t) + \mu A_2(\psi(t), \theta^{(1)}(t), x^{(2)}(t), \mu)\, x^{(2)}(t).$$

The first equation of this system can be rewritten as follows

$$\frac{dx^{(2)}(t)}{dt} = \left[b^0\left(\psi,\psi_\Delta\right) + \left(B\left(\psi,\psi_\Delta,\theta^{(1)},\theta^{(1)}_\Delta,x^{(1)},x^{(1)}_\Delta,\mu\right) - b^0\left(\psi,\psi_\Delta\right)\right) \right] x^{(2)}(t)$$
$$(4.27)$$

$$+ \left[b_1^0\left(\psi,\psi_\Delta\right) + \left(B_1\left(\psi,\psi_\Delta,\theta^{(1)},\theta^{(1)}_\Delta,x^{(1)},x^{(1)}_\Delta,\mu\right) - b_1^0\left(\psi,\psi_\Delta\right)\right) \right] x^{(2)}(t-\Delta).$$

According to Lemma 2.2 and the inequality (4.25), we can conclude that the equation (4.27) possesses a Green's function for the problem of bounded solutions which satisfies the inequality

$$\| G_1(t, s) \| \leq c_2 e^{-(\gamma-\varepsilon)(t-s)} \qquad (4.28)$$

with a positive constant c_2 for all $t > s$. Rewriting the second equation of (4.26) as

$$\frac{d\theta^{(2)}(t)}{dt} = \left[A_1^0(\psi) + \left(A_1\left(\psi,\theta^{(1)},\mu\right) - A_1^0(\psi)\right) \right] \theta^{(2)}(t) + \mu A_2(\psi, \theta^{(1)}, x^{(2)})x^{(2)}(t), \quad (4.29)$$

and employing the inequalities (4.13) and (4.25), we obtain the estimate

$$\min_{\|\xi\|=1} \left\langle A_1\left(\psi,\theta^{(1)},\mu\right)\xi,\xi\right\rangle \geq \alpha-\varepsilon, \qquad (4.30)$$

which yields

$$\left\| \Omega_s^t \left(A_1\left(\psi, \theta^{(1)}, \mu \right) \right) \theta_0 \right\| \le e^{(\alpha - \varepsilon)(t - s)} \| \theta_0 \|$$
(4.31)

for all $o \le t \le s$ and arbitrary θ_0. The solution of (4.26), bounded for $t \ge 0$, we define by the relations

$$\theta^{(2)}(t) = -\mu \int_t^\infty \Omega_s^t \left(A_1\left(\psi, \theta^{(1)}, \mu \right) \right) A_2\left(\psi, \theta^{(1)}, x^{(2)}, \mu \right) x^2(s) \, ds,$$

$$x^{(2)}(t) = G_1(t,0) \, w(0) + \int_{-\Delta}^0 G_1(t, s + \Delta) B_1\left(\psi(s + \Delta), \psi(s), \right.$$
(4.32)

$$\theta^{(1)}(s + \Delta), \theta^{(1)}(s), x^{(1)}(s + \Delta), x^{(1)}(s), \mu \right) w(s) \, ds;$$

where $\theta^{(2)}(t) = v(t)$ and $x^{(2)}(t) = w(t)$ for $-\Delta \le t \le 0$.

By using the inequalities (4.10) and (4.28), we get (just as for $x^{(1)}(t)$)

$$\left\| x^{(2)}(t) \right\| = c_2 e^{-(\gamma - \varepsilon)t} \delta + \int_{-\Delta}^0 c_2 e^{-(\gamma - \varepsilon)(t - s - \Delta)} M_3 \delta \, ds \le c_2' e^{-(\gamma - \varepsilon)t} \delta, t \ge 0, \quad (4.33)$$

where

$$c_2' = c_2 \left(1 + \frac{M_3}{\gamma - \varepsilon} \left(e^{-(\gamma - \varepsilon)\Delta} - 1 \right) \right).$$

We set $c_3 = \max(1, c_2')$ and obtain as a result

$$\| x^{(2)}(t) \| \le c_3 e^{-(\gamma - \varepsilon)t} \delta$$
(4.34)

for all $t \ge -\Delta$.

Assume that the function $w(t)$ is such that the inequality $\delta c_3 < 1$ holds. Then for $t \ge 0$ and $\mu \le \mu_0$, the function $x^{(2)}(t)$ belongs to the region (4.9). Using the first relation of (4.32), we get

$$\left\| \theta^{(2)}(t) \right\| \le \mu \int_t^\infty e^{(\alpha - \varepsilon)(t - s)} M_1 c_3 e^{-(\gamma - \varepsilon)s} \delta \, ds \le \frac{\mu M_1}{\alpha + \gamma - 2\varepsilon} e^{-(\gamma - \varepsilon)t} \delta, \ t \ge 0. \quad (4.35)$$

Hence, if $w(t)$ is chosen as indicated above, then for all $t \geq 0$, the functions $\theta^{(2)}(t)$ and $x^{(2)}(t)$ belong to the region (4.9) and satisfy the inequalities

$$\| \theta^{(2)}(t) \| \leq e^{-(\gamma-\varepsilon)t}, \quad \| x^{(2)}(t) \| \leq e^{-(\gamma-\varepsilon)t}.$$

We set

$$\rho = \min\left(\frac{1}{c_1},\frac{1}{c_3}\right); \quad c_0 = \max(c_1, c_3);$$

$$A_1^{(n-1)} = A_1\left(\psi, \theta^{(n-1)}, \mu\right);$$

$$B_1^{(n-1)}(t, t-\Delta, \mu) = B_1\left(\psi, \psi_\Delta, \theta^{(n-1)}, \theta_\Delta^{(n-1)}, y^{(n-1)}, y_\Delta^{(n-1)}, \mu\right); \tag{4.36}$$

$$B^{(n-1)}(t, t-\Delta, \mu) = B\left(\psi, \psi_\Delta, \theta^{(n-1)}, \theta_\Delta^{(n-1)}, y^{(n-1)}, y_\Delta^{(n-1)}, \mu\right),$$

and assume that the functions $x^{(n-1)}(t)$ and $\theta^{(n-1)}(t)$ satisfy the inequalities

$$\| \theta^{(n-1)}(t) \| \leq c_0 e^{-(\gamma-\varepsilon)\delta}, \quad \| x^{(n-1)}(t) \| \leq c_0 e^{-(\gamma-\varepsilon)\delta} \tag{4.37}$$

for $t \geq 0$ and $\mu < \mu^0$. Let us now show that these inequalities are valid for $x^{(n)}(t)$ and $\theta^{(n)}(t)$. In fact, the nth approximation $x^{(n)}(t)$, $\theta^{(n)}(t)$ is defined by the relations

$$\theta^{(n)}(t) = -\mu \int_t^\infty \Omega_s^t\left(A_1^{(n-1)}\right) A_2\left(\psi, \theta^{(n-1)}, x^{(n)}\right) x^{(n)}(s)\, ds;$$

$$\tag{4.38}$$

$$x^{(n)}(t) = G_{n-1}(t,0)w(0) + \int_{-\Delta}^0 G_{n-1}(t,s+\Delta) B_1^{(n-1)}(s+\Delta, s, \mu)\, w(s)\, ds;$$

where $x^{(n)}(t) = w(t)$ and $\theta^{(n)}(t) = v(t)$ for $t \in [-\Delta, 0]$; moreover, the functions $w(t)$ and $v(t)$ satisfy the inequalities $\delta < \rho$ and $\max\limits_{t\in[-\Delta,0]} \| v(t) \| \leq e^{-\lambda\Delta}$. Taking (4.10), (4.24), and (4.37) into account, we obtain

$$\left\| A_1^{(n-1)} - A_1^0 \right\| \leq \mu K_1 \left\| \theta^{(n-1)}(t) \right\| \leq \mu K_1 < \varepsilon;$$

$$\left\| B^{(n-1)} - b^0 \right\| \leq \mu K_2 \left(\left\| \theta^{(n-1)}(t) \right\| + \left\| \theta^{(n-1)}(t-\Delta) \right\| + \left\| y^{(n-1)}(t) \right\| \right.$$

$$\left. + \left\| y^{(n-1)}(t-\Delta) \right\| \right) \leq 4\mu K_2 < \varepsilon;$$

$$\left\| B_1^{(n-1)} - b_1^0 \right\| \le \mu K_3 \left(\left\| \theta^{(n-1)}(t) \right\| + \left\| \theta^{(n-1)}(t-\Delta) \right\| + \left\| y^{(n-1)}(t) \right\| \right.$$

$$\left. + \left\| y^{(n-1)}(t-\Delta) \right\| \right) \le 4\mu K_3 < \varepsilon \tag{4.39}$$

for all $t \ge 0$. By analogy with (4.30), using the first inequality in (4.9), we find that

$$\min_{|\xi|=1} \left\langle A_1^{(n-1)} \xi, \xi \right\rangle \ge \alpha - \varepsilon.$$

This implies that

$$\left\| \Omega_s^t \left(A_1^{(n-1)} \right) \theta_0 \right\| \le e^{(\alpha-\varepsilon)(t-s)} \left\| \theta_0 \right\| \tag{4.40}$$

for all $0 \le t \le s$ and arbitrary θ_0. The last two inequalities of (4.39) guarantee the existence of the Green's function $G_{n-1}(t, s)$ for the problem of bounded solutions of (4.18); this function   satisfies the inequality

$$\left\| G_{n-1}(t,s) \right\| \le c_2 e^{-(\alpha-\varepsilon)(t-s)}, \; t > s. \tag{4.41}$$

Estimating the relations (4.38) with the help of the inequalities (4.40) and (4.41), we get

$$\left\| x^{(n)}(t) \right\| \le c_2 e^{-(\gamma-\varepsilon)} \delta + \int_{-\Delta}^{0} c_2 e^{-(\gamma-\varepsilon)(t-s-\Delta)} M_3 \delta ds \le c_3 e^{-(\gamma-\varepsilon)t} \delta \le c_0 e^{-(\gamma-\varepsilon)t} \delta;$$

$$\left\| \theta^{(n)}(t) \right\| \le \mu \int_{t}^{\infty} M_1 c_0 e^{-(\gamma-\varepsilon)s} e^{-(\alpha-\varepsilon)(t-s)} M_3 \delta ds \tag{4.42}$$

$$\le \frac{\mu M_1}{\alpha + \gamma - 2\varepsilon} e^{-(\gamma-\varepsilon)t} \delta \le c_0 e^{-(\gamma-\varepsilon)t} \delta.$$

This completes the proof.

By induction one can conclude that the successive approximations $x^{(n)}(t)$, $\theta^n(t)$ satisfy the inequalities (4.37) for all $n = 1, 2, 3, ...,$ $t \ge 0$, and $0 < \mu \le \mu^0$. Moreover, these functions continuously depend on the parameter $\psi_0 \in \mathcal{C}_m$. This is so, because the function $a(\varphi, u(\varphi))$ belongs to the space $C^1_{\mathrm{Lip}}(D)$, and therefore, $\psi(t) = \psi(t, \psi_0)$ is a continuous function of $\psi_0 \in \mathcal{C}_m$ as a solution of the equation (4.4). The functions

$\Omega_s^t\left(A_1^{(n-1)}\right)$, $x^{(n)}(t)$, and $\theta^{(n)}(t)$ are also continuous in $\psi \in \mathcal{T}_m$.

We now prove that the sequences $\{x^{(n)}(t)\}$ and $\{\theta^{(n)}(t)\}$ are convergent. Denoting $r_{n+1}(t) = x^{(n+1)}(t) - x^{(n)}(t)$, $\eta_{n+1}(t) = \theta^{(n+1)}(t) - \theta^{(n)}(t)$, and taking (4.18) into account, we find that $r_{n+1}(t)$ and $\eta_{n+1}(t)$ satisfy the equations

$$\frac{dr_{n+1}(t)}{dt} = B^{(n)}r_{n+1}(t) + B_1^{(n)}r_{n+1}(t-\Delta) + \left(B^{(n)} - B^{(n-1)}\right)x^{(n)}(t)$$

$$+ \left(B_1^{(n)} - B_1^{(n-1)}\right)x^{(n-1)}(t-\Delta);$$

$$\eta_{n+1}(t) = -\mu \int_t^\infty \left[\Omega_s^t\left(A_1^{(n)}\right)A_2\left(\psi(s),\theta^{(n)}(s),x^{(n+1)}(s),\mu\right)x^{(n+1)}(s)\right.$$

$$\left. - \Omega_s^t\left(A_1^{(n-1)}\right)A_2\left(\psi(s),\theta^{(n-1)}(s),x^{(n)}(s),\mu\right)x^{(n)}(s)\right]ds.$$

$$(4.43)$$

Since $r_{n+1}(t) = 0$ for $t \in [-\Delta, 0]$, the bounded solution $r_{n+1}(t)$ of the first equation of (4.43) is defined, for $t \geq 0$, by

$$r_{n+1}(t) = \int_0^t G(t,s)\left[\left(B^{(n)} - B^{(n-1)}\right)x^{(n)}(s) + \left(B_1^{(n)} - B_1^{(n-1)}\right)x^{(n)}(s-\Delta)\right]ds \qquad (4.44)$$

Estimating (4.44) with the help of (4.10), (4.37), and (4.41), we obtain

$$\left\|r_{n+1}(t)\right\| \leq \int_0^t c_2 e^{-(\gamma-\varepsilon)(t-s)}\mu K_2\left(\left\|r_n(s)\right\| + \left\|r_n(s-\Delta)\right\| + \left\|\eta_n(s)\right\|\right)c_0 e^{-(\gamma-\varepsilon)s}\,ds$$

$$+ \int_0^t c_2 e^{-(\gamma-\varepsilon)(t-s)}\mu K_3\left(\left\|r_n(s)\right\| + \left\|r_n(s-\Delta)\right\| + \left\|\eta_n(s)\right\|\right.$$

$$+ \left.\left\|\eta_n(s-\Delta)\right\|\right)\left\|x_n(s-\Delta)\right\|ds. \qquad (4.45)$$

Since $\eta_n(t) = r_n(t)$ for $t \in [-\Delta, 0]$, we have $\|r_n(t-\Delta)\| = \|r_n(t)\|$ and $\|\eta_n(t-\Delta)\| = \|\eta_n(t)\|$ for $t \geq 0$. Changing the variable $s - \Delta = s_1$ in (4.45), we find

$$\left\|r_{n+1}(t)\right\| \leq 2\mu K_2 c_2 c_0 \delta e^{-(\gamma-\varepsilon)t}\int_0^t\left(\left\|r_n(s)\right\| + \left\|\eta_n(s)\right\|\right)ds + 2\mu K_3 c_2$$

$$\times \int_{-\Delta}^{t-\Delta} e^{-(\gamma-\varepsilon)(t-s_1-\Delta)}\left(\left\|r_n(s_1+\Delta)\right\|+\left\|\eta_n(s_1+\Delta)\right\|\right)\left\|x_n(s_1)\right\|ds_1$$

$$\leq 2\mu K_2 c_2 c_0 \delta \left(\left\|r_n(t)\right\|_0 +\left\|\eta_n(t)\right\|_0\right) e^{-(\gamma-\varepsilon)t}+\frac{2\mu K_3 c_2}{\gamma-\varepsilon}$$

$$\times \left(\left\|r_n(t)\right\|_0 +\left\|\eta_n(t)\right\|_0\right)\delta e^{-(\gamma-\varepsilon)t}\, e^{(\gamma-\varepsilon)\Delta}\left(1-e^{-(\gamma-\varepsilon)\Delta}\right)$$

$$+2\mu K_3 c_2 \left(\left\|r_n(t)\right\|_0 +\left\|\eta_n(t)\right\|_0\right) e^{-(\gamma-\varepsilon)t}\, e^{(\gamma-\varepsilon)\Delta}\int_0^t e^{(\gamma-\varepsilon)s_1}c_0\delta e^{-(\gamma-\varepsilon)s_1}\,ds_1$$

$$\leq 2\mu c_2 c_0 e^{-(\gamma-\varepsilon)t}\, t\left(\left\|r_n(t)\right\|_0 +\left\|\eta_n(t)\right\|_0\right)\left(K_2+K_3 e^{\gamma\Delta}\right)\delta$$

$$+\frac{2\mu K_3 c_2}{\gamma-\varepsilon}\,\delta e^{(\gamma-\varepsilon)\Delta}\, e^{-(\gamma-\varepsilon)t}\left(\left\|r_n(t)\right\|_0 +\left\|\eta_n(t)\right\|_0\right). \tag{4.46}$$

The second relation in (4.43) can be transformed as follows

$$\eta_{n+1}(t)=-\mu\int_t^\infty\left\{\left[\Omega_s^t\!\left(A_1^{(n)}\right)-\Omega_s^t\!\left(A_1^{(n-1)}\right)\right]A_2\!\left(\psi(s),\theta^{(n)}(s),x^{(n+1)}(s),\mu\right)x^{(n+1)}(s)\right.$$

$$+\Omega_s^t\!\left(A_1^{(n-1)}\right)A_2\!\left(\psi(s),\theta^{(n)}(s),x^{(n+1)}(s),\mu\right)\left[x^{(n+1)}(s)-x^{(n)}(s)\right]$$

$$+\Omega_s^t\!\left(A_1^{(n-1)}\right)\left[A_2\!\left(\psi(s),\theta^{(n)}(s),x^{(n+1)}(s),\mu\right)\right.$$

$$\left.\left.-A_2\!\left(\psi(s),\theta^{(n)}(s),x^{(n)}(s),\mu\right)\right]x^{(n)}(s)\right\}ds. \tag{4.47}$$

Denoting $\psi_n = \Omega_s^t\!\left(A_1^{(n-1)}\right)\theta_0$, one can easily show that the difference $\psi_{n+1}-\psi_n$ is a solution of the differential equation

$$\frac{d\left(\psi_{n+1}-\psi_n\right)}{dt}=A_1^{(n)}\left(\psi_{n+1}-\psi_n\right)+\left(A_1^{(n)}-A_1^{(n-1)}\right)\psi_n,$$

which can be presented in the form

$$\psi_{n+1}-\psi_n =\int_s^t\Omega_\tau^t\!\left(A_1^{(n)}\right)\left(A_1^{(n)}-A_1^{(n-1)}\right)\Omega_s^\tau\!\left(A_1^{(n-1)}\right)\theta_0\,d\tau, \tag{4.48}$$

if we take into account that $\psi_{n+1} - \psi_n = 0$ for $t = s$. Emplying this relation and the inequalities (4.10) and (4.40), we obtain

$$\left\| \psi_{n+1} - \psi_n \right\| \leq \int_s^t e^{(\alpha-\varepsilon)(t-\tau)} \mu K_1 \left\| \theta^{(n)} - \theta^{(n-1)} \right\| e^{(\alpha-\varepsilon)(\tau-s)} \left\| \theta_0 \right\| d\tau$$

$$\leq \mu K_1 e^{(\alpha-\varepsilon)(t-s)} (s-t) \left\| \theta^{(n)} - \theta^{(n-1)} \right\|_0 \left\| \theta_0 \right\| \tag{4.49}$$

for $s \geq t \geq 0$. Estimating (4.47) with the help of (4.49), we get

$$\left\| \eta_{n+1}(t) \right\| \leq \mu \int_t^\infty \mu K_1 e^{(\alpha-\varepsilon)(t-s)} (s-t) \left\| \eta_n \right\|_0 M_1 c_0 e^{-(\gamma-\varepsilon)s} \delta \, ds$$

$$+ \mu \int_t^\infty e^{(\alpha-\varepsilon)(t-s)} \mu K_1 \left(\left\| \eta_n \right\| + \left\| r_{n+1} \right\| \right) c_0 e^{-(\gamma-\varepsilon)s} \delta \, ds$$

$$\leq \frac{\mu^2 K_1 M_1 c_0 \gamma}{(\gamma-\varepsilon)^2} e^{-(\gamma-\varepsilon)t} \left\| \eta_n(t) \right\|_0 + \frac{\mu^2 K_1 c_0 \delta}{\gamma-\varepsilon} e^{-(\gamma-\varepsilon)t} \left\| \eta_n(t) \right\|_0 \tag{4.50}$$

$$+ \mu M_1 \int_t^\infty e^{(\alpha-\varepsilon)(t-s)} \left\| r_{n+1}(s) \right\| ds + \mu^2 K_1 c_0 \delta \int_t^\infty e^{(\alpha-\varepsilon)(t-s)} e^{-(\gamma-\varepsilon)s} \left\| r_{n+1}(s) \right\| ds.$$

Let

$$\vartheta_1 = \int_t^\infty e^{(\alpha-\varepsilon)(t-s)} \left\| r_{n+1}(s) \right\| ds,$$

$$\vartheta_2 = \int_t^\infty e^{(\alpha-\varepsilon)(t-s)} e^{-(\gamma-\varepsilon)s} \left\| r_{n+1}(s) \right\| ds.$$

By using the inequalities (4.46), we find that

$$\vartheta_1 \leq \int_t^\infty e^{(\alpha-\varepsilon)(t-s)} e^{-(\gamma-\varepsilon)t} (As + B) \delta \left(\left\| r_n(s) \right\|_0 + \left\| \eta_n(s) \right\|_0 \right) ds$$

$$\leq \left[\frac{A}{\gamma+\alpha-2\varepsilon} t + \frac{A}{(\gamma+\alpha-2\varepsilon)^2} + \frac{B}{\gamma+\alpha-2\varepsilon} \right] e^{-(\gamma-\varepsilon)t} \delta \left(\left\| r_n \right\|_0 + \left\| \eta_n \right\|_0 \right);$$

$$\mathscr{I}_2 \leq \int_t^\infty e^{(\alpha-\varepsilon)(t-s)} e^{-(\gamma-\varepsilon)s} (As + B)\delta\left(\|r_n(s)\|_0 + \|\eta_n(s)\|_0\right) ds$$

$$\leq \left[\frac{A}{\gamma+\alpha-2\varepsilon} t + \frac{A}{(\gamma+\alpha-2\varepsilon)^2} + \frac{B}{\gamma+\alpha-2\varepsilon}\right] e^{-(\gamma-\varepsilon)t} \delta\left(\|r_n\|_0 + \|\eta_n\|_0\right), \qquad (4.51)$$

where

$$A = 2\mu c_2 c_0 (K_2 + e^{\gamma\Delta}K_3), \quad B = \frac{2\mu K_3 c_2}{\gamma-\varepsilon} e^{(\gamma-\varepsilon)\Delta}.$$

Taking (4.51) into account, we can rewrite the inequality (4.50) in the form

$$\|\eta_{n+1}(t)\| \leq \frac{\mu^2 K c_0}{\gamma+\alpha-2\varepsilon}\left(\frac{M_1}{\gamma+\alpha-2\varepsilon} + 1\right)\delta e^{-(\gamma-\varepsilon)t}\|\eta_n\|_0 + \left(\mu M_1 + \mu^2 K_1 c_0 \delta\right)$$

$$\times \left(\frac{A}{\gamma+\alpha-2\varepsilon} t + \frac{A}{(\gamma+\alpha-2\varepsilon)^2} + \frac{B}{\gamma+\alpha-2\varepsilon}\right) e^{-(\gamma+\alpha-2\varepsilon)t} \delta\left(\|r_n\|_0 + \|\eta_n\|_0\right),$$

or in another form

$$\|\eta_{n+1}(t)\|_0 \leq \mu H \delta e^{-(\gamma-\varepsilon)t}(t+1)\left(\|\xi_n(t)\|_0 + \|\eta_n(t)\|_0\right), \qquad (4.52)$$

where H is some positive constant independent of μ. Using the inequalities (4.46) and (4.52), we find

$$\|\eta_{n+1}(t)\|_0 + \|r_{n+1}(t)\|_0 \leq \mu N S \delta(\|\eta_n(t)\|_0 + \|r_n(t)\|_0). \qquad (4.53)$$

This estimate involves the uniform convergence of the sequences of functions $\{\theta^{(n)}(t)\}$ and $\{x^{(n)}(t)\}$ for all

$$t \geq 0, \quad \psi_0 \in \mathscr{C}_m, \quad 0 < \mu \leq \mu_0 \qquad (4.54)$$

if μ_0 is chosen sufficiently small such that the inequality

$$\mu_0 N S < \frac{1}{2} \qquad (4.55)$$

holds, where

$$S = \sup_{t \geq 0}(t+1)e^{-(\gamma-\varepsilon)t}, \ N = \text{const}.$$

Proceeding to the limit as $n \to \infty$ in the relations (4.37) and (4.38), we obtain the limiting functions $\theta(t)$ and $x(t)$ which are solutions of the system of differential equations (4.6) and satisfy the inequalities

$$\| \theta(t) \| \leq c_0 e^{-(\gamma-\varepsilon)t}\delta, \ \| x(t) \| c_0 e^{-(\gamma-\varepsilon)t}\delta, \ (t \geq 0) \tag{4.56}$$

These inequalities imply that the solutions of (4.6) decrease exponentially as $t \to \infty$.

If we return to the initial system of equations (4.1) and denote its solution $y_t = y_t(y_0(t), \varphi_0(t))$, $\varphi(t) = \varphi_t(y_0(t), \varphi_0(t))$ on the initial interval $t \in [-\Delta, 0]$ by $y_0(t)$ and $\varphi_0(t)$, respectively, then we can make the following conclusion:

For every fixed ψ_0 and sufficiently small μ, there exist solutions

$$\varphi_t = \psi(t) + \mu\theta(t),$$

$$y_t = u(\varphi_t) + \mu^2 x(t) = u(\psi(t)) + [u(\varphi_t) - u(\psi(t))] + \mu^2 x(t)$$

of this system with the initial conditions $y_0(t)$, $\varphi_0(t)$, for $t \in [-\Delta, 0]$, which satisfy the inequalities

$$\| y_t - u(\psi(t)) \| \leq K\mu\| \theta(t) \| + \mu^2\| x(t) \| \leq \mu \overline{K} c_0 \delta \ e^{-(\gamma-\varepsilon)t},$$

$$\| \varphi_t - \psi(t) \| \leq \mu \ c_0 \delta \ e^{-(\gamma-\varepsilon)t}$$

provided that

$$\| \varphi_0(t) - \psi(t) \| \leq \mu \ \text{ and } \ \| y_0(t) - u(\varphi_0(t)) \| \leq \mu^2.$$

This completes the proof of the theorem.

5. REDUCIBILITY OF LINEAR SYSTEMS OF DIFFERENCE EQUATIONS WITH QUASIPERIODIC COEFFICIENTS

§1. Statement of the Problem and Auxiliary Statements

Consider a linear system of difference equations

$$x_{n+1} = A\, x_n + P(\varphi_n)\, x_n, \quad \Delta\varphi_n = \omega, \; n = \pm 1, 2, 3, \ldots, \tag{1.1}$$

where A is a constant, $P(\varphi)$ is a quasiperiodic $(s \times s)$-dimensional matrix, $\omega = (\omega_1, \omega_2, \ldots, \omega_n)$ is a frequency basis of $P(\varphi)$, $x = (x^1, x^2, \ldots, x^s)$, and $\varphi = (\varphi^1, \varphi^2, \ldots, \varphi^m)$ are vectors whose dimensionality is s and m, respectively.

Let us state the problem : We want to find a change of variables

$$x_n = \Phi(\varphi_n)\, y_n \tag{1.2}$$

with a non-singular matrix $\Phi(\varphi)$ quasiperiodic in φ which reduces (1.1) to a system with constant coefficients. Under certain assumptions, this reduction is always possible. Below, we present a method which enables us to construct the matrix $\Phi(\varphi_n)$ and, consequently, to find the general solution of (1.1).

Before we start proving the theorem on the reducibility of the system (1.1), let us formulate and prove some auxiliary statements which will be necessary in what follows.

Given a matrix $P(\varphi)$, we suppose it to be analytic with respect to $\varphi = (\varphi^1, \varphi^2, \ldots, \varphi^m)$ in the region $|\operatorname{Im} \varphi| \le \rho_0$, periodic in φ with period 2π, and bounded by a constant M, i. e.,

$$|P(\varphi)| = \sum_{i,j} P_{i,j}(\varphi) \le M.$$

We also suppose that the Fourier series of this matrix is

$$P(\varphi) = \sum_k P^{(k)} e^{i(r,\varphi)}, \tag{1.3}$$

where $P^{(k)}$ are constants of the matrix and $k = (k_1, k_2, \ldots, k_m)$ is a vector with integral components. It is well-known that the coefficients $P^{(k)}$ of this Fourier series satisfy the inequality

$$|P^{(k)}| \le M e^{-\rho_0|k|}. \tag{1.4}$$

Denote by $\overline{P}(\varphi)$ the average value of the matrix $P(\varphi)$

$$P^0 = \overline{P}(\varphi) = \frac{1}{(2\pi)}m \int_0^{2\pi} \ldots \int_0^{2\pi} P(\varphi) \, d\varphi^1 \ldots d\varphi^m$$

and by S_N, as before, the operator

$$S_N P(\varphi) = \sum_{|k| \le N} P^{(k)} e^{i(k,\varphi)}. \tag{1.5}$$

Henceforth, we shall need the following statement (Bogolyubov, 1964).

Lemma 5.1. *If* $0 < \delta < 1$ *and* $\nu > 1$, *then*

$$\sum_{|k| \neq 0} |k|^\nu e^{-2\delta|k|} = \left(\frac{\nu}{e}\right)^\nu \frac{1}{\delta^{\nu+m}} \left(1 + e^m\right). \tag{1.6}$$

Let us now consider the problem of periodic solutions of the equation

$$U(\varphi + \omega) A = A\, U(\varphi) + P(\varphi) \tag{1.7}$$

where A is a constant $(s \times s)$–dimensional matrix. Suppose that the eigenvalues $\lambda_1, \lambda_2, \ldots, \lambda_s$ of the matrix A are real, different and nonzero. Therefore, without loss of generality, we can assume that A is a diagonal matrix. Taking into account that the matrix $P(\varphi)$ can be represented in the form of a series the series (1.3), we try to find the solutions of the equation (1.7) in the form

$$U(\varphi) = \sum_k U^{(k)} e^{i(k,\varphi)} \tag{1.8}$$

where $U^{(k)}$ are unknown matrices. Inserting (1.8) and (1.3) in the equation (1.7), we obtain the coefficients $u^{(k)}$

$$U^{(k)}e^{i(k,\omega)}A = A\,U^{(k)} + P^{(k)}. \tag{1.9}$$

Let us clarify under what conditions this equation is solvable. Denoting elements of the matrices $U^{(k)}$ and $P^{(k)}$ by

$$U^{(k)} = \left\{ U^{(k)}_{\alpha\beta} \right\},\; P^{(k)} = \left\{ P^{(k)}_{\alpha\beta} \right\},\; \alpha,\beta=1,2,\ldots,s, \tag{1.10}$$

we can rewrite the equation (4.19) as a system

$$U^{(k)}_{\alpha\beta}\, e^{i(r,\omega)}\lambda_\beta = \lambda_\alpha U^{(k)}_{\alpha\beta} + P^{(k)}_{\alpha\beta}, \tag{1.11}$$

by solving which, one can find

$$U^{(k)}_{\alpha\beta} = \frac{P^{(k)}_{\alpha\beta}}{\left(\lambda_\beta e^{i(k,\omega)} - \lambda_\alpha\right)}. \tag{1.12}$$

For $|k| = 0$, we have

$$U^{(0)}_{\alpha\beta} = \frac{P^{(0)}_{\alpha\beta}}{\left(\lambda_\beta - \lambda_\alpha\right)} \tag{1.13}$$

and this implies that the original system is solvable only if

$$P^{(0)}_{\alpha\alpha} = 0,\, \alpha=1,2,\ldots,s. \tag{1.14}$$

Under conditions (1.14), i. e., for average values of the diagonal elements of $P(\varphi)$ equal to zero, the equations (1.12) are always solvable for all $k = (k_1, k_2, \ldots, k_m), |k| \neq 0$.

If all $U^{(k)}$ are determined, then (1.8) gives a formal solution of the equation (1.7). The series (1.8) should be convergent in order to be a real solution of the equation (1.7). To prove this, we find the upper bound of the matrix $U^{(k)}$. Taking (1.12), (1.13), and (1.14) into account, we find the following estimates for the elements $U^{(k)}_{\alpha\beta}$ of $U^{(k)}$

$$\left|U_{\alpha\beta}^{(0)}\right| \leq \frac{\left|P_{\alpha\beta}^{(0)}\right|}{\min\limits_{\alpha\neq\beta}\left|\lambda_\beta - \lambda_\alpha\right|}, \quad \left|U_{\alpha\beta}^{(k)}\right| \leq \frac{\left|P_{\alpha\beta}^{(k)}\right|}{2\sin\dfrac{(k,\omega)}{2}\min\limits_{\alpha}\left|\lambda_\alpha\right|}. \tag{1.15}$$

Assume that for some positive ε and d the inequality

$$\left|\sin\frac{(k,\omega)}{2}\right| \leq \varepsilon |k|^{-d}, \quad |k| \neq 0 \tag{1.16}$$

holds for all vectors $k = (k_1, k_2, ..., k_m)$ with integer components. Taking (1.16) into account, we transform the inequalities (1.15) as follows

$$\left|U^{(0)}\right| \leq \frac{\left|P^{(0)}\right|}{\min\limits_{\alpha\neq\beta}\left|\lambda_\beta - \lambda_\alpha\right|}, \quad \left|U^{(k)}\right| \leq \frac{\left|P^{(k)}\right|}{2\varepsilon\min\limits_{\alpha}\left|\lambda_\alpha\right|}|k|^{-d}. \tag{1.17}$$

The problem of the convergence and differentiability of the series (1.8), for an analytical matrix $P(\varphi)$, is solved by the following statement (Bogolyubov, Mitropolsky, and Samoilenko, 1969; Martinyuk and Perestyuk, 1974).

Lemma 5.2. *Let the matrix $P(\varphi)$ be periodic in $\varphi = (\varphi_1, \varphi_2, ..., \varphi_m)$ with period 2π, analytic in the region*

$$|\operatorname{Im}\varphi| \leq \rho_0, \tag{1.18}$$

real for $\operatorname{Im}\varphi = 0$, and satisfying the conditions

$$|P(\varphi)| \leq M_0 \tag{1.19}$$

and (1.14). Furthermore, let the eigenvalues $\lambda_1, \lambda_2, ..., \lambda_s$ of the matrix A be real, different and nonzero, and let the vector $\omega = (\omega_1, \omega_2, ..., \omega_m)$ satisfy the inequality (1.16). Then the equation (1.7) has a periodic solution $U(\varphi)$ with period 2π which is analytic in the region

$$|\operatorname{Im}\varphi| \leq \rho_0 - 2\delta, \ 0 < 2\delta < \rho_0 \tag{1.20}$$

real for $\operatorname{Im}\varphi = 0$ and satisfies the inequalities

$$\left|U - U^{(0)}\right| \leq \frac{1}{2\varepsilon \min_{\alpha}\left|\lambda_{\alpha}\right|}\left(\frac{d}{e}\right)^{d}\left(1 + e^{m}\right)\frac{M_0}{\delta^{\alpha+m}} \leq c_1(\varepsilon)\frac{M_0}{\delta^{\alpha+m}},$$

$$\left|\frac{\partial U}{\partial \varphi}\right| = \sum_{\alpha=1}^{m}\left|\frac{\partial U}{\partial \varphi_{\alpha}}\right| \leq c_1(\varepsilon)\frac{M_0}{\delta^{\alpha+m+1}}.$$

Proof. Conditions (1.14) and (1.16) ensure the existence of the formal solution of the equation (1.17)

$$U = U^{(0)} + \sum_{|k|\neq 0} U^{(k)} e^{i(k,\varphi)} \tag{1.21}$$

and the validity of (1.17) for the coefficients of (1.21). Therefore, the inequality

$$\left|U - U^{(0)}\right| \leq \sum_{|k|\neq 0}\left|U^{(k)}\right|e^{|\operatorname{Im}\varphi\|k|} \leq \frac{1}{2\varepsilon \min_{\alpha}\left|\lambda_{\alpha}\right|}\sum_{|k|\neq 0}\left|P^{(k)}\right|\left|k\right|^{d}e^{|\operatorname{Im}\varphi\|k|} \tag{1.22}$$

holds. Since the matrix $P(\varphi)$ is analytic in the region (1.18), we have

$$\left|P^{(k)}\right| < M_0 e^{-\rho_0|k|} \tag{1.23}$$

and thus,

$$\left|U - U^{(0)}\right| \leq \frac{1}{2\varepsilon \min_{\alpha}\left|\lambda_{\alpha}\right|}\sum_{|k|\neq 0}M_0 e^{-\rho_0|k|}|k|^{d}e^{|\operatorname{Im}\varphi\|k|}$$

$$\leq \frac{M_0}{2\varepsilon \min_{\alpha}\left|\lambda_{\alpha}\right|}\sum_{|k|\neq 0}|k|^{d}e^{(|\operatorname{Im}\varphi|-\rho_0)|k|} \leq \frac{M_0}{2\varepsilon \min_{\alpha}\left|\lambda_{\alpha}\right|}\sum_{|k|\neq 0}|k|^{d}e^{-2\delta|k|}. \tag{1.24}$$

Using now (1.6), we can write (1.24) in the form

$$\left|U - U^{(0)}\right| \leq \frac{M_0}{2\varepsilon \min_{\alpha}\left|\lambda_{\alpha}\right|}\left(\frac{d}{e}\right)^{d}\left(1 + e^{m}\right)\delta^{-(d+m)} \leq c_1(\varepsilon)\frac{M_0}{\delta^{\alpha+m}}. \tag{1.25}$$

Differentiating the series (1.21) and estimating its coefficients, we find

$$\left|\frac{\partial U(\varphi)}{\partial\varphi}\right| \le \sum_{\alpha=1}^{m}\left|\frac{\partial U}{\partial\varphi_{\alpha}}\right| \le \sum_{\alpha=1}^{m}\sum_{|k|\ne0}\left|k_{\alpha}\right|\left|U^{(k)}\right|e^{|\mathrm{Im}\,\varphi|\|k\|} \le m\sum_{|k|\ne0}|k|\left|U^{(k)}\right|e^{|\mathrm{Im}\,\varphi|\|k\|}$$

$$\le \frac{mM_0}{2\varepsilon\min\limits_{\alpha}\left|\lambda_{\alpha}\right|}\sum_{|k|\ne0}|k|^{d+1}e^{-2\delta|k|} \le c_1(\varepsilon)\frac{M_0}{\delta^{\alpha+m+1}} \qquad (1.26)$$

for $|\,\mathrm{Im}\,\varphi\,| \le \rho_0 - 2\delta$. It follows from (1.25) and (1.26) that the matrix $U(\varphi)$ exists and is analytic and periodic in φ with period 2π for all φ from the region (1.20). The fact that it is real for $\mathrm{Im}\,\varphi = 0$ follows from the complex conjugacy of the coefficients $U^{(k)}$ and $U^{(-k)}$, because for the matrix $P(\varphi)$ real for $\mathrm{Im}\,\varphi = 0$, the coefficients $P^{(k)}$ and $P^{(-k)}$ are complex conjugate.

§2. Theorem on Reducibility

The main result, obtained when solving the problem formulated above, is given by the following theorem.

Theorem 5.1. (Martinyuk and Perestyuk, 1974). *Suppose that the system* (1.1) *satisfies the following conditions:*

(i) the matrix $P(\varphi)$ is periodic in $\varphi = (\varphi^1, \ldots, \varphi^m)$ with period 2π, analytic in the region

$$|\,\mathrm{Im}\,\varphi\,| = \sup_{\alpha}\left|\mathrm{Im}\,\varphi_{\alpha}\right| \le \rho_0, \ \rho_0 > 0 \qquad (2.1)$$

and real for $\mathrm{Im}\,\varphi = 0$;

(ii) for some positive ε and d, the inequality

$$\left|\sin\frac{(k,\omega)}{2}\right| > \varepsilon|k|^{\lceil-d}, \ |k| \ne 0, \qquad (2.2)$$

holds for all vectors $k = (k_1, \ldots, k_m)$ with integer components;

(iii) the eigenvalues $\lambda = (\lambda_1, \dots, \lambda_s)$ *of the matrix* A *are real, different, and nonzero.*

Then one can always find a sufficiently small positive constant M_0, *such that for*

$$|P(\varphi)| = \sum_{i,j=1}^{s} |P_{i,j}(\varphi)| \le M_0 \qquad (2.3)$$

the system of equations (1.1) *is reduced to the form* (A_0 *is a constant matrix*)

$$y_{n+1} = A_0 y_n, \quad \Delta \varphi_n = \omega \qquad (2.4)$$

by a nondegenerate change of variables

$$x_n = \Phi(\varphi_n) y_n, \qquad (2.5)$$

with a matrix $\Phi(\varphi)$ *periodic in* φ *with period* 2π, *analytic with analytic reciprocal in the region*

$$|\operatorname{Im}\varphi| \le \frac{\rho_0}{2}, \qquad (2.6)$$

and real for $\operatorname{Im}\varphi = 0$.

This theorem implies that the fundamental matrix of (1.1) has the form

$$x_n = \Phi(\varphi_n) A_0^n, \qquad (2.7)$$

where $\Phi(\varphi_n)$ is a non-singular quasiperiodic real matrix with the same frequency basis as $P(\varphi)$.

   Proof. We prove Theorem 5.1 by constructing the reducing matrix $\Phi(\varphi)$. Let us represent the matrix $P(\varphi)$ as the sum of two matrices: $P(\varphi) = D + Q$, where D is a diagonal matrix ($D = \{d_1, \dots, d\}$) with the diagonal elements equal to the elements of the matrix P. Let us make a change of variables in the system (1.1) according to

$$x_n = \left(E + U_1(\varphi_n) \right) y_n^{(1)} \qquad (2.8)$$

and choose a periodic solution of the equation

$$U_1(\varphi + \omega) A = A U_1(\varphi) + P(\varphi) - \overline{D} \qquad (2.9)$$

to be $U_1(\varphi)$; here,

$$\overline{D} = \frac{1}{(2\pi)^m} \int\limits_0^{2\pi} \cdots \int\limits_0^{2\pi} D(\varphi)\, d\varphi^1 \ldots d\varphi^m.$$

Since A is the diagonal matrix with real, different, and nonzero elements $\lambda_1, \ldots, \lambda_s$, the equation (2.9) is always solvable; moreover, the function $U_1(\varphi)$ is analytic in the region

$$|\operatorname{Im}\varphi| \le \rho_0 - 2\delta_0 = \rho_1 \qquad (2.10)$$

real for $\operatorname{Im}\varphi = 0$ and satisfies the inequality

$$|U_1(\varphi)| \le \left(\frac{1}{\min\limits_{\alpha \ne \beta}|\lambda_\beta - \lambda_\alpha|} + \frac{1}{2\varepsilon\min\limits_{\alpha}|\lambda_\alpha|}\left(\frac{d}{e}\right)^d (1+e)^m \frac{1}{\delta^{d+m}}\right) M_0. \qquad (2.11)$$

Inserting (2.8) in (1.1) and taking (2.9) into account, we obtain

$$y_{n+1}^{(1)} = \left(A + \overline{D}\right) y_n^{(1)} + P_1(\varphi_n)\, y_n^{(1)}, \quad \Delta\varphi_n = \omega, \qquad (2.12)$$

where

$$P_1(\varphi_n) = (E + U_1(\varphi_n + \omega))^{-1}\, (P(\varphi_n)\, U_1(\varphi_n) - U_1(\varphi_n + \omega)\, \overline{D}). \qquad (2.13)$$

We set $r_0 = \min\limits_{\alpha \ne \beta}|\lambda_\beta - \lambda_\alpha|$ and choose so small an $M_0 = M_0(r_0, \delta_0)$ that the inequality

$$4s\left[\frac{1}{r_0} + \frac{1}{2\varepsilon\min\limits_{\alpha}|\lambda_\alpha|}\left(\frac{d}{e}\right)^d (1+e)^m \frac{1}{\delta^{d+m}}\right] M_0^{2-\gamma} < 1, \quad 1 < \gamma < 2, \qquad (2.14)$$

holds. Then for the function $U_1(\varphi)$, we find

$$|U_1(\varphi)| < M_0^{\gamma-1}/4s. \qquad (2.15)$$

Estimating now $P_1(\varphi)$, we get

$$|P_1(\varphi)| \le \left(|E| + \sum_{v=1}^{\infty} |U_1(\varphi)|^v \right) (|P(\varphi_n)| |U_1(\varphi_n)| + |U_1(\varphi_n + \omega)| |\overline{D}|)$$

$$\le \left(s + \frac{M_0^{\gamma-1}}{4s - M_0^{\gamma-1}} \right) \frac{M_0^{\gamma}}{2s} \le M_0^{\gamma} = M_1. \tag{2.16}$$

Let $A_1 = A + \overline{D}$ and let $\lambda_1^{(1)}, \lambda_2^{(1)}, \ldots, \lambda_s^{(1)}$ be the diagonal elements of the matrix A_1. We have

$$\min_{\alpha \ne \beta} \left| \lambda_\beta^{(1)} - \lambda_\alpha^{(1)} \right| \ge \min_{\alpha \ne \beta} \left| \lambda_\beta - \lambda_\alpha \right| - \min_{\alpha \ne \beta} \left| d_\beta - d_\alpha \right| \ge r_0 - 2M_0 = r_1,$$

$$\min_{\alpha} \left| \lambda_\alpha \right| \ge \min_{\alpha} \left| \lambda_\alpha \right| - \max_{\alpha} \left| d_\alpha \right| \ge \sigma_0 - M_0 = \sigma_1. \tag{2.17}$$

For small M_0, e. g., for $M_0 < \frac{r_0}{4}$ and $M_0 < \frac{\sigma_0}{2}$, i. e., for $M_0 < \min\left\{\frac{r_0}{4}, \frac{\sigma_0}{2}\right\}$, the constants r_1 and σ_1 are positive and, consequently, $\lambda_1^{(1)}, \lambda_2^{(1)}, \ldots, \lambda_s^{(1)}$ are different. Clearly, the system

$$y_{n+1}^{(1)} = A_1 y_n^{(1)} + P_1(\varphi) y_n^{(1)}, \quad \Delta\varphi_n = \omega \tag{2.18}$$

possesses the same properties as the original system (1.1), and therefore, one can transform (2.18) just as the system (1.1). Let us show that the process of the transformation of the system (1.1) by the changes of variables such as (2.5) can be infinitely repeated if ρ_0 and δ are properly chosen. As follows from the above argument, the function $U_1(\varphi)$ is analytic in the region

$$|\operatorname{Im} \varphi| \le \rho_0 - 2\delta_0 = \rho_1,$$

and the function $U_2(\varphi)$, obtained as a result of the second iteration, is analytic in the region

$$|\operatorname{Im} \varphi| \le \rho_1 - 2\delta_1 = \rho_0 - 2\delta_0 - 2\delta_1 = \rho_2,$$

etc. Hence, all ρ_i $(i = 1, 2, \ldots)$ should be bounded by a certain number, e. g., by $\rho_0/2$. We set $\delta_0 = q$, $\delta_1 = q^2, \ldots$, $\delta_i = q^{i+1}$ and choose q such that the following equality holds

$$2(q + q^2 + \ldots + q^i + \ldots) = \frac{\rho_0}{2} \tag{2.19}$$

i. e.,

$$q = \rho_0 / (4 + \rho_0).$$ (2.20)

Then by successive changes of variables such as (2.5), we obtain a set of difference equations for $U_1(\varphi)$, $U_2(\varphi)$, ..., $U_p(\varphi)$ with the constants M satisfying the relation $M_{p+1} = M_p^{\gamma}$, $1 < \gamma < 2$.

Assume that, having changed the variables in (1.1) p times (according to (2.5)), we obtain the system of difference equations

$$y_{n+1}^{(p)} = A_p y_n^{(p)} + P_p(\varphi_n) y_n^{(p)}, \quad \Delta \varphi_n = \omega$$ (2.21)

which satisfy the following conditions:

(i) the matrix $P_p(\varphi)$ is periodic in φ with period 2π, analytic in φ in the region

$$|\operatorname{Im} \varphi| \leq \rho_p,$$ (2.22)

real for real φ, and bounded

$$|P_p(\varphi)| \leq M_p;$$ (2.23)

(ii) diagonal elements $\lambda_1^{(1)}, \lambda_2^{(1)}, ..., \lambda_s^{(1)}$ of the real matrix A_p are nonzero and satisfy the inequalities

$$\min_{\alpha \neq \beta} \left| \lambda_\beta^{(p)} - \lambda_\alpha^{(p)} \right| \geq r_p, \quad \min_\alpha \left| \lambda_\alpha^{(p)} \right| \geq \sigma_p.$$ (2.24)

We show that under these assumptions, one can always find a positive constant M^0 independent of p, such that for all $M_0 < M^0$, there exists a change of variables

$$y_n^{(p)} = \left(E + U_{p+1}(\varphi_n) \right) y_n^{(p+1)},$$ (2.25)

which reduces the system of difference equations (2.21) to the form

$$y_n^{(p+1)} = A_{p+1} y_n^{(p+1)} + P_{p+1}(\varphi_n) y_n^{(p+1)}, \quad \Delta \varphi_n = \omega;$$ (2.26)

and moreover,

(a) the matrices $P_{p+1}(\varphi)$ and $U_{p+1}(\varphi)$ are periodic in φ with period 2π, analytic in the region

$$|\operatorname{Im}\varphi| \le \rho_{p+1} \qquad (2.27)$$

real for real φ and such that

$$|P_{p+1}(\varphi)| \le M_{p+1}, \quad |U_{p+1}(\varphi)| \le M^{\gamma-1}/4s; \qquad (2.28)$$

(b) the diagonal elements $\lambda_1^{(p+1)}, \lambda_2^{(p+1)}, \dots, \lambda_s^{(p+1)}$ of the real diagonal matrix A_{p+1} are distinct nonzero and such that

$$\min_{\alpha \ne \beta}\left|\lambda_\beta^{(p+1)} - \lambda_\alpha^{(p+1)}\right| \ge r_{p+1}, \quad \min_\alpha\left|\lambda_\alpha^{(p+1)}\right| \ge \sigma_{p+1}, \qquad (2.29)$$

where the constants $M_{p+1}, \rho_{p+1}, r_{p+1}$, and σ_{p+1} can be obtained from M_p, ρ_p, r_p, and σ_p as follows

$$M_{p+1} = M_p^\gamma, \quad \rho_{p+1} = \rho_p - 2\delta_0^p, \quad r_{p+1} = r_p - M_p^{\gamma-1},$$

$$\sigma_{p+1} = \sigma_p - M_p^{\gamma-1}, \quad 1 < \gamma < 2, \quad \delta_0 < \rho_0 / (4 + \rho_0). \qquad (2.30)$$

Indeed, after the change of variables (2.25), the system (1.1) turns into (2.26), where $U_{p+1}(\varphi)$ is a solution of the system

$$U_{p+1}(\varphi + \omega)\, A_{p+1} = A_{p+1}\, U_{p+1}(\varphi) + P_p(\varphi) - \overline{D}_p \qquad (2.31)$$

which satisfies conditions of Lemma 5.2 by virtue of the assumptions (i) and (ii). Hence, the inequality

$$\left|U_{p+1}(\varphi)\right| \le \left(\frac{1}{r_p} + \frac{1}{2\varepsilon\sigma_p}\left(\frac{d}{e}\right)^d \frac{(1+e)^m}{\delta_0^{(d+m)p}}\right) M_p \qquad (2.32)$$

holds for $|\operatorname{Im}\varphi| \le \rho_{p+1}$.

We choose M_0 so small that, together with conditions (2.14), the inequalities

$$r_0 - \sum_{v=0}^{\infty} M_0^{(\gamma-1)\gamma^v} > 0; \quad \sigma_0 - \sum_{v=0}^{\infty} M_0^{(\gamma-1)\gamma^v} > 0;$$

$$(2.33)$$

$$4s\left[\frac{1}{r_0 - \sum\limits_{v=0}^{\infty} M_0^{(\gamma-1)\gamma^v}} + \frac{1}{2\varepsilon\left(\sigma_0 - \sum\limits_{v=0}^{\infty} M_0^{(\gamma-1)\gamma^v}\right)}\left(\frac{d}{e}\right)^d \frac{(1+e)^m}{\delta_0^{(d+m)p}}\right] M_0^{2-\gamma} M_0^{(2-\gamma)\gamma^p} < 1$$

should be valid for all integer $p > 0$. In this case (i. e., when the inequalities (2.33) hold), the matrix $U_{p+1}(\varphi)$ satisfies the second inequality in (2.28).

Since

$$P_{p+1}(\varphi) = (E + U_{p+1}(\varphi + \omega))^{-1} (P_p(\varphi) U_{p+1}(\varphi) - U_{p+1}(\varphi) \overline{D}_p),$$

we have

$$|P_{p+1}(\varphi)| \le |(E + U_{p+1}(\varphi + \omega))^{-1}| (|P_p(\varphi)||U_{p+1}(\varphi)| - |U_{p+1}(\varphi)||\overline{D}_p|)$$

$$\le \left(s + \frac{M_p^{\gamma-1}}{4s - M_p^{\gamma-1}}\right)\left(M_p \frac{M_p^{\gamma-1}}{4s} + \frac{M_p^{\gamma-1}}{4s}\right) \le M_p^\gamma = M_{p+1},$$

Taking into account that A_{p+1} is a diagonal matrix, we obtain

$$\min_{\alpha \ne \beta}\left|\lambda_\alpha^{(p+1)} - \lambda_\beta^{(p+1)}\right| \ge \min_{\alpha \ne \beta}\left|\lambda_\alpha^{(p)} - \lambda_\beta^{(p)}\right| - \max_{\alpha \ne \beta}\left|\overline{P}_{p\alpha\alpha} - \overline{P}_{p\beta\beta}\right|$$

$$> r_p - 2M_p > r_p - 2M_p^{\gamma-1};$$

$$\min_{\alpha \ne \beta}\left|\lambda_\alpha^{(p+1)} + \overline{P}_{p\alpha\alpha}\right| \ge \min_\alpha\left|\lambda_\alpha^{(p+1)}\right| - \max_\alpha\left|\overline{P}_{p\alpha\alpha}\right| > \sigma_p - M_p > \sigma_p - M_p^{\gamma-1}.$$

Therefore, if the value M_0 sartisfying the inequalities (2.14) and (2.33), is taken to play the role of M^0, then the inequalities (2.27)–(2.29) can always be satisfied. This implies that we can infinitely repeat the process of the transformation of the system (1.1).

Denote

$$\Phi_k(\varphi) = (E + U_1(\varphi)) (E + U_2(\varphi)) \dots (E + U_p(\varphi)) = \prod_{\alpha=1}^{p}\left(E + U_\alpha(\varphi)\right). \qquad (2.34)$$

Let us show that the sequence (2.34) converges uniformly to a matrix $\Phi(\varphi)$ analytic in the region

$$| \operatorname{Im} \varphi | < \frac{\rho_0}{2}. \tag{2.35}$$

The analyticity of $\Phi(\varphi)$ in (2.35) follows from the choice of $\rho_1, \rho_2, \ldots, \rho_p, \ldots$
    By virtue of (2.25), we have the estimate

$$\left| \Phi_{p+1}(\varphi) - \Phi_p(\varphi) \right| \leq \prod_{\alpha=1}^{p} \left(E + \frac{M_{\alpha-1}^{\gamma-1}}{4s} \right) \frac{M_p^{\gamma-1}}{4} \leq \frac{s}{4} \sum_{\alpha=1}^{\infty} \left(1 + \frac{M_{\alpha-1}^{\gamma-1}}{4} \right) M_p^{\gamma-1} \leq c M_p^{\gamma-1}.$$

Therefore,

$$\left| \Phi_{p+s}(\varphi) - \Phi_p(\varphi) \right| \leq c \sum_{\alpha=p}^{p+s-1} M_{\alpha}^{\gamma-1} < c \sum_{\alpha=p}^{\infty} M_{\alpha}^{\gamma-1}, \tag{2.36}$$

and this means that the sequence (2.34) is uniformly convergent. Denoting the limiting matrix of the sequence (2.34) by $\Phi(\varphi)$, we find that it is periodic in φ with period 2π, real for $\operatorname{Im} \varphi = 0$, and analytic for $| \operatorname{Im} \varphi | < \rho_0 / 2$. Thus, passing to the limit as $p \to \infty$ in the expression

$$x_n = (E + U_1(\varphi)) (E + U_2(\varphi)) \ldots (E + U_p(\varphi)) y^{(p)} \tag{2.37}$$

we find, that by the change of variables

$$x_n = \Phi(\varphi) y_n$$

the system (1.1) is transformed into the system

$$y_{n+1} = A_0 y_n, \Delta \varphi_n = \omega, \tag{2.38}$$

where

$$A_0 = \lim_{p \to \infty} \left(A + \overline{D}_1 + \overline{D}_2 + \ldots + \overline{D}_p \right).$$

    We now prove that $\Phi(\varphi)$ is a non-singular matrix. Estimating the difference $\Phi(\varphi) - E$, we find

$$|\Phi(\varphi) - E| = \left|\prod_{\alpha=1}^{p}\big(E + U_\alpha(\varphi) - E\big)\right| \le \left|\left(\frac{M_0^{\gamma-1}}{4s} + \frac{M_{e-1}^{\gamma-1}}{4s}\right)I\right.$$

$$+ \left(\frac{M_0^{\gamma-1}}{4s}\frac{M_1^{\gamma-1}}{4s} + \frac{M_{p-2}^{\gamma-1}}{4s}\frac{M_{p-1}^{\gamma-1}}{4s}s\right)I + \ldots + \frac{M_0^{\gamma-1}}{4s}$$

$$\left.\times\frac{M_1^{\gamma-1}}{4s}\ldots\frac{M_{p-2}^{\gamma-1}}{4s}s^{p-1}I\right| \le \left(\prod_{\alpha=1}^{p}\left(1 + \frac{M_{\alpha-1}}{4}\right) - 1\right)|I|, \tag{2.39}$$

where I is a matrix with unit elements.

If M_0 is sufficiently small, then (2.39) implies the inequality

$$|\Phi_p(\varphi) - E| \le c < 1,$$

which yields

$$|\Phi(\varphi) - E| \le c < 1.$$

Consequently, the series $\sum_{i=0}^{\infty}\Phi\big(E - \Phi(\varphi)\big)^i$ converges and defines a matrix $\Phi^{-1}(\varphi)$ reciprocal to $\Phi(\varphi)$. We finally note that the method of construction of the reducing matrix $\Phi(\varphi)$ can be employed when constructing a fundamental matrix of solutions of the system (1.1).

§3. Reducibility of Linear Systems of Difference Equations with a Smooth Right-Hand Side

Consider the case when the matrix $P(\varphi)$ has a finitely many derivatives. Following the paper by Mitropolsky and Samoilenko, (1965), we use the so-called smoothing operator (see §1 in Chapter 6). Let us denote this operator by S_N. For a function $f(\varphi)$, periodic in each of its arguments φ^1, φ^2, ..., φ^m with period 2π, which, consequently, can be expanded in the Fourier series

$$f(\varphi) = \sum_k f^{(k)} e^{i(k,\varphi)}$$

(here, k is a vector with integer components and (k, φ) is the scalar product), we define the smoothing operator $S_N f(\varphi)$ as an operator which truncates the Fourier series, i. e.,

$$S_N f(\varphi) = \sum_{|k| \le N} f^{(k)} e^{i(k,\varphi)}, \ |k| = \sum_{v=1}^{m} |k_v|.$$

For this operator, we have the following lemma.

Lemma 5.3. (Mitropolsky and Samoilenko, 1965) *We have*

$$|f(\varphi) - S_N f(\varphi)|_\lambda \le CN^{-l + \lambda + \delta} |f(\varphi)|_l,$$

for $\lambda \ge 0$, *and*

$$|S_N f(\varphi)|_\lambda \le CN^{\lambda + \delta} |f(\varphi)|_b,$$

for $0 \le \lambda \le l - m - 1$ *and* $f(\varphi) \in C^l(E_m)$, *where* $m \le \delta \le m + 1$.

Consider a system of difference equations

$$x_{n+1} = A x_n + P(\varphi) x_n, \ \Delta\varphi_n = \omega, \tag{3.1}$$

where the matrix $P(\varphi)$ is l times continuously differentiable and periodic in φ with period 2π. Assume that A is a real diagonal matrix

$$A = \{\lambda_1, ..., \lambda_s\}.$$

Theorem 5.2. (Martinyuk and Perestyuk, 1975b) *Suppose that the right-hand side of* (3.1) *satisfies the following conditions:*

(i) *the matrix* $P(\varphi)$ *is periodic in* $\varphi = (\varphi^1, \varphi^2, ..., \varphi^m)$ *with period* 2π *and* l *times continuously differentiable;*

(ii) *for some positive* ε *and* d, *the inequality*

$$\left| \sin\frac{(k,\omega)}{2} \right| \ge \varepsilon |k|^{-d}, \ |k| \ne 0 \tag{3.2}$$

holds for all vectors $k = (k_1, ..., k_m)$ *with integer components;*

(iii) *the eigenvalues* $\lambda = (\lambda_1, ..., \lambda_s)$ *of the matrix* A *are real, different, and nonzero.*

Then one can find a sufficiently small constant M *and an integer* $l = l(k_0)$, *such that for*

$$|P(\varphi)| \le M_0 < M^0 \ \text{and} \ |P(\varphi)|_l \le c, \tag{3.3}$$

the system of equations (3.1) *is reduced to the form*

$$y_{n+1} = A_0 y_n, \Delta\varphi_n = \omega \tag{3.4}$$

where $\Phi(\varphi)$ *is a constant matrix, by a non-degenerate change of variables*

$$x_n = \Phi(\varphi_n) y_n \tag{3.5}$$

where the matrix $\Phi(\varphi)$ *is periodic in* φ *with period* 2π *and* k_0 *times continuously differentiable.*

Proof. Let us transform (3.1), by using the change of variables

$$x_n = (E + U_1(\varphi_n, N_0)) y_n^{(1)}, \tag{3.6}$$

where $U_1(\varphi_n, N_0)$ is a solution (periodic in $\varphi = (\varphi^1, \varphi^2, ..., \varphi^m)$) of the matrix equation

$$U(\varphi + \omega) A = A U_1(\varphi) + S_{N_0}(P(\varphi)) - D. \tag{3.7}$$

Here, $S_{N_0}(P)$ is a finite sum of the Fourier expansion of $P(\varphi)$, i. e.,

$$S_{N_0}(P) = \sum_{|k| \le N_0} P^{(k)} e^{i(k,\varphi)};$$

$$\overline{P} = \frac{1}{(2\pi)^m} \int_0^{2\pi} ... \int_0^{2\pi} P(\varphi) d\varphi^1 d\varphi^2 ... d\varphi^m; \tag{3.8}$$

$$U_1(\varphi_n, N_0) = \sum_{|k| \le N_0} U_1^{(k)} e^{i(k,\varphi)}.$$

By taking the results of the previous section into account, one can find that the equation (3.7) possesses a periodic solution $U_1(\varphi_n, N_0)$ for which the following estimateholds.

$$\left|U_1(\varphi, N_0)\right|_\lambda = \max_{|\rho|=\rho_1+...+\rho_m=\lambda} \left|D_\varphi^{|\rho|} U_1(\varphi, N_0)\right| \le CN_0^{\lambda+m+1} \left|U_1(\varphi, N_0)\right|_0$$

$$\le CN_0^{\lambda+m+1} \sum_{|k|\le N_0} \left|U_1^{(k)}\right| \le CN_0^{\lambda+m+1} \left[\frac{\left|P^{(0)}\right|}{\min_{\alpha\ne\beta}\left|\lambda_\alpha-\lambda_\beta\right|} + \sum_{|k|\le N_0} \frac{\left|P^{(k)}\right|\left|k\right|^d}{\varepsilon} \right]$$

$$\le \frac{C_1 N^{\lambda+m+1}}{\max(r_0,\varepsilon)} \sum_{|k|\le N_0} N_0^d |P|_0 \le \frac{2^m C_1 N_0^{1+\lambda+d+2m+1}}{\bar{r}} |P|_0, \tag{3.9}$$

where $\bar{r} = \max(r_0, \varepsilon)$. As a result of the change (3.6), we obtain the following system of equations for $y_n^{(1)}$

$$y_{n+1}^{(1)} = (A + \bar{D})y_n^{(1)} + P_1(\varphi)y_n^{(1)}, \tag{3.10}$$

where

$$P_1(\varphi) = P_1(\varphi, N_0) = E + U_1(\varphi+\omega, N_0)^{-1}(P(\varphi)U_1(\varphi, N_0)$$

$$-U_1(\varphi+\omega)N_0)\bar{D} + R_{N_0};$$

$$R_{N_0} = P - S_{N_0}(P). \tag{3.11}$$

By virtue of the fact that $R_{N_0} \to 0$ as $N_0 \to \infty$ for a sufficiently smooth function $P(\varphi)$, one can always choose N_0 such that the variable R_{N_0} has the second order of smallness with respect to M. Moreover, the constant M can be chosen so small that the matrix $U_1(\varphi)$ becomes a value of the first order of smallness with respect to M_0. Thus, one can easily find that the matrix $P_1(\varphi)$ is a value of the second order of smallness with respect to M_0.

By denoting $A_1 = A + \bar{D}$, we rewrite the system (3.10) as

$$y_{n+1}^{(1)} = A_1 y_n^{(1)} + P_1(\varphi_n)y_n^{(1)} \tag{3.12}$$

where the right–hand side satisfies the conditions (i) – (iii). Hence, we can change the variables in (3.12) as follows

$$y_n^{(1)} = \left(E + U_2(\varphi)\right)y_n^{(2)}, \tag{3.13}$$

where $U_2(\varphi_n)$ is the periodic solution of the equation

$$U_2(\varphi + \omega)A_1 = A_1 U_2(\varphi) + S_{N_1}\left(P_1(\varphi) - \overline{P_1(\varphi)}\right). \tag{3.14}$$

Inserting (3.13) in the system (3.12) and taking (3.14) into account, we obtain the system of equations for $y_n^{(2)}$

$$y_{n+1}^{(2)} = \left(A_1 + \overline{P_1}(\varphi)\right)y_n^{(2)} + \left(E + U_2(\varphi + \omega)\right)^{-1}\left(P_1(\varphi)U_2(\varphi)\right.$$

$$\left. - U_2(\varphi + \omega)\overline{P_1}(\varphi) + R_{N_1}\right). \tag{3.15}$$

We define N_1 by the relation $N_1 = N_0^\gamma, 1 < \gamma < 2$. Having fixed N_1, we can show that the matrix

$$P_2(\varphi) = \left(E + U_2(\varphi + \omega)\right)^{-1}\left(P_1(\varphi)U_2(\varphi) - U_2(\varphi + \omega)\overline{P_1}(\varphi) + R_{N_1}\right)$$

is a variable of the second order of smallness with respect to $|P_1(\varphi)|_0$ and, therefore, of the fourth order of smallness with respect to M_0.

If the process of transformation of the initial equation is continued, then on the pth step we reduce (3.1) to the following system of equations

$$y_{n+1}^{(p)} = \left(A_{p-1} + \overline{P}_{p-1}\right)y_n^{(p)} + P_p \tag{3.16}$$

by the change of variables

$$y_{n+1}^{(p-1)} = \left(E + U_p(\varphi)\right)y_n^{(p)}. \tag{3.17}$$

Here, the matrix $U_p(\varphi)$ is a periodic solution of the equation

$$U_p(\varphi + \omega)A_{p-1} = A_{p-1}U_p(\varphi) + S_{N_{p-1}}\left(P_{p-1}\right) - \overline{P}_{p-1}, \tag{3.18}$$

where

$$A_{p-1} = A + \overline{P} + \sum_{j=1}^{p-2}\overline{P}_j;$$

$$P_p(\varphi) = \left(E + U_p(\varphi + \omega)\right)^{-1}\left(P_{p-1}(\varphi)U_p(\varphi) - U_p(\varphi + \omega)\overline{P}_{p-1}(\varphi) + R_{N_{p-1}}^{(p-1)}\right); \quad (3.19)$$

$$N_p = N_{p-1}^\gamma, \; p \geq 2.$$

Clearly, the matrix $P_p(\varphi)$ has the (2^p)–th order of smallness with respect to M_0, because $P_p(\varphi)$ has the second order of smallness with respect to P_{p-1}.

We now estimate the matrix $P_1(\varphi)$. By using the inequality (3.9) and he properties of the operator S_N we find

$$\left|P_1(\varphi)\right| \leq \left|\left(E + U_1(\varphi + \omega, N_0)\right)^{-1}\right|\left(2^{m+1}C_1 N_0^{d+2m+1}\left|P(\varphi)\right|_0^2\right.$$

$$+ CN^{-l+m+1}\left|P(\varphi)_l\right|\right). \qquad (3.20)$$

Differentiating $P_1(\varphi)$, we obtain

$$\left|D_\varphi^l P_1(\varphi)\right| \leq c_2\left[\left|\left(E + U_1(\varphi + \omega, N_0)\right)^{-1}\right|\left(\left|P(\varphi)U_1(\varphi, N_0)\right|_l\right.\right.$$

$$+ \left|U_1(\varphi, N_0)\right|_l\left|\overline{D}\right| + \left|P(\varphi)\right|_l + \left|S_N P(\varphi)\right|_l\right)$$

$$+ \max_{\alpha=1,\ldots,l}\left\{\left|\left(E + U_1(\varphi + \omega, N_0)\right)^{-1}\right|_\alpha \times \left(\left|P(\varphi)U_1(\varphi, N_0)\right|_{l-\alpha}\right.\right.$$

$$+ \left|U_1(\varphi, N_0)\right|_{l-\alpha}\left|\overline{D}\right| + \left|P(\varphi)\right|_{l-\alpha} + \left|S_N P(\varphi)\right|_{l-\alpha}\right\}\right]$$

$$\leq c_3\left[\left|\left(E + U_1(\varphi + \omega, N_0)\right)^{-1}\right|\left(\max_{\alpha=1,\ldots,l}\left\{\left|P(\varphi)\right|_\alpha \frac{N_0^{l-\alpha+d+2m+1}}{\overline{r}}\right.\right.\right.$$

$$\times \left|P(\varphi)\right|_0 + \frac{N_0^{l+d+2m-1}}{\overline{r}}\left|P(\varphi)\right|_0^2 + \left|P(\varphi)\right|_l + N_0^{l+m+1}\left|P(\varphi)\right|_0\right)$$

$$+ \left|\left(E + U_1(\varphi + \omega, N_0)\right)^{-1}\right|_0^{l+1}\frac{N_0^{d+2m+1}}{\overline{r}}\left|P(\varphi)\right|_0$$

$$\times \max_{\alpha=1,\ldots,l}\left\{N_0^\alpha \max_\beta\left(\left|P(\varphi)\right|_\beta\left|U_1(\varphi, N_0)\right|_{l-\alpha-\beta}\right.\right.$$

$$+ \frac{N_0^{l+d+2m+1}}{\overline{r}}\left|P(\varphi)\right|_0^2 + N^{l+m+1}\left|P(\varphi)\right|_0\right\}\right].$$

$$(3.21)$$

According to Lemma 11 in (Bogolyubov, Mitropolsky, and Samoilenko, 1969), we have

$$l > \frac{\gamma}{\gamma - 1}(\beta + m + 1);$$

$$1 < \gamma < 2; \ \beta > \frac{\gamma(k_0 + d + 2m + 1)}{2 - \gamma},$$

where k_0 is a natural number and, consequently, one can always find a positive M^0, such that the inequalities

$$\left| U_1(\varphi, N_1) \right|_{k_0} < M_0^{\gamma - 1}/4s; \ \left| P_1(\varphi, N_1) \right| < M_0^{\gamma} \leq M_1;$$

$$\left| D_\varphi^l P_1(\varphi, N_1) \right| \leq N_1^l; \ N_1^{-\beta} = M_1 = N_0^{-\gamma\beta},$$

and the following generalization of the Landau–Hadamard inequality

$$|f|_\lambda \leq c|f|_0 \left(\frac{|f|_r}{|f|_0} \right)^{\frac{\lambda}{r}}, \ 0 \leq \lambda \leq r$$

hold for $M_0 < M^0$. Hence, one can easily get

$$|P|_\alpha \leq c|P|_0 \left(\frac{|P|_l}{|P|_0} \right)^{\alpha/l} \leq cM_0 \left(\frac{N_0^l}{M_0} \right)^{\alpha/l} = cM_0 \frac{l - \alpha}{l} N_0^\alpha,$$

and this yields

$$\max_\alpha \left\{ |P|_\alpha \frac{N_1^{l - \alpha + d + 2m + 1}}{\bar{r}} \right\} \leq \frac{c}{\bar{r}} N_1^{l + d + 2m + 1};$$

$$\max_\alpha \left\{ N^\alpha \max_\beta \left(|P|_\beta \left| U_1 \right|_{l - \alpha - \beta} \right) \right\} \leq \max \left\{ cM_0^{\frac{l - \beta}{l}} \right.$$

$$\left. \times \frac{2^m c_1}{\bar{r}} N_1^{l + d + 2m + 1} M_0 \right\} \leq \frac{2^m c c_1}{\bar{r}} N_1^{l + d + 2m + 1} M_0. \tag{3.22}$$

It follows from (3.21) and (3.22) that

$$\left|D_\varphi^l P_1(\varphi,N_1)\right| \le c_4\left[(s+1)\left(\frac{N_1^{l+d+2m+2}M_0 + N_1^{l+d+2m+1}M_0^2}{\overline{r}}\right)\right.$$

$$+N_0^l + N_1^{l+m+1}M_0\Big) + (s+1)^{l+1}\frac{N_1^{d+2m+1}}{\overline{r}}$$

$$\times\left(\frac{N_1^{l+d+2m+1} + N_1^{l+d+2m+1}M_0^2}{\overline{r}} + N_1^{l+m+1}M_0\right)\Bigg],$$

and for small M_0 this implies that

$$\left|D_\varphi^l P_1(\varphi,N_1)\right| \le N_1^l.$$

Let us continue the process of transformation of the initial system (3.1) by changes of variables such as (3.6). If we set $M_j = M_{j-1}^\gamma = N_j^{-\beta}$ and $M_j = N_{j-1}^\gamma$, then for sufficiently small M_0, we have

$$\left|U_p(\varphi,N_p)\right| \le \frac{M_{p-1}^{\gamma-1}}{4s}; \quad \left|P_p(\varphi)\right| \le M_p;$$

$$\left|A_p - A\right| \le \sum_{\alpha=0}^{p-1} M_\alpha < \frac{r}{2}. \tag{3.23}$$

These inequalities yield

$$\left|\prod_{j=1}^{p+1}\left(E + U_j(\varphi,N_j)\right) - \prod_{j=1}^{p}\left(E + U_j(\varphi,N_j)\right)\right|$$

$$\le \left|\prod_{j=1}^{p}\left(E + U_j(\varphi,N_j)\right)\right|\left|U_{p+1}(\varphi,N_{p+1})\right|$$

$$\le \left|\prod_{j=1}^{p}\left(E + \frac{M_{j-1}^{\gamma-1}}{4s}I\right)\right|\frac{M_p^{\gamma-1}}{4s} \le \frac{s}{4}\prod_{j=1}^{\infty}\left(1 + \frac{M_{j-1}^{\gamma-1}}{4}\right)M_p^{\gamma-1} \le \frac{s}{2}M_p^{\gamma-1}.$$

In fact, this yield the criterion of the uniform convergence for the sequence

$$\prod_{j=1}^{p}\left(E + U_j(\varphi,N_j)\right),$$

namely,

$$\left| \prod_{j=1}^{p+k_0} \left(E + U_j(\varphi, N_j) \right) - \prod_{j=1}^{p} \left(E + U_j(\varphi, N_j) \right) \right| \le \frac{s}{2} \sum_{j=p}^{p+k-1} M_j^{\gamma-1} < s N_p^{\gamma-1}.$$

By introducing the notations

$$\Phi(\varphi) = \lim_{p \to \infty} \prod_{j=1}^{p} \left(E + U_j(\varphi, N_j) \right),$$

$$A_0 = A + \overline{D} + \sum_{j=1}^{\infty} \overline{D}_j,$$

one can easily find that the matrix $\Phi(\varphi)$ is periodic in φ with period 2π. Furthermore, the inequality

$$\left| \prod_{j=1}^{p} \left(E + U_j(\varphi, N_j) \right) - E \right| \le \left[\prod_{j=1}^{p} \left(1 + \frac{M_{j-1}^{\gamma-1}}{4} \right) - 1 \right] |I| < 1$$

implies the convergence of the series $\displaystyle\sum_{j=0}^{\infty} (E - \Phi(\varphi))^j$, and hence, the matrix $\Phi(\varphi)$ is non–singular.

We also have the inequality

$$\left| D_\varphi^{k_0} \left[\prod_{j=1}^{p+1} \left(E + U_j(\varphi, N_j) \right) - \prod_{j=1}^{p} \left(E + U_j(\varphi, N_j) \right) \right] \right|$$

$$\le \left| D_\varphi^{k_0} \left[\prod_{j=1}^{p} \left(E + U_j(\varphi, N_j) \right) U_{p+1}(\varphi, N_{p+1}) \right] \right|$$

$$\le c \max_{\alpha=1,\dots,k_0} \left\{ \prod_{j=1}^{p} \left| \left(E + U_j(\varphi, N_0) \right) \right|_\alpha \left| U_{p+1}(\varphi, N_{p+1}) \right|_{k_0-\alpha} \right\}$$

$$\le c_5 \left| U_{p+1}(\varphi, N_{p+1}) \right|_{k_0} \le c_5 \frac{M_p^{\gamma-1}}{4s}.$$

This inequality implies that the matrix $\Phi(\varphi)$ is k_0 times differentiable.

6. INVARIANT TOROIDAL SETS FOR SYSTEMS OF DIFFERENCE EQUATIONS. INVESTIGATION OF THE BEHAVIOR OF TRAJECTORIES ON TOROIDAL SETS AND IN THEIR VICINITIES

§1. Auxiliary Lemmas

Consider the canonical form of a real $(s \times s)$-dimensional matrix A. Let $\lambda_1, \lambda_2, \ldots, \lambda_3$ be eigenvalues of the matrix A and let $\mathcal{J} = \mathcal{J}_{\rho_1}(\lambda_1), \mathcal{J}_{\rho_2}(\lambda_2), \ldots, \mathcal{J}_{\rho_p}(\lambda_p))$ be its Jordan form. If λ_j is a real eigenvalue, then the real $(\rho_j \times \rho_j)$-dimensional Jordan cell

$$
\mathcal{J}_{\rho_j}(\lambda_j) = \begin{pmatrix} \lambda_j & 0 & 0 & \ldots & 0 & 0 \\ \varepsilon_1 & \lambda_j & 0 & \ldots & 0 & 0 \\ 0 & \varepsilon_1 & \lambda_j & \ldots & 0 & 0 \\ \ldots & \ldots & \ldots & \ldots & \ldots & \ldots \\ 0 & 0 & 0 & \ldots & \varepsilon_1 & \lambda_j \end{pmatrix},
\tag{1.1}
$$

where ε_1 is an arbitrary nonzero real number, corresponds to it in $\mathcal{J}$. If $\lambda_j = \alpha_j + i\beta_j$ and $\overline{\lambda}_j = \alpha_j - i\beta_j$ are two complex conjugate eigenvalues, then the pair of $(\rho_j \times \rho_j)$-dimensional Jordan cells $\mathcal{J}_{\rho_j}(\lambda_j)$ and $\mathcal{J}_{\rho_j}(\overline{\lambda}_j)$ of the type (1.1) corresponds to $\lambda_j, \overline{\lambda}_j$ in the matrix $\mathcal{J}$. We can establish a correspondence between this pair and a $(2\rho_j \times 2\rho_j)$-dimensional matrix

$$
\mathcal{J}_{2\rho_j}\{\lambda_j, \overline{\lambda}_j\} = \begin{pmatrix} S_j & 0 & 0 & \ldots & 0 & 0 \\ \varepsilon_1 E_2 & S_j & 0 & \ldots & 0 & 0 \\ 0 & \varepsilon_1 E_2 & S_j & \ldots & 0 & 0 \\ \ldots & \ldots & \ldots & \ldots & \ldots & \ldots \\ 0 & 0 & 0 & \ldots & \varepsilon_1 E_2 & S_j \end{pmatrix},
\tag{1.2}
$$

223

where E_2 is the unit square matrix of the second order

$$S_j = \begin{pmatrix} \alpha_j & -\beta_j \\ \beta_j & \alpha_j \end{pmatrix}.$$

The matrix

$$B = \{D_1, \dots, D_p\} = \begin{pmatrix} D_1 & 0 & \dots & 0 \\ 0 & D_2 & \dots & 0 \\ \dots & \dots & \dots & \dots \\ 0 & 0 & \dots & D_p \end{pmatrix},$$

where

$$D_j = \begin{cases} \mathcal{I}_{P_j}(\lambda_j), & \text{if } \lambda_j \text{ is a real number} \\ \mathcal{I}_{2P_j}(\lambda_j, \overline{\lambda}_j), & \text{if } \lambda_j \text{ and } \overline{\lambda}_j \text{ are complex conjugate numbers} \end{cases}$$

is called the canonical form of the real matrix A. It is well-known that for every real matrix A there exists a real non-singular matrix C which reduces A to the canonical form $B : CA\,C^{-1} = B$.

Lemma 6.1. (Martinyuk, 1975). *Let A be a matrix in the real canonical form and let $\lambda_1, \dots, \lambda_s$ be its eigenvalues. Suppose also that $\varepsilon_1 > 0$ and $\gamma_1 = \max |\lambda_j|$. Then*

$$\| A^n x \| \le (\gamma_1 + \varepsilon_1)^n \| x \|. \tag{1.3}$$

for all integer $n \ge 0$.

Proof. Since $A = \{D_1, \dots, D_p\}$, we have $A^n = \{D_1^n, \dots, D_p^n\}$. For $D_j = \mathcal{I}_{P_j}(\lambda_j)$ (here $\mathcal{I}_{P_j}(\lambda_j)$ is defined by (1.1)) the structure of the Jordan cell

$$\mathcal{I}_{P_j}(\lambda_j) = \lambda_j E_{P_j} + \varepsilon_1 Z_{P_j},$$

where

$$Z_{\rho_j} = \begin{pmatrix} 0 & 0 & \ldots & 0 & 0 \\ 1 & 0 & \ldots & 0 & 0 \\ \ldots & \ldots & \ldots & \ldots & \ldots \\ 0 & 0 & \ldots & 1 & 0 \end{pmatrix}$$

yields

$$D_j^n = \lambda_j^n E_{\rho_j} + n\lambda_j^{n-1}\varepsilon_1 Z_{\rho_j} + \frac{n(n-1)}{2!}\lambda_j^{n-2}\varepsilon_1^2 Z_{\rho_j}^{\rho_j-1}$$

$$+ \ldots + \frac{n(n-1)\ldots(n-\rho_j+2)}{(\rho_j-1)!}\lambda_j^{n-\rho_j+1}\varepsilon_1^{\rho_j-1}Z_{\rho_j}^{\rho_j-1} =$$

(1.4)

$$= \begin{pmatrix} \lambda_j^n & 0 & 0 & \ldots & 0 \\ n\varepsilon_1\lambda_j^{n-1} & \lambda_j^n & 0 & \ldots & 0 \\ \frac{n(n-1)}{2!}\varepsilon_1^2\lambda_j^{n-2} & n\varepsilon_1\lambda_j^{n-1} & \lambda_j^n & \ldots & 0 \\ \ldots & \ldots & \ldots & \ldots & \ldots \\ \frac{n(n-1)(n-2)\ldots(n-\rho_j+2)}{\rho_j-1}\varepsilon_1^{\rho_j-1}\lambda_j^{n-\rho_j+1} & & \ldots & n\varepsilon_1\lambda_j^{n-1} & \ldots & \lambda_j^n \end{pmatrix}.$$

Therefore, if $x_j = \left\{ x^{j_1}, \ldots, x^{j_{\rho_j}} \right\}$ is a ρ_j-dimensional vector, then the vector $D_j^n x_j$ has the form

$$D_j^n x_j = \left\{ \lambda_j^n, x^{j_1}, (n\varepsilon_1\lambda_j^{n-1}x^{j_1} + \lambda_j^n x^{j_2}), \ldots \right.$$

$$\left. \ldots, \sum_{v=1}^{\rho_j} \frac{n(n-1)\ldots(n-(\rho_j-v)+1)}{(\rho_j-v)!}\lambda_j^{n-(\rho_j-v)}\varepsilon_1^{(\rho_j-v)}x^{j_v} \right\}.$$

(1.5)

This implies that

$$\left\| D_j^n x_j \right\|^2 = \lambda_j^{2n}x^{2j_1} + (\lambda_j^n x^{j_2} + n\varepsilon_1\lambda_j^{n-1}x^{j_1})^2 + \ldots$$

$$\ldots + (\lambda_j^n x^{\rho_j} + n\varepsilon_1\lambda_j^{n-1}x^{\rho_j-1} + \ldots + \frac{n(n-1)\ldots(n-\rho_j-v)}{(\rho_j-1)!}\lambda_j^{n-\rho_j+1}\varepsilon_1^{\rho_j-1}x^{j_1})^2.$$

Moreover, the following relations hold

$$\left(\lambda_j^n x^{j_2} + n\varepsilon_1 \lambda_j^{n-1} x^{j_1}\right)^2 = \lambda_j^{2n} x^{2j_2} + 2n\varepsilon_1 \lambda_j^{2n-1} x^{j_1} x^{j_2} + n^2 \varepsilon_1^2 \lambda_j^{2n-2} x^{2j_1}$$

$$\le \left(\gamma_1^n + n\varepsilon_1 \gamma_1^{n-1}\right)\left[\gamma_1^n x^{2j_2} + n\varepsilon_1 \gamma_1^{n-1} x^{2j_1}\right] \le \left(\gamma_1 + \varepsilon_1\right)^n \left[\gamma_1^n x^{2j_2} + n\varepsilon_1 \gamma_1^{n-1} x^{2j_1}\right],$$

$$(1.6)$$

$$\left(\lambda_j^n x^{j_3} + n\varepsilon_1 \lambda_j^{n-1} x^{j_2} + \frac{n(n-1)}{2!} \lambda_j^{n-2} x^{j_1}\right)^2 = \lambda_j^{2n} x^{2j_3} + n^2 \varepsilon_1^2 \lambda_j^{2n-2} x^{2j_2}$$

$$+ \frac{n^2(n-1)^2}{4} \lambda_j^{2n-4} x^{2j_1} + 2n\varepsilon_1 \lambda_j^{2n-1} x^{j_3} x^{j_2} + n(n-1)\lambda_j^{2n-2} x^{j_3} x^{j_2}$$

$$+ n^2(n-1)\varepsilon_1 \lambda_j^{2n-3} x^{j_1} x^{j_2} \le \left(\gamma_1^n + n\varepsilon_1 \gamma_1^{n-1} + \frac{n(n-1)}{2!} \varepsilon_1^2 \gamma_1^{n-2}\right)$$

$$\times \left(\gamma_1^n x^{2j_3} + n\varepsilon_1 \gamma_1^{n-1} x^{2j_2} + \frac{n(n-1)}{2!} \varepsilon_1^2 \gamma_1^{n-2} x^{2j_1}\right)$$

$$\le (\gamma_1 + \varepsilon_1)^n \left(\gamma_1^n x^{2j_3} + n\varepsilon_1 \gamma_1^{n-1} x^{2j_3} + \frac{n(n-1)}{2!} \varepsilon_1^2 \gamma_1^{n-2} x^{2j_1}\right)$$

etc., where $\gamma_1 = |\lambda_j|$. Taking the inequalities (1.6) into account, we find

$$\left\|D_j^n x_j\right\|^2 \le \gamma_1^{2n} x^{2j_1} + (\gamma_1 + \varepsilon_1)^n \left(\gamma_1^n x^{2j_2} + n\varepsilon_1 \gamma_1^{n-1} x^{2j_1}\right) + (\gamma_1 + \varepsilon_1)^n (\gamma_1^n x^{2j_3}$$

$$+ n\varepsilon_1 \gamma_1^{n-1} x^{2j_3} + \frac{n(n-1)}{2!} \varepsilon_1^2 \gamma_1^{n-2} x^{2j_2}\right) + (\gamma_1 + \varepsilon_1)^n (\gamma_1^n x^{2j_4} + n\varepsilon_1 \gamma_1^{n-1} x^{2j_3}$$

$$+ \frac{n(n-1)}{2!} \varepsilon_1^2 \gamma_1^{n-2} x^{2j_2} + \frac{n(n-1)(n-2)}{3!} \varepsilon_1^3 \gamma_1^{n-3} x^{2j_1}\right) + \ldots \le [\gamma_1^{2n} + (\gamma_1$$

$$+ \varepsilon_1)^n n\varepsilon_1 \gamma_1^{n-1} + (\gamma_1 + \varepsilon_1)^n \frac{n(n-1)}{2!} \varepsilon_1^2 \gamma_1^{n-2} + \frac{n(n-1)(n-2)}{3!} \varepsilon_1^3 \gamma_1^{n-3} + \ldots\right] x^{2j_1}$$

$$+ \left[(\gamma_1 + \varepsilon_1)^n \gamma_1^n + \gamma_1 + \varepsilon_1)^n n\varepsilon_1 \gamma_1^{n-1} + (\gamma_1 + \varepsilon_1)^n \frac{n(n-1)}{2!} \varepsilon_1^2 \gamma_1^{n-2} + \ldots\right] x^{2j_2} + \ldots$$

$$\le (\gamma_1 + \varepsilon_1)^{2n} \left\|x_j\right\|^2$$

$$(1.7)$$

And this yields

$$\left\|A^n x\right\|^2 = \left\|\left\{D_1^n x^1, D_2^n x^2, \ldots, D_p^n x^p\right\}\right\|^2 = \left\|D_1^n x^1\right\|^2$$

$$+ \left\|D_2^n x^2\right\|^2 + \ldots + \left\|D_p^n x^p\right\|^2 \leq (\gamma_1 + \varepsilon_1)^{2n} \left\|x^1\right\|^2$$

$$+ (\gamma_1 + \varepsilon_1)^{2n} \left\|x^2\right\|^2 + \ldots + (\gamma_1 + \varepsilon_1)^{2n} \left\|x^p\right\|^2$$

$$\leq (\gamma_1 + \varepsilon_1)^{2n} \left(\left\|x^1\right\|^2 + \left\|x^2\right\|^2 + \ldots + \left\|x^p\right\|^2 \right) = (\gamma_1 + \varepsilon_1)^{2n} \|x\|^2.$$

The last estimate completes the proof of Lemma 6.1. In the case of the complex conjugate roots $\lambda_j = \alpha_j + i\beta_j$ and $\lambda_j = \alpha_j - i\beta_j$, the proof is similar.

Consider a function $f(x^1, x^2, \ldots, x^s)$ which is defined for x from the bounded closed region D. Let $r_1, r_2, \ldots, r_s$ be nonnegative integers. We say that $f(x)$ belongs to the space $C_0^{r_1, r_2, \ldots, r_s}(D)$, if it possesses all derivatives whose order with respect to each argument x^i does not exceed r_i, and these derivatives are continuous in D. Then

$$C^r(D) = \bigcap_{r_1 + \ldots + r_s = r} C_0^{r_1, r_2, \ldots, r_s}(D)$$

is a space of functions r times continuously differentiable in D.

Lemma 6.2. (Samoilenko, 1966a). *Suppose that a function* $F(\varphi^1, \ldots, \varphi^m, x^1, \ldots, x^s)$ *is periodic in* $\varphi^1, \ldots, \varphi^m$ *with period* 2π *and belongs to the space* $C_0^{l_1, \ldots, l_m, r_1, r_2, \ldots, r_s}(D)$. *Let* $F^{(k)}(x) = F^{(k_1 \ldots k_m)}(x^1, \ldots, x^s)$ *be its Fourier coefficients, and let* $k_{\beta_1}, \ldots, k_{\beta_j}$ *be nonzero coordinates of the vector* $k = (k_1, \ldots, k_m)$. *Then* $F^{(k)}(x) \in C_0^{r_1, \ldots, r_s}(D)$ *and*

$$\left| \frac{\partial^\alpha F^{(k)}(x)}{(\partial x^1)^{\alpha_1} \ldots (\partial x^s)^{\alpha_s}} \right| \leq \min_{\substack{0 \leq l_1^i \leq l_{\beta_1} \\ \cdots \cdots \cdots \\ 0 \leq l_j^i \leq l_{\beta_j}}} \left[\left| k_{\beta_1} \right|^{-l_1^i} \ldots \left| k_{\beta_j} \right|^{-l_j^i} \right]$$

$$\times \max_{\varphi, x} \left| \frac{\partial^{l_1^i + \ldots + l_j^i + \alpha}}{\left(\partial \varphi^{\beta_1} \right)^{l_1^i} \ldots \left(\partial \varphi^{\beta_j} \right)^{l_j^i} (\partial x^1)^{\alpha_1} \ldots (\partial x^s)^{\alpha_s}} \right| \tag{1.8}$$

where $\alpha = \alpha_1 + \ldots + \alpha_s, 0 \leq \alpha_\nu \leq r_j, \nu = 1, 2, \ldots, s.$

Henceforth. we shall use some smoothing operators and their main properties. Recall that an operator T_θ, depending on the parameter θ, is called smoothing, if, as a result of its action, a function $f(x)$, which is r times continuously differentiable, turns into a function $T_\theta f(x)$ which is infinitely differentiable, and moreover,

$$|f(x) - T_\theta f(x)|_r \to 0, \quad \text{as } \theta \to \theta_0. \tag{1.9}$$

Consider the function $h(\varphi, x)$ defined for $-a < x < a$. Following the paper by Samoilenko, (1966a), we introduce the operator T_{NM} by

$$T_{NM}h(\varphi, x) = \iint\limits_{|x'|<a} \kappa_{NM}\left(\varphi - \varphi^1, x - x^1\right) h\left(\varphi^1, x^1\right) d\varphi^1 dx^1 \tag{1.10}$$

where $\kappa_{NM} = N\kappa(N\varphi)M\,\kappa(Mx)$ and $\kappa(z)$ is a function, all the derivatives of which exist and are continuous; furthermore, it is such that

$$\kappa(z) = 0 \quad \text{for } |z| > 1,$$

$$\int\limits_{-\infty}^{\infty} z^k \kappa(z)dz = \begin{cases} 1 & \text{for} \quad k=0, \\ 0 & \text{for} \quad 0 \leq k < l \end{cases}$$

(l is a fixed integer).

We also consider the operators T_{NM}^o and T_{NM}^1 , which approximate smooth functions $h(\varphi, x)$ such that $h(\varphi, 0) = 0$ and $\dfrac{\partial h(\varphi, 0)}{\partial x} = 0$. They are given by

$$T_{NM}^o h(\varphi, x) = \int\limits_{|x_i^1|<a}\!\!\!\dots\int \kappa_N\left(\varphi - \varphi^1\right)\left[\kappa_M\left(x - x^1\right) - \kappa_M\left(-x^1\right)\right]h\left(\varphi^1, x^1\right)d\varphi^1 dx^1; \tag{1.11}$$

$$T_{NM}^1 h(\varphi, x) = \int\limits_{|x_i^1|<a}\!\!\!\dots\int \kappa_N\left(\varphi - \varphi^1\right)\left[\kappa_M\left(x - x^1\right) - \kappa_M\left(-x^1\right)\right.$$

$$\left. + \frac{\partial \kappa_M\left(-x^1\right)}{\partial x^1}x\right]h\left(\varphi^1, x^1\right)d\varphi^1 dx^1; \quad -a < x_i < a, \quad i = 1, \dots, s, \tag{1.12}$$

where $\kappa_p(z) = p\kappa(pz_1) \dots p\kappa(pz_s)$.

The following statements are valid for the operators under consideration (Bogolyubov, Mitropolsky, and Samoilenko, 1969).

Lemma 6.3. *For every integer nonnegative* $\rho = (\rho_1, \dots, \rho_m), r = (r_1, \dots, r_s)$

and for all $x \in D_1$, *where the domain* D_1 *is given by*

$$-a + M^{-1} \le x^i \le a - M^{-1}, \quad M^{-1} < a, \quad i = 1, \dots, s,$$

we have

$$\left| \frac{\partial^{|p|+|r|}}{\left(\partial\varphi^1\right)^{p_1} \dots \left(\partial\varphi^m\right)^{p_m} \left(\partial x^1\right)^{r_1} \dots \left(\partial x^s\right)^{r_s}} T_{NM} f \right| \le C N^{|p|} M^{|r|+\delta} |f(\varphi, x)|_0 \qquad (1.13)$$

where

$$\delta = \begin{cases} 0 & \text{for} \quad T_{NM} = T_{NM}^0, \\ 1 & \text{for} \quad T_{NM} = T_{NM}^1; \end{cases}$$

and the constant C *is independent of* $N, M, $ *and* f.

Lemma 6.4. *If* $f(\varphi, x) \in C^l(D)$, *then*

$$\left| f(\varphi, x) - T_{NM} f(\varphi, x) \right| \le c \sup_{|p| + |r| = l} N^{-|p|} M^{-|r| + \delta} \left| \frac{\partial^{|p|+|r|} f(\varphi, x)}{\left(\partial\varphi^1\right)^{p_1} \dots \left(\partial x^s\right)^{r_s}} \right|,$$

where $T_{NM} = \left\{ T_{NM}^0, T_{NM}^1 \right\}$.

§2. Periodic Solutions of Some Classes of Functional Equations

Let us investigate the problem of the periodic solutions of the equations

$$u(\varphi + \omega, Ax) = Au(\varphi, x) + F(\varphi, x); \qquad (2.1)$$

$$w(\varphi + \omega, Ax) = w(\varphi, x) + f(\varphi, x), \qquad (2.2)$$

where the vector functions $F(\varphi, x) = (F^1, \dots, F^s)$ and $f(x) = (f^1, \dots, f^m)$ are periodic in $\varphi = (\varphi^1, \dots, \varphi^m)$ with period 2π; they are defined for $\|x\| \le \eta$; and A is an $(s \times s)$-

dimensional matrix in the real canonical form, $\lambda = (\lambda_1, \dots, \lambda_s)$ being its eigenvalues and $\omega = (\omega_1, \dots, \omega_m)$.

Assume that the functions $F(\varphi, x)$ and $f(\varphi, x)$ belong to the space

$$C_0^{l,\tau}(\|x\| \leq \eta) = C_0^{l,\dots,l,\tau,\dots,\tau}(\|x\| \leq \eta), \quad l \geq m + 2, \quad \tau \geq 2,$$

and satisfy the condition

$$F(\varphi, 0) = \left.\frac{\partial F(\varphi, x)}{\partial x}\right|_{x=0} = f(\varphi, 0) = 0. \tag{2.3}$$

We now clarify under what restrictions on the eigenvalues of A and on ω, the systems (2.1) and (2.2) possess the solutions $u(\varphi, x) = (u^1, \dots, u^m)$, $w(\varphi, x) = (w^1, \dots, w^m)$ periodic in φ with period 2π and such that

$$u(\varphi, 0) = \left.\frac{\partial u(\varphi, x)}{\partial x}\right|_{x=0} = w(\varphi, 0) = 0 \tag{2.4}$$

These restrictions are established by the following lemma.

Lemma 6.5 (Martinyuk, 1977). *Suppose that the functions $F(\varphi, x)$ and $f(\varphi, x)$ are periodic with period 2π, belong to the space $C_0^{l,\tau}(\|x\| \leq \eta)$, (where $\tau \geq 2$, $l = m + 1 + \sigma$), and satisfy the conditions (2.3). Let the eigenvalues of the matrix A be such that*

$$0 < |\lambda_j| < 1,$$

$$\tag{2.5}$$

$$|\lambda_\alpha| \, |\lambda_\beta| \, |\lambda_j|^{l-1} \leq \gamma_1 < 1$$

for any $\alpha, \beta, j = 1, \dots, s$ and for some constant $\gamma_1 > 0$. Then the system of equations

$$u(\varphi + \omega, Ax) = A^\delta u^\delta(\varphi, x) w^{1-\delta}(\varphi, x) + \delta F(\varphi, x) +$$

$$+ (1 - \delta) f(\varphi, x), \quad \delta = 0, 1, \tag{2.6}$$

has the solution

$$u^{(\delta)}(\varphi, x) = L^{(\delta)}\{\delta F(\varphi, x) + (1 - \delta) f(\varphi, x)\} = \sum_{|k|=-\infty}^{\infty} u_k^{(\delta)}(x) e^{i(k, \varphi)}, \tag{2.7}$$

periodic in φ *with period* 2π. *This solution belongs to the space* $C_0^{\sigma,\tau}(\|x\|\leq\eta)$ *and satisfies* (2.4) *and the following inequalities*

$$\left|u^{(\delta)}(\varphi,x)\right|\leq a_0\left[\max_{\substack{i,|r_1|=1+\delta,\\|x|\leq\eta}}\left|D_{\varphi^i}^l D_x^{r_1}F^{(\delta)}(\varphi,x)\right|+\max_{\substack{|r_1|=1+\delta,\\|x|\leq\eta}}\left|D_x^{r_1}F^{(\delta)}(\varphi,x)\right|\right],$$

(2.8)

$$\left|D_{\varphi}^{\rho}D_x^r u^{(\delta)}(\varphi,x)\right|\leq a_1\left[\max_{\substack{i,|r_1|=1+\delta,\\|x|\leq\eta}}\left|D_{\varphi^i}^l D_x^{r_1}F^{(\delta)}(\varphi,x)\right|+\max_{\substack{|r_1|=1+\delta,\\|x|\leq\eta}}\left|D_x^{r_1}F^{(\delta)}(\varphi,x)\right|\right]$$

for $|r|=1$, $|\rho|\leq\sigma$ *and*

$$\left|D_{\varphi}^{\rho}D_x^r u^{(\delta)}(\varphi,x)\right|\leq a_2\left[\max_{\substack{i,|r_1|=|r|,\\|x|\leq\eta}}\left|D_{\varphi^i}^l F^{(\delta)}(\varphi,x)\right|+\max_{\substack{|r_1|=|r|,\\|x|\leq\eta}}\left|D_x^{r_1}F^{(\delta)}(\varphi,x)\right|\right]$$

(2.9)

for $2\leq|r|\leq\tau$, $|\rho|\leq\sigma$, *where* a_0, a_1 *and* a_2 *are positive constants, which do not depend on* $F^{(\delta)}(\varphi,x)$.

Obviously, the system of equations (2.6) coincides with (2.1) for $\delta=0$ and with (2.2) for $\delta=1$.

Proof. We give the proof for the case $\delta=1$, i.e., for the system (2.1). Since the matrix A has the real canonical form, one of its parameters is an arbitrary nonzero number ε_1. Assume that ε_1 is a positive fixed number. Then, according to Lemma 6.1, we have for the function $A^n x$

$$\left\|A^n x\right\|\leq\left(\varepsilon_1+\max_j|\lambda_j|\right)^n\|x\|.$$

This implies that

$$\left\|A^n x\right\|\leq\left(\varepsilon_1+\gamma_1\right)^n\|x\|\leq\|x\|$$

(2.10)

for all $n\geq 0$ provided that $\varepsilon_1+\gamma_1<1$.

Taking (2.10) into account, we substitute $A^n x$ and $\varphi_n = n\omega + \varphi$ for x and φ, respectively in thew equation (2.1). As a result, we obtain the equation

$$u((n+1)\omega + \varphi, A^{n+1}x) = Au(n\omega + \varphi, A^n x) + F(n\omega + \varphi, A^n x), \qquad (2.11)$$

the solution of which satisfies the initial equation (2.1) at $n = 0$.

The solution of eqn.(2.11) can be represented as

$$u(\varphi_n, A^n x) = -\sum_{v=1}^{\infty} A^{n-v} F\left(\varphi_{n+v-1}, A^{n+v-1}x\right), n \ge 0. \qquad (2.12)$$

We set $n = 0$ in (2.12) and obtain

$$u(\varphi, x) = -\sum_{v=1}^{\infty} A^{-v} F\left(\varphi_{v-1}, A^{v-1}x\right). \qquad (2.13)$$

Let us show that the function $u(\varphi, x)$ defined by (2.13) exists, belongs to the space $C_0^{l,\tau}(\|x\| \le \eta)$, and satisfies (2.4), (2.8) and (2.9). Taking into account that $F(\varphi, x) = $ $\in C_0^{l,2}(\|x\| \le \eta)$ and $F(\varphi, 0) = \dfrac{\partial F(\varphi, x)}{\partial x}\bigg|_{x=0} = 0$ and expanding this function at $x = 0$ in the Taylor series with a remainder defined by the second derivatives, we can write

$$F\left(\varphi_{v-1}, A^{v-1}x\right) = -\frac{1}{2}\sum_{\alpha,\beta} \frac{\partial^2 F\left(\varphi_{v-1}, A^{v-1}x\right)}{\partial x^\alpha \partial x^\beta} \left\{A^{v-1}x\right\}_\alpha \left\{A^{v-1}x\right\}_\beta \qquad (2.14)$$

where $\left\{A^{v-1}x\right\}_\alpha$ is the α-th coordinate of the vector $A^{v-1}x$. Inserting (2.14) into (2.13), we find

$$u_j(\varphi, x) = -\frac{1}{2}\sum_{v=1}^{\infty}\sum_{\mu,\alpha,\beta} \left\{A^{-v}x\right\}_{j\mu} \frac{\partial^2 F\left(\varphi_v, A^{v-1}x\right)}{\partial x^\alpha \partial x^\beta} \left\{A^{v-1}x\right\}_\alpha \left\{A^{v-1}x\right\}_\beta, \qquad (2.15)$$

$$j = 1, \dots, s.$$

For the $j\mu$-th element of the matrix A^v, we have

$$\left|\left\{A^v\right\}_{j\mu}\right| \le P_{j\mu}(v)\left|\lambda_j\right|^v,$$

where $P_{j\mu}(v)$ is a polynomial in v whose degree does not exceed $s - 1$. Therefore, (2.15) implies that

$$
|u_j(\varphi,x)| \leq \frac{1}{2}\max_{\substack{\mu,\alpha,\beta \\ |x|\leq\eta}} \left| \frac{\partial^2 F^j(\varphi,x)}{\partial x^\alpha \partial x^\beta} \right| \sum_{v=1}^{\infty} \sum_{\mu,\alpha,\beta} \left| \left\{ A^{-v}x \right\}_{j\mu} \left\{ A^{v-1}x \right\}_\alpha \left\{ A^{v-1}x \right\}_\beta \right|
$$

$$
\leq \frac{1}{2}\max_{\substack{\alpha,\beta \\ |x|\leq\eta}} \left| \frac{\partial^2 F(\varphi,x)}{\partial x^\alpha \partial x^\beta} \right| \sum_{v=1}^{\infty} \sum_{\alpha,\beta} \left[\sum_{\mu=1}^{s} P_{j\mu}(v) \sum_{\mu=1}^{s} P_{\alpha\mu}(v) \sum_{\mu=1}^{s} P_{\beta\mu}(v) \right]
$$

$$
\times \left| \lambda_j \right|^{-v} \left| \lambda_\alpha \right|^{v-1} \left| \lambda_\beta \right|^{v-1} \eta^2 \leq a_0 \max_{\substack{|r|=2 \\ \|x\|\leq\eta}} \left| D_x^r F(\varphi,x) \right|. \tag{2.16}
$$

This inequality means that the function $u(\varphi,x)$ defined by (2.13) exists; moreover, it is continuous in φ and x and satisfies the relation (2.4) for $|p| + |r| = 0$. Furthermore, it follows from (2.13) that the function $u(\varphi,x)$ is periodic in φ with period 2π and satisfies the condition $u(\varphi, 0) = 0$.

One can easily show that $u(\varphi,x)$ belongs to the space $C_0^{l,1}(\|x\|\leq\eta)$ and that the inequalities (2.8) and (2.9) hold for its derivatives.

The expression

$$
\frac{\partial u(\varphi,x)}{\partial x} = -\sum_{v=1}^{\infty} A^{-v} \frac{\partial F\left(\varphi_{v-1}, A^{v-1}x\right)}{\partial y} A^{v-1}
$$

involves the equality $\left. \dfrac{\partial u(\varphi,x)}{\partial x} \right|_{x=0} = 0$, and since the sum

$$
\sum_{|v|=1}^{\infty} A^{-v} D_\varphi^p D_x^r \left\{ F\left(\varphi_v, A^{v-1}x\right) \right\}
$$

exists, we have

$$
\left| D_\varphi^p D_x^r u(\varphi,x) \right| \leq a_0 \max_{\substack{|r|=r \\ \|x\|\leq\eta}} \left| D_\varphi^p D_x^{r_1} F(\varphi,x) \right|.
$$

Clearly, for $|p| \leq l$ and $2\leq |r| \leq \tau$, the function $u(\varphi,x)$ belongs to the space $C_0^{l,\tau}(\|x\| \leq \eta)$ and satisfies the inequality (2.9). Let us now show that under conditions (2.5) the system (2.1) possesses, in the space $C_0^{l,2}$, the unique periodic solution which satisfies (2.4). Assume the opposite, i.e. that the system (2.1) has two solutions with these properties in $C_0^{l,2}(\|x\|\leq\eta)$. Then the difference between these solutions is the solution of the homogeneous system (2.1) which satisfies (2.4). But for every solution of the homogeneous system of equations (2.1), the relation

$$u(\varphi_n, A^n x) = A^n u(\varphi, x), \quad n \geq 0, \tag{2.17}$$

holds. By differentiating (2.17), we find

$$A^{-n} \frac{\partial u(\varphi_n, A^n x)}{\partial y} A^n = \frac{\partial u(\varphi, x)}{\partial x}, \quad n \geq 0.$$

By using Taylor's formula, we obtain

$$A^{-n} \sum_{\alpha=1}^{s} \frac{\partial u_y'(\varphi_n, A^n \bar{x})}{\partial y_\alpha} \{A^n x\}_\alpha A^n = \frac{\partial u(\varphi, x)}{\partial x}.$$

By passing to the limit as $n \to \infty$, we obtain in the last expression

$$\frac{\partial u(\varphi, x)}{\partial x} = 0,$$

i.e., $u(\varphi, x) = u(\varphi, 0)$, and this contradicts the assumption of the existence of two solutions.

Now consider the problem of the existence of periodic solutions of the equation

$$u(\varphi + \omega) = u(\varphi) + F(\varphi) - \bar{F}(\varphi) \tag{2.18}$$

where the function $F(\varphi)$ is periodic in φ with period 2π. Let us represent $F(\varphi)$ and $u(\varphi)$ in the form of the Fourier series

$$F(\varphi) = \sum_k F_k e^{i(k,\varphi)}; \tag{2.19}$$

$$u(\varphi) = \sum_k u_k e^{i(k,\varphi)}. \tag{2.20}$$

Inserting (2.19) and (2.20) in (2.18), we obtain the following relation

$$u_k e^{i(k,\omega)} = u_k + F_k - F_0 \tag{2.21}$$

for the coefficients u_k. Equation (2.21) implies that the equation (2.18) possesses a periodic solution only in the case when $u_0 = 0$ and

$$e^{i(k,\omega)} - 1 \neq 0 \tag{2.22}$$

for $|k| \neq 0$. Solving eqn.(2.21), we get

$$u_k = \frac{F^k}{e^{i(k,\omega)} - 1}.$$

Hence, the series

$$u(\varphi) = \sum_{k \neq 0} \frac{F^k}{e^{i(k,\omega)} - 1} e^{i(k,\varphi)} \tag{2.23}$$

is a formal solution of eqn.(2.18).

We now prove that the series (2.22) converges under some additional conditions which are established by the following lemma.

Lemma 6.6. (Martinyuk, 1975). *Suppose that the function $F(\varphi)$ is $r = 2m + \sigma + 1$ times continuously differentiable, and that the constant vector $\omega = (\omega_1, \ldots, \omega_m)$ with incommensurable components satisfies the inequality*

$$\left| \frac{\sin(k,\omega)}{2} \right| \geq \frac{c}{|k|^m} \tag{2.24}$$

for an arbitrary vector $k = (k_1, \ldots, k_m)$ with integer components. Then the series (2.23) is a σ-times continuously differentiable solution of the equation (2.18); it is periodic with period 2π and satisfies the inequality

$$|u(\varphi)|_\sigma \leq c_1 \max_i \left| D_{\varphi^i}^\tau F(\varphi) \right|_0, \tag{2.25}$$

where c_1 is a constant depending on c, m and τ.

Proof. Since $F(\varphi)$ is τ times continuously differentiable, the inequality

$$|F_k| \leq \left(\max_i |k_i| \right)^{-\tau} \max_i \left| D_{\varphi^i}^\tau F(\varphi) \right| \tag{2.26}$$

holds for $|k| \neq 0$. By using (2.23), (2.24), and (2.26), we obtain

$$\left|D_\varphi^\rho u(\varphi)\right| = \left|\sum_{|k|\neq 0} \frac{(ik_1)^{\rho_1}\ldots(ik_m)^{\rho_m}}{e^{i(\omega,k)}-1} F_k e^{i(k,\varphi)}\right| \leq c^{-1}\sum_{|k|\neq 0}\max_i \left|k_i\right|^{|\rho|}|k|^m |F_k|$$

$$\leq c^{-1}\max_i\left|D_\varphi^\tau i F(\varphi)\right|_0 \sum_{|k|\neq 0}\max_i\left|k_i\right|^{-\tau+|\rho|}|k|^m \leq c^{-1}\max_i\left|D_\varphi^\tau i F(\varphi)\right|_0$$

$$\times\sum_{l=1}^{\infty} m^\tau \sum_{|k|=l}|k|^{-\tau+m+|\rho|} \leq c^{-1}m^\tau\max_i\left|D_\varphi^\tau i F(\varphi)\right|_0 \sum_{l=1}^{\infty}e^{-\tau+m+|\rho|}\sum_{|k|=l}1$$

$$\leq c^{-1}m^\tau\max_i\left|D_\varphi^\tau i F(\varphi)\right|_0 \sum_{i=1}^{\infty}e^{-\tau+2m+|\rho|-1} \leq c^{-1}\max_i\left|D_\varphi^\tau i F(\varphi)\right|_0.$$

This inequality implies that the series (2.23) converges and is σ times differentiable. The estimate (2.25) also follows from this inequality.

§3. Existence of Invariant Toroidal Sets for Nonlinear Systems of Difference Equations

Consider the system of difference equations

$$\varphi_{n+1} = \varphi_n + a\,(\varphi_n, y_n, \mu),$$

$$y_{n+1} = [E + b(\varphi_n, y_n, \mu)]\,y_n + c(\varphi_n, \mu),$$

(3.1)

where the functions $y = (y^1, y^2, \ldots, y^s)$, $\varphi = (\varphi^1, \varphi^2, \ldots, \varphi^m)$, $a(\varphi, y, \mu)$, $b(\varphi, y, \mu)$, and $c(\varphi,\mu)$ are periodic in φ with period 2π; they are defined in the region

$$\|y\| = \left(\sum_{i=1}^{m}(y^i)^2\right)^{\frac{1}{2}} \leq d,\ \mu \in [0, \mu_0].$$

(3.2)

We shall investigate the problem of the existence of invariant tori for the system (3.1) under the assumption that $c(\varphi,0) = 0$, i.e., that for $\mu = 0$ the system (3.1) possesses the trivial invariant torus $y \equiv 0$. We say that a surface is invariant if it possesses the following property:

If at the initial moment $n = k$, a solution of (3.1) lies on the surface, then it remains on this surface for all n.

We seek an invariant manifold $\mathfrak{I}(\mu)$ of the system (3.1) in the form

$$y = u\,(\varphi, \mu), \tag{3.3}$$

where $u(\varphi, \mu)$ is a continuous function periodic in φ with period 2π. The expression (3.3) defines the invariant torus $\mathfrak{I}(\mu)$, if for every integer n we have

$$u(\varphi_{n+1}(\varphi), \mu) \equiv [E + b(\varphi_n(\varphi), u(\varphi_n(\varphi), \mu), \mu)]\, u\,(\varphi_n(\varphi), \mu) + c(\varphi_n(\varphi), \mu), \tag{3.4}$$

where $\varphi_n(\varphi)$ is a solution of the system

$$\varphi_{n+1} = \varphi_n + a\,(\varphi_n,\, u(\varphi_n(\varphi), \mu),\, \mu), \varphi_k(\varphi) = \varphi, \tag{3.5}$$

k is an arbitrary integer, and φ is an arbitrary constant.

We construct the invariant toroidal set $\mathfrak{T}(\mu)$ as the limit of a sequence of sets $\mathfrak{T}^0(\mu), \mathfrak{T}^1(\mu), \ldots, \mathfrak{T}^i(\mu)$; each of which is the invariant torus

$$\mathfrak{T}^i(\mu): \ y = u^i\,(\varphi, \mu), \ i = 0,1, \ldots, \tag{3.6}$$

of the system of equations

$$\varphi_{n+1} = \varphi_n + a(\varphi_n,\, u^{i-1}(\varphi_n, \mu),\, \mu),$$

$$y_{n+1} = [E + b(\varphi_n,\, u^{i-1}(\varphi_n, \mu),\, \mu)]\, y_n + c(\varphi_n, \mu). \tag{3.7}$$

This method for finding $\mathfrak{T}(\mu)$ is justified by the following lemma.

Lemma 6.7. (Samoilenko, Martinyuk, and Perestyuk, 1973). *Suppose that the functions* $a(\varphi, y, \mu)$, $b(\varphi, y, \mu)$ *and* $c(\varphi, \mu)$ *are continuous in* φ *and* y *for* $\| y \| \leq d$ *and* $\mu \in [0, \mu_0]$. *If the sequence* (3.6) *converges uniformly for every* $\mu \in [0, \mu_0]$, *i.e.,*

$$\lim_{i \to \infty} u^i(\varphi, \mu) = u(\varphi, \mu) \tag{3.8}$$

then the limiting function $u(\varphi, \mu)$ *defines the invariant torus* $\mathfrak{T}(\mu): y = u(\varphi, \mu)$ *of the system* (3.1).

In fact, since $y = u^i(\varphi, \mu)$ is the invariant torus of (3.7), the relations

$$\varphi_n^i = \varphi + \sum_{j=k}^{n-1} a\left(\varphi_j^i, u^{i-1}\left(\varphi_j^i, \mu\right), \mu\right),$$

(3.9)

$$u^i\left(\varphi_n^i, \mu\right) = u^i(\varphi, \mu) + \sum_{j=k}^{n-1}\left[b\left(\varphi_j^i, u^{i-1}\left(\varphi_j^i, \mu\right), \mu\right)u^i\left(\varphi_j^i, \mu\right) + c\left(\varphi_j^i, \mu\right)\right]$$

hold for the solutions φ_n^i and y_n^i lying on this torus. We introduce an auxiliary sequence of piecewise linear continuous functions defined by

$$f^i(t) = \varphi_n^i + (t - n)(\varphi_{n+1}^i - \varphi_n^i), \quad n \le t \le n + 1.$$

(3.10)

Note that $f^i(n) = \varphi_n^i$ for all integers n. Taking into account the inequality $|u^{i-1}| \le d$, and the periodicity and continuity of the function $a(\varphi, y, \mu)$ for $\|y\| \le d$, we find that the sequence of functions $f^i(t)$, $i = 0, 1, \ldots$, is uniformly bounded and equicontinuous for t from an arbitrary finite interval T of the real axis $R : -\infty < t < \infty$. Hence, the sequence $f^i(t)$ contains a subsequence $f^{i_v}(t)$, $t = 0, 1, \ldots$, which is uniformly convergent on T. We set

$$f^i(t) = \lim_{v \to \infty} f^{i_v}(t), \quad t \in T.$$

(3.11)

Taking into account that $f(n) = \varphi_n$, where $\varphi_n = \lim_{v \to \infty} \varphi^{i_v}$, we pass to the limit in the equalities obtained from (3.9) for $i = i_v$. The assumption concerning the continuity of the functions a, b, c and u^{i-1} ensures the validity of all the changes of the limit sign position which are necessary to obtain the identities showing that the continuous periodic function $y = u(\varphi, \mu)$ is the invariant torus of the system (3.1); these identities are to be obtained from (3.8), (3.9), and (3.11).

Hence, the solution of the problem concerning the existence of the invariant torus (3.1) is connected with the solution of the same problem for the system (3.7).

We construct the set $\mathcal{T}^{i+1}(\mu)$ by using the Green's function for the problem of bounded solutions of the linear system of equations

$$y_{n+1} = [E + b(\varphi_n(\varphi), u^i(\varphi_n(\varphi), \mu), \mu)]y + c(\varphi_n(\varphi), \mu),$$

(3.12)

which can be obtained from the system (3.7), if we replace φ_n by the general solution $\varphi_n(\varphi)$, $\varphi_k(\varphi) = \varphi$ of the first equation of (3.7). Let $G(n, k, \varphi, \mu)$ be this Green's function. Then

$$y_n(\varphi, \mu) = \sum_{k=-\infty}^{\infty} G(n-1, k, \varphi, \mu)\, c(\varphi_k(\varphi), \mu) \qquad (3.13)$$

is a family of bounded solutions of (3.12) which depend on φ and μ as on parameters. This family covers the invariant toroidal set

$$\mathcal{T}^{i+1}(\mu): y = u^{i+1}(\varphi, \mu) = \sum_{k=-\infty}^{\infty} G(0, k, \varphi, \mu)\, c(\varphi_k, \mu), \qquad (3.14)$$

if the function $G(0, k, \varphi, \mu)\, c(\varphi_k(\varphi))$ is periodic in φ with period 2π and such that

$$\sum_{k=-\infty}^{\infty} G(0, k, \varphi_n(\varphi), \mu)\, c(\varphi_k(\varphi_n(\varphi)), \mu) = \sum_{k=-\infty}^{\infty} G(n-1, k, \varphi, \mu)\, c(\varphi_k(\varphi), \mu). \quad (3.15)$$

If the function $G(n, k, \varphi, \mu)$ satisfies the above-mentioned conditions, then system (3.7) possesses the invariant toroidal set $\mathcal{T}^{i+1}(\mu): y = u^i(\varphi, \mu)$ defined by

$$u^i(\varphi, \mu) = \sum_{k=-\infty}^{\infty} G(0, k, \varphi, \mu)\, c(\varphi_k(\varphi), \mu). \qquad (3.16)$$

We call the function $G(0, k, \varphi, \mu)$ which defines the invariant torus $y = u^i(\varphi, \mu)$ of the system (3.7), according to (3.16), the Green's function for the problem of the invariant tori of (3.7). Let us find the general form of this function.

Consider a system of equations of the form

$$\varphi_{n=1} = \varphi_n + a(\varphi_n, \mu), \quad y_{n+1} = [E + b(\varphi_n, \mu)]\, y_n. \qquad (3.17)$$

Denote by $\Omega_n^k(\varphi, \mu)$ a fundamental matrix of solutions of the linear system

$$y_{n+1} = [E + b(\varphi_n(\varphi, \mu), \mu)]\, y_n, \qquad (3.18)$$

in which $\varphi_n(\varphi, \mu)$ is the solution of the first equation of (3.17) and $\varphi_k(\varphi, \mu) = \varphi$.
  We introduce the function

$$G_0(k, \varphi, \mu) = \begin{cases} \Omega_0^k(\varphi, \mu) c(\varphi_k(\varphi, \mu), \mu) & \text{for } k<0, \\ E & \text{for } k=0, \\ \Omega_o^k(\varphi,\mu)\left(c(\varphi_k(\varphi, \mu), \mu) - E\right) & \text{for } k>0, \end{cases} \qquad (3.19)$$

where $C(\varphi, \mu)$ is a matrix periodic in φ with period 2π. Assume that $G_0(k, \varphi, \mu)$ is such that the inequality

$$\sum_{k=-\infty}^{\infty} |G_o(k,\varphi,\mu)| < K < \infty \qquad (3.20)$$

holds. Taking into account the identity $\varphi_n(\varphi+2\pi, \mu) = \varphi_n(\varphi, \mu) + 2\pi$, we find that

$$G(n, k, \varphi, \mu) = \begin{cases} \Omega_n^k(\varphi, \mu) c(\varphi_k(\varphi, \mu), \mu) & \text{for } k<n, \\ E & \text{for } k=n, \\ \Omega_n^k(\varphi,\mu)\left(c(\varphi_k(\varphi,\mu), \mu) - E\right) & \text{for } k>n \end{cases} \qquad (3.21)$$

is the Green's function for the problem of the bounded solutions of the system

$$y_{n+1} = (E + b(\varphi_n(\varphi, \mu), \mu))y_n + c(\varphi_n(\varphi, \mu), \mu). \qquad (3.22)$$

Consequently, $G_0(k, \varphi, \mu) = G(0, k, \varphi, \mu)$ is the Green's function for the problem of the invariant tori of the system (3.17), provided that the function $G(n, k, \varphi, \mu)$ satisfies (3.15). The last statement can be easily verified, if we take into account that

$$\Omega_n^k(\varphi_p(\varphi, \mu), \mu) = \Omega_{n+p}^{k+p}(\varphi, \mu) \qquad (3.23)$$

for arbitrary integers n, k, and p.

Now consider a system of difference equations

$$\varphi_{n+1} = \varphi_n + a_0(\varphi_n) + a_1(\varphi_n)$$

$$y_{n+1} = [E + b_0(\varphi_n) + b_1(\varphi_n)] y_n + c(\varphi_n) \qquad (3.24)$$

where $a_0(\varphi), b_0(\varphi), a_1(\varphi), b_1(\varphi)$, and $c(\varphi)$ are continuously differentiable functions; the functions $a_0(\varphi)$ and $b_0(\varphi)$ are fixed whereas $a_1(\varphi), b_1(\varphi)$, and $c(\varphi)$ are arbitrary but small in the sense of the norm $c'(\varphi)$. Denote by $\varphi_n(\varphi)$ ($\varphi_k(\varphi) = \varphi$) the solution of the first equation of (3.24). We have

Lemma 6.8. (Samoilenko, Martinyuk, and Perestyuk, 1973). *Assume that the system of equations* (3.24) *is such that for all* $a_1(\varphi)$ *and* $b_1(\varphi)$ *satisfying the inequality*

$$\max\left\{\left|a_1(\varphi)\right|_1, \left|b_1(\varphi)\right|_1\right\} \leq M, \tag{3.25}$$

there exists a Green's function $G_0(k,\varphi)$ *for the problem of invariant tori such that*

$$| G_0(k, \varphi)\, c(\varphi_k(\varphi)) |_1 \leq K \lambda^{|k|} |c(\varphi)|_1 \tag{3.26}$$

for all integer k *; moreover* $M > 0$ *and* $0 < \varphi < 1$. *Then* (3.24) *possesses the invariant torus* $\mathfrak{C} : y = u(\varphi)$ *with* $u(\varphi)$ *being a continuously differentiable function which satisfies the inequality*

$$| u(\varphi) |_1 \leq \frac{2K}{1-\lambda} | c(\varphi) |_1. \tag{3.27}$$

The statement of Lemma 6.8 follows from the representation

$$u(\varphi) = \sum_{k=-\infty}^{\infty} G_0(k, \varphi)c(\varphi_k(\varphi)) \tag{3.28}$$

and condition (3.26).

We employ Lemma 6.8 to construct the sequence of invariant tori (3.6). For this purpose, we set

$$a_0(\varphi) = a(\varphi, 0, 0), \quad b(\varphi, 0, 0) = b_0(\varphi),$$

$$a_1(\varphi, y, \mu) = a(\varphi, y, \mu) - a(\varphi, 0, 0), \tag{3.29}$$

$$b_1(\varphi, y, \mu) = b(\varphi, y, \mu) - b(\varphi, 0, 0),$$

and assume that $c(\varphi, \mu)$, $a(\varphi, y, \mu)$ and $b(\varphi, y, \mu)$ are continuously differentiable for $\|y\| \leq d$ and $0 \leq \mu \leq \mu_0$. Let

$$\max \{| a_1(\varphi, y, \mu) |_1, | b_1(\varphi, y, \mu) |_1, | c(\varphi, \mu) |_1\} \leq L(d, \mu_0), \tag{3.30}$$

where $L(d, \mu_0) \to 0$ as $d \to 0$ and $\mu_0 \to 0$. Then the following theorem is valid.

Theorem 6.1 (Samoilenko, Martinyuk, and Perestyuk, 1973). *Suppose that the* *function $a_0(\varphi)$ and $b_0(\varphi)$ are such that the system (3.24) satisfies the conditions of Lemma 6.8. Then one can always find a number μ^0 $(0 \le \mu^0 \le \mu)$ such that for all $\mu<\mu^0$ the sequence of systems (3.7) determines the sequence of the invariant tori (3.6), where each torus is continuously differentiable and satisfies the inequality*

$$| u^i(\varphi, \mu)|_1 \le \frac{2K}{1-\lambda} L(0, \mu^0). \tag{3.31}$$

The proof of this theorem can be carried out by analogy with the proof of Theorem 4.1.

We now prove that the sequence of invariant tori (3.6) is convergent. Since $y = u^{i+1}(\varphi, \mu)$ is the invariant torus of the system (3.7), the function $u^{i+1}(\varphi, \mu)$ satisfies the equation

$$u^{i+1}(\varphi_{n+1}, \mu) = (E + b_0(\varphi_n) + b_1(\varphi_n, u^i(\varphi_n, \mu), \mu))u^{i+1}(\varphi_n, \mu) + c(\varphi_n, \mu) \tag{3.32}$$

where $\varphi_n = \varphi_n(\varphi, \mu)$ is the general solution of the equation

$$\varphi_{n+1} = \varphi_n + a_0(\varphi_n) + a_1(\varphi_n, u^i(\varphi_n, \mu), \mu). \tag{3.33}$$

The functions $a_0(\varphi), b_0(\varphi), a_1(\varphi, y, \mu)$ and $b_1(\varphi, y, \mu)$ are defined by (3.29). The equality (3.32) holds for any n and, therefore, for $n = 0$, the function $u^{n+1}(\varphi_n, \mu)$ is periodic in φ with period 2π and solves the functional equation

$$u^{i+1}(\varphi + a_0(\varphi) + a_1(\varphi, u^i(\varphi, \mu), \mu)$$
$$= [E + b_0(\varphi) + b_1(\varphi, u^i(\varphi, \mu), \mu)] u^{i+1}(\varphi, \mu) + c(\varphi, \mu). \tag{3.34}$$

We set

$$u^{i+1}(\varphi, \mu) - u^i(\varphi, \mu) = w^{i+1}(\varphi, \mu). \tag{3.35}$$

Then the function $w^{i+1}(\varphi, \mu)$ is periodic in φ with period 2π and satisfies the equation

$$w^{i+1}(\varphi + a_0(\varphi) + a_1(\varphi, u^i(\varphi, \mu), \mu), \mu) - [E + b_0(\varphi) + b_1(\varphi, u^i(\varphi, \mu), \mu)] w^{i+1}(\varphi, \mu)$$
$$= u^i(\varphi + a_0(\varphi) + a_1(\varphi, u^{i-1}(\varphi, \mu), \mu), \mu) - u^i(\varphi + a_0(\varphi) + a_1(\varphi, u^i(\varphi, \mu), \mu), \mu)$$
$$+ b_1(\varphi, u^i(\varphi, \mu), \mu) u^i(\varphi, \mu) - b_1(\varphi, u^{i-1}(\varphi, \mu), \mu) u^i(\varphi, \mu). \tag{3.36}$$

This means that $y = w^{i+1}(\varphi, \mu)$ defines the invariant torus of the system of difference equations

$$\varphi_{n+1} = \varphi_n + a_0(\varphi_n) = a_1(\varphi_n, u^i(\varphi_n, \mu), \mu),$$

$$y_{n+1} = [E + b_0(\varphi_n) + b_1(\varphi_n, u^i(\varphi_n, \mu), \mu)]\, y_n + c_i(\varphi_n, \mu), \tag{3.37}$$

where $c_i(\varphi, \mu)$ denotes the right-hand side of eqn.(3.36). The system (3.37) has the same form as (3.24), and consequently, we can represent $w^{i+1}(\varphi, \mu)$ in terms of the Green's function for the problem of the invariant tori of the system (3.6):

$$w^{i+1}(\varphi, \mu) = \sum_{k=-\infty}^{\infty} G_0(k, \varphi)\, c_i(\varphi_k, \mu). \tag{3.38}$$

Then, taking (3.27) into account, we obtain

$$|\,w^{i+1}(\varphi, \mu)\,|_0 \le \frac{2K}{1-\lambda}\,|\,c_i(\varphi, \mu)\,|_0$$

$$\le \frac{2K}{1-\lambda}\,|\,u^k|_1\,(|\,a_1\,|_1 + |\,b_1\,|_1)\,w^i(\varphi, \mu)|_0. \tag{3.39}$$

By virtue of the inequalities (3.27) and (3.30), we find

$$|\,w^{i+1}(\varphi, \mu)\,|_0 \le \rho_0|w^i(\varphi, \mu)\,|_0 \le \rho_0^{i-1}|\,u^1(\varphi, \mu)\,|_0 \le \rho_0^{i-1}\frac{2K}{1-\lambda}\,|\,c(\varphi, \mu)\,|_0 \tag{3.40}$$

where ρ_0 is a positive constant less than unity for small μ_0. The inequality (3.40) proves that the sequence of invariant tori (3.6) is uniformly convergent and that the limiting function is continuous in μ at $\mu = 0$.

The above argument yields:

Lemma 6.9. (Samoilenko, Martinyuk, and Perestyuk, 1973). *Suppose that the conditions of Theorem 6.1 are satisfied. Then the sequence (3.6) of invariant tori of the system (3.7) converges uniformly to the function $u(\varphi, \mu)$, i.e.,*

$$\lim_{i\to\infty} u^i(\varphi, \mu) = u(\varphi, \mu)$$

moreover,

$$\lim_{\mu\to 0} |\,u(\varphi, \mu)\,|_0 = 0. \tag{3.41}$$

Then main theorem concerning the existence of the invariant tori of the perturbed system of difference equations (3.1) follows from Lemmas 6.7 and 6.9.

Theorem 6.2 (Samoilenko, Martinyuk, and Perestyuk, 1973). *Suppose that the right-hand side of the system*

$$\varphi_{n+1} = \varphi_n + a_0(\varphi_n, y_n, \mu)$$

$$y_{n+1} = [E + b(\varphi_n, y_n, \mu)] y_n + c(\varphi_n, \mu),$$

(3.42)

satisfies the following conditions:

(i)   the functions $a(\varphi,y,\mu)$, $b(\varphi,y,\mu)$, and $c(\varphi,\mu)$ are continuously differentiable for $\| y \| \le d$ and $0 \le \mu \le \mu_0$;

(ii)   max $\{| a(\varphi, y, \mu) - a(\varphi, 0, 0) |_1 , | b(\varphi, y, \mu) - b(\varphi, 0, 0) |_1 , | c(\varphi, \mu) |_1\} = L(d,\mu_0)$, where $L(d, \mu_0) \to 0$ as $d \to 0$ and $\mu_0 \to 0$;

(iii)   for arbitrary functions $b_1(\varphi)$, $a_1(\varphi)$, and $c_1(\varphi)$, which are sufficiently small in the norm $c'(\varphi)$, the perturbed linearized equations

$$\varphi_{n+1} = \varphi_n + a(\varphi_n, 0, 0) + a_1(\varphi_n),$$

$$y_{n+1} = [E + b(\varphi_n, 0, 0) + b_1(\varphi_n)] y_n + c(\varphi_n)$$

(3.43)

possess a Green's function $G_0(k, \varphi)$ for the problem of the invariant tori such that

$$| G_0(k, \varphi) c_1(\varphi_k(\varphi)) |_1 \le K \lambda^{|k|} | c_1(\varphi) |_1$$

(3.44)

for all integer k, where K is a positive constant and $0 < \lambda < 1$.

Then one can find a number μ^0 such that for all $0 < \mu < \mu^0$ the system (3.1) possesses the invariant torus $\mathcal{T}(\mu)$: $y = u(\varphi, \mu)$, for which

$$\lim_{\mu \to 0} | u(\varphi,\mu) |_0 = 0.$$

(3.45)

In order to employ in practice the scheme presented above, i.e., for the construction of the invariant tori for concrete systems of the form (3.1), it is necessary to learn the properties of Green's function for the problem of the invariant tori of the linearized system (3.43). Let us point out some systems for which a Green's function for the problem of the invariant tori exists and satisfies the inequality (3.44).

Consider a system of the form (3.32) such that the trivial solution of the system

$$y_{n+1} = [E + b(\varphi_n(\varphi), 0, 0) + b_1(\varphi_n(\varphi))] y_n$$

(3.46)

is asymptotically stable for all the matrices $b_1(\varphi)$ sufficiently small in the norm. For these systems, a Green's function for the problem under consideration exists and has the form

$$G_0(k, \varphi) = \begin{cases} \Omega_0^k(\varphi) & \text{for} \quad k<0, \\ E & \text{for} \quad k=0, \\ 0 & \text{for} \quad k>0, \end{cases} \qquad (3.47)$$

where $\Omega_n^k(\varphi)$ is the fundamental matrix of solutions of the system (3.46).

Let the matrix $b(\varphi, 0, 0)$ be such that

$$\max_{<\eta,\eta>=1} \left\langle [E+b(\varphi,0,0)][E+b(\varphi,0,0)]^{-1}\eta,\eta \right\rangle \leq \lambda, \qquad (3.48)$$

where $0 < \lambda < 1$ and $[E+b(\varphi,0,0)]^{-1}$ is the matrix transposed with respect to the matrix $[E+b(\varphi,0,0)]$. If the inequality (3.48) holds, then one can easily show that the function $G_0(k, \varphi)$ satisfies the inequality (3.44). Thus we have

Theorem 6.3. (Samoilenko, Martinyuk, and Perestyuk, 1973). *Suppose that the system (3.1) satisfies conditions (i) and (ii) of Theorem 6.2, and let the matrix $b(\varphi, 0, 0)$ be such that the inequality (3.4) holds. Then one can find a number $\mu^0 > 0$, such that for all $0 < \mu < \mu^0$, the system of equations (3.1) possesses the invariant toroidal set $\mathcal{T}(\mu)$: $y = u(\varphi, \mu)$ for which $\lim_{\mu \to 0} |u(\varphi, \mu)|_0 = 0$.*

§4. Reducibility of Systems of Difference Equations on the Toroidal Set

Consider a system of difference equations

$$\varphi_{n+1} = \varphi_n + \omega + f(\varphi_n), \qquad (4.1)$$

where $\varphi = (\varphi^1, \varphi^2, \dots, \varphi^m)$ and $f(\varphi) = (f^1(\varphi), \dots, f^m(\varphi))$; the function $f(\varphi)$ is periodic in φ with period 2π. We assume, that $\varphi = (\varphi^1, \varphi^2, \dots, \varphi^m)$ are angle coordinates on the torus $\mathfrak{F}_m$, and interpret (4.1) as a dynamical system given on this torus.

Along with (4.1), let us consider the system

$$\varphi_{n+1} = \varphi_n + \omega + \Delta + f(\varphi_n, \Delta), \tag{4.2}$$

where $f(\varphi, \Delta)$ and Δ are variables of the first order of smallness. Show that one can always find a change of variables $\varphi \to \theta$ and $\Delta \to \Delta^{(1)}$ which transforms (4.2) to the form

$$\theta_{n+1} = \theta_n + \omega + \Delta^{(1)} + f^{(1)}(\theta, \Delta^{(1)}), \tag{4.3}$$

where $\Delta^{(1)}$ and $f^{(1)}(\theta, \Delta^{(1)})$ are variables of the second order of smallness. The existence of this change is established by the following theorem

Theorem 6.4. (Martinyuk and Perestyuk, 1975a). *Suppose that the right-hand side of (4.2) satisfies the conditions:*

(i)   the function $f(\varphi, \Delta)$ is periodic in φ with period 2π and $l = l(s_0)$ times continuously differentiable with respect to φ and Δ for

$$|\Delta| = \max_i |\Delta| \leq M_0^{-1}; \tag{4.4}$$

moreover, $f(\varphi, \Delta)$ satisfies the inequalities

$$|f(\varphi, \Delta)| = \max_i |f_i(\varphi, \Delta)| < \delta_0,$$

$$\left| D_\varphi^p D_\Delta^r f(\varphi, \Delta) \right| \leq N_0^{|p|} M_0^{|r|} \quad \text{for } |p| + |r| = l; \tag{4.5}$$

(ii)   the number $\omega = (\omega_1, \omega_2, \dots, \omega_m)$ satisfies the inequality

$$\left| \sin\left(\frac{k, \omega}{2} \right) \right| \geq \frac{c}{|k|^m}. \tag{4.6}$$

Then one can find a number δ^0 which does not depend on N_0, M_0, and $f(\varphi, \Delta)$, such that for all $\delta_0 \leq \delta^0$ there exists a change of variables

$$\varphi_n = \theta_n + u(\theta_n, \Delta^{(1)}), \quad \Delta = \Delta^{(1)} + v(\Delta^{(1)}), \tag{4.7}$$

which transforms the system (4.2) into

$$\theta_{n+1} = \theta_n + \omega + \Delta^{(1)} + f^{(1)}(\theta_n, \Delta^{(1)}). \tag{4.8}$$

Furthermore, the functions $u(\theta, \Delta^{(1)})$, $v(\Delta^{(1)})$ *and* $f(\theta, \Delta^{(1)})$ *defined in the region* $|\Delta^{(1)}| \leq M^{-1}$ *are periodic in* θ *with period* 2π *and* l *times continuously differentiable. They satisfy the inequalities*

$$|v(\Delta^{(1)})| < M^{-1}, \quad |v(\Delta^{(1)})|_{s_0+1} < N^{-1}, \quad |u(\theta, \Delta^{(1)})|_{s_0+1} < N^{-1},$$

$$(4.9)$$

$$|f^{(1)}(\theta, \Delta^{(1)})| < \delta, \quad \left| D_\varphi^p D_{\Delta^{(1)}}^r f^{(1)}(\theta, \Delta^{(1)}) \right| \leq N^{|p|} M^{|r|} \text{ for } |p| + |r| = l,$$

where

$$N = N_0^\alpha, \quad M = N^\nu, \quad \delta = M^{-\beta}, \quad \alpha = 1 + \frac{1}{2(s_0+1)}, \quad \beta = s_0 + 2,$$

$$\nu = 4(m+1) + \frac{1}{4(s_0+2)^2}, \quad l(s_0) = 1 + 8(s_0+2)(m+1).$$

Proof. For functions $F(\varphi)$ which can be represented as the Fourier series (2.18), we denote

$$\overline{F(\varphi)} = \frac{1}{(2\pi)^m} \int_0^{2\pi} \cdots \int_0^{2\pi} F(\varphi) d\varphi^1 \ldots d\varphi^m,$$

$$(4.10)$$

$$\widetilde{F} = \sum_{|k|\neq 0} \frac{F_k}{e^{i(k,\omega)} - 1} e^{i(k,\varphi)}$$

and set

$$\varphi = \theta + \widetilde{Tf(\theta, \Delta)}, \quad \Delta^{(1)} = \Delta + \overline{Tf(\theta, \Delta)},$$

$$(4.11)$$

where $T = T_{NM}$ is a smoothing operator and $\overline{Tf(\theta, \Delta)}$ is the average value (in θ) of the function $Tf(\theta, \Delta)$, i.e.,

$$\overline{Tf(\theta, \Delta)} = \frac{1}{(2\pi)^m} \int_0^{2\pi} \cdots \int_0^{2\pi} Tf(\theta, \Delta) d\theta^1 \ldots d\theta^m.$$

After the change (4.11), the system of equations (4.2) transforms into

$$\theta_{n+1} + \widetilde{Tf(\theta, \Delta)} = \theta_n + \widetilde{Tf(\theta, \Delta)} + \omega + \Delta + f(\theta + \widetilde{Tf(\theta, \Delta)}, \Delta)$$

$$(4.12)$$

The last system can be represented as follows

$$\theta_{n+1} - \theta_n - \omega = \overline{Tf(\theta_n, \Delta)} + \Delta - \overline{Tf(\theta_{n+1}, \Delta)} + f(\theta_n + \overline{Tf(\theta_n, \Delta)}, \Delta)$$

$$- \overline{Tf(\theta_n + \omega, \Delta)} + \overline{Tf(\theta_n + \omega, \Delta)} + Tf(\theta_n, \Delta) - Tf(\theta_n, \Delta)$$

$$- \overline{Tf(\theta_n, \Delta)} + \overline{Tf(\theta_n, \Delta)}. \tag{4.13}$$

Let us choose $\overline{Tf(\theta_n, \Delta)}$ as a periodic solution of eqn. (2.17), i.e., of the equation

$$\overline{Tf(\theta_n + \omega, \Delta)} = \overline{Tf(\theta_n, \Delta)} + Tf(\theta_n, \Delta) - \overline{Tf(\theta_n, \Delta)}.$$

By virtue of this choice, one can rewrite (4.13) as

$$\theta_{n+1} - \theta_n - \omega = \Delta^{(1)} - [\overline{Tf(\theta_{n+1} - \theta_n - \omega + \theta_n + \omega, \Delta)}$$

$$\tag{4.14}$$

$$- \overline{Tf(\theta_n + \omega, \Delta)}] + f(\theta_n + \overline{Tf(\theta_n, \Delta)}, \Delta) - Tf(\theta_n, \Delta).$$

Solving (4.14) with respect to $\theta_{n+1} - \theta_n - \omega$, we obtain the equation

$$\theta_{n+1} = \theta_n + \omega + \Delta^{(1)} + f_1^{(1)}(\theta_n, \Delta, \Delta_1), \tag{4.15}$$

where

$$f_1^{(1)}(\theta_n, \Delta, \Delta_1) = \left[\left(E + \frac{\partial \overline{Tf}}{\partial \theta} \right)^{-1} - E \right] \Delta^{(1)} + \left(E - \frac{\partial \overline{Tf}}{\partial \theta} \right)^{-1}$$

$$\times \left[f(\theta_n + \overline{Tf}, \Delta) - Tf(\theta_n, \Delta) \right].$$

By solving the second equation of (4.11) with respect to Δ: $\Delta = \Delta(\Delta^{(1)})$ and substituting this value of Δ into the first equation of (4.11), we arrive at the systems (4.8) in which

$$f^{(1)}(\theta_n, \Delta^{(1)}) = f_1^{(1)}(\theta_n, \Delta(\Delta^{(1)}), \Delta^{(1)}). \tag{4.16}$$

Consequently, we can set

$$u = \overline{Tf(\theta_n, \Delta(\Delta^{(1)}))}, \quad v = \Delta(\Delta^{(1)}) - \Delta^{(1)}.$$

In order to complete the proof of Theorem 6.4, it suffices to verify the validity of (4.9). For the function $v(\Delta^{(1)})$ and $u(\theta, \Delta^{(1)})$, these estimates can be carried out by analogy with the approach of Bogolyubov, Mitropolsky, and Samoilenko (1969).

We estimate the function $f^{(1)}(\theta_n, \Delta^{(1)})$. By using the estimates for $v(\Delta^{(1)})$ and $u(\theta, \Delta^{(1)})$, one can easily find

$$\left| \left(E + \frac{\partial \overline{Tf}}{\partial \theta} \right)^{-1} \right| < c_3, \quad \left| \left(E + \frac{\partial \overline{Tf}}{\partial \theta} \right)^{-1} - E \right|$$

$$= \left| \frac{\partial \overline{Tf}}{\partial \theta} \left(E + \frac{\partial \overline{Tf}}{\partial \theta} \right)^{-1} \right| < c_3, \quad \left| \frac{\partial \overline{Tf}}{\partial \theta} \right| < c_3 N^{2(m+1)} \delta_0. \tag{4.17}$$

This implies that

$$|f^{(1)}(\theta, \Delta^{(1)})| < c_3 N^{2(m+1)} \delta_0 M^{-1} + c_3 [|f(\theta_n + \overline{Tf(\theta_n, \Delta)}, \Delta)$$

$$- Tf(\theta_n + \overline{Tf(\theta_n, \Delta)}, \Delta)|_0 + |Tf(\theta_n + \overline{Tf(\theta_n, \Delta)}, \Delta) - Tf(\theta_n, \Delta)|_0]. \tag{4.18}$$

By employing the properties of smoothing operators, given by the inequalities (1.13) and (1.14), we get

$$|f(\theta, \Delta) - f(\theta_n + Tf(\theta_n, \Delta)|_0 \le cN^{-l} N_0^l = c \left(\frac{N_0}{N} \right)^l = cN_0^{(1-\alpha)l}; \tag{4.19}$$

$$|Tf(\theta + \overline{Tf}, \Delta) - Tf(\theta, \Delta)|_0 < m \left| \frac{\partial Tf(\theta, \Delta)}{\partial \theta} \right|_0 |\overline{Tf}|_0$$

$$\le c_4 N \delta_0 |\overline{Tf}|_0 \le \bar{c}_4 N^{2(m+1)} \delta_0^2 \tag{4.20}$$

By virtue of (4.19) and (4.20), the inequality (4.18) yields

$$|f^{(1)}(\theta, \Delta^{(1)})| \le c_5 \left(N^{\alpha(m+1)} M_0^{\left(1-\frac{1}{\alpha}\right)\beta - \alpha} + N_0^{(1-\alpha)l + \alpha\beta\nu} \right.$$

$$+ N^{2(m+1)} M_0^{-\beta(2-\alpha)} \big) \delta_0^2 \leq \delta^0.$$

The estimates of the lth derivatives of $f^{(1)}(\theta_n, \Delta^{(1)})$ can be performed just as in Bogolyubov, Mitropolsky, and Samoilenko (1969).

We now apply Theorem 6.4. to prove the theorem on the reducibility of the system of difference equations on a torus.

Theorem 6.5. (Martinyuk and Perestyuk, 1975a). *Suppose that the right-hand side of the system difference equations*

$$\varphi_{n+1} = \varphi_n + \omega + \Delta + f(\varphi_n) \tag{4.21}$$

is such that $f(\varphi) = (f^1(\varphi), \dots, f^m(\varphi))$ *is a vector function periodic in* φ *with period* 2π, *and* $\omega = (\omega_1, \dots, \omega_m)$ *is a vector with incommensurable components, such that the inequality*

$$\left| \sin\left(\frac{k, \omega}{2}\right) \right| \geq \frac{c}{|k|^m}, \quad |k| \neq 0 \tag{4.22}$$

holds for all $k = (k_1, k_2, \dots, k_m)$ *with integer components. Then for given positive constants* ε *and* c_0 *and integer* $s_0 \geq 1$, *one can find* $\delta_0 = \delta_0(c_0, \varepsilon, s)$ *and the integer* $l = l(s_0)$ *such that, for the* l *times continuously differentiable function* $f(\varphi)$, *satisfying the inequalities*

$$|f(\varphi)|_0 < \delta_0, \quad |f(\varphi)|_l < c_0,$$

there exist a constant vector $\Delta = (\Delta_1, \dots, \Delta_m)$ *for which*

$$|\Delta|_0 < \varepsilon,$$

and an s_0 *times continuously differentiable function* $\Phi = (\Phi^1, \dots, \Phi^m)$ *periodic in* $\theta = (\theta^1, \dots \theta^m)$ *with period* 2π *for which*

$$|\Phi(\theta)|_{s_0} < \varepsilon,$$

such that by the change of variables

$$\varphi = \theta + \Phi(\theta) \tag{4.23}$$

the system of equations (4.21) *is reduced to the form*

$$\theta_{n+1} = \theta_n + \omega. \tag{4.24}$$

Proof. Assuming δ_0 to be so small that $\delta_0 < \delta^0 = \min\left(e^\beta, c_0^{-\beta v}\right)$, we can easily find that the conditions of Theorem 6.5 can be made consistent with the conditions (4.5). Consequently, one can apply Theorem 6.4 to eqn.(4.21) and find the following change of variables

$$\varphi_n = \varphi_n^{(1)} + u^{(1)}\left(\varphi_n^{(1)}, \Delta^{(1)}\right) = \varphi_n^{(1)} + T_{N_1 M_1} f\left(\varphi_n^{(1)}, \Delta\left(\Delta^{(1)}\right)\right),$$

$$\Delta = \Delta\left(\Delta^{(1)}\right) = \Delta^{(1)} + v^{(1)}\left(\Delta^{(1)}\right) \tag{4.25}$$

reducing the equation (4.21) to

$$\varphi_{n+1}^{(1)} = + \varphi_n^{(1)} + \omega + \Delta^{(1)} + f^{(1)}\left(\varphi_n^{(1)}, \Delta^{(1)}\right), \tag{4.26}$$

where the functions $u^{(1)}\left(\varphi_n^{(1)}, \Delta^{(1)}\right)$, $v^{(1)}(\Delta^{(1)})$, and $f^{(1)}\left(\varphi_n^{(1)}, \Delta^{(1)}\right)$ are periodic in $\varphi^{(1)}$ with period 2π and such that

$$|v^{(1)}(\Delta^{(1)})| \le M_1^{-1} = M_0^{-\alpha}, \quad |v^{(1)}(\Delta^{(1)})|_{s_0+1} \le N_1^{-1} = N_0^{-\alpha},$$

$$|u^{(1)}(\varphi^{(1)}, \Delta^{(1)})|_{s_0+1} \le N_1^{-1}, \quad |f^{(1)}(\varphi^{(1)}, \Delta^{(1)})| \le \delta_1 \le \delta_0^{-\alpha} \tag{4.27}$$

$$\left| D_{\varphi^{(1)}}^r D_{\Delta^{(1)}}^\rho f^{(1)}\left(\varphi^{(1)}, \Delta^{(1)}\right) \right| \le N_1^{|r|} M_1^{|\rho|}$$

for $|r| + |\rho| = l$ and $|\Delta^{(1)}| \le M_1^{-1}$.

The inequalities (4.37) ensure the validity of Theorem 6.4 for the system (4.36), and thus, for arbitrary s, $s = 1, 2, \ldots$, one can find the change of variables

$$\varphi_n^{(s-1)} = \varphi_n^s + u^{(s)}\left(\varphi_n^{(s)}, \Delta^{(s)}\right) = \varphi_n^s + T_{N_s M_s} f^{(s-1)}\left(\varphi^{(s)}, \Delta^{(s-1)}\left(\Delta^{(s)}\right)\right),$$

$$\tag{4.28}$$

$$\Delta^{(s-1)} = \Delta^{(s-1)}(\Delta^{(s)}) = \Delta^{(s)} + v^{(s)}(\Delta^{(s)}),$$

reducing the equation for $\varphi_n^{(s-1)}$ to the form

$$\varphi_{n+1}^{(s)} = \varphi_n^{(s)} + \omega + \Delta^{(s)} + f^{(s)}\left(\varphi_n^s, \Delta^{(s)}\right); \tag{4.29}$$

where the functions $u^{(s)}(\varphi^{(s)}, \Delta^{(s)})$, $v^{(s)}(\Delta^{(s)})$, and $f^{(s)}(\varphi^{(s)}, \Delta^{(s)})$ are periodic in $\varphi^{(s)}$ with period 2π and such that

$$|v^{(s)}(\Delta^{(s)})| \leq M_s^{-1} = M_{s-1}^{-\alpha},$$

$$|v^{(s)}(\Delta^{(s)})|_{s_0+1} \leq N_s^{-1} = M_{s-1}^{-\alpha},$$

$$|u^{(s)}(\varphi_n^s, \Delta^{(s)})|_{s_0+1} \leq N_s^{-1}, \tag{4.30}$$

$$|f^{(s)}(\varphi_n^s, \Delta^{(s)})| \leq \delta_3 = \delta_{s-1}^{-\alpha},$$

$$\left| D_{\varphi^{(s)}}^r D_{\Delta^{(s)}}^\rho f^{(s)}(\varphi^{(s)}, \Delta^{(s)}) \right| \leq N_s^{|r|} M_s^{|\rho|}$$

for $|r| + |\rho| = l$ and $\left|\Delta^{(s)}\right| \leq M_s^{-1}$.

Clearly, a superposition of s changes (4.28) leads to the change of variables

$$\varphi_n = \varphi_n^{(s)} + \Phi^{(s)}\left(\varphi_n^{(s)}, \Delta^{(s)}\right) = \theta_n + \Phi^{(s)}\left(\varphi_n^{(s)}, \Delta^{(s)}\right),$$

$$\Delta = A^{(s)}(\Delta^{(s)}) = \Delta^{(s)} + \Psi^{(s)}(\Delta^{(s)}) \tag{4.31}$$

which links φ and Δ with $\varphi^{(s)} = \theta$ and $\Delta^{(s)}$ and transforms the initial equation (4.21) into the equation (4.29).

We have $\left|\Delta^{(s)}\right| \leq M_s^{-1} \to 0$ and $|f^{(s)}(\theta, \Delta^{(s)})| \leq \delta_3 \to 0$ as $s \to \infty$, and thus, in order to complete the proof of the theorem on reducibility, it remains to show that the number sequence $A^{(s)}(0)$ converges and that the sequence of functions $\Phi^{(s)}(\theta, 0)$ (and of their derivatives up to the s_0-th order) converges uniformly (see, Bogolyubov, Mitropolsky, and Samoilenko (1969)). This, in fact, completes the proof of the theorem, because by means of the change of variables

$$\varphi_n = \theta_n + \Phi^{(\infty)}(\theta_n, 0),$$

the equation (4.21) with $\Delta = A^{(\infty)}(0)$ can be transformed into

$$\theta_{n+1} = \theta_n + \omega$$

and this change of variables, together with the constant $A^{(\infty)}(0)$, satisfies all conditions of Theorem 6.4.

§5. Reducibility of Nonlinear Systems of Difference Equations in the Vicinity of a Toroidal Set

Consider a system of difference equations

$$x_{n+1} = Ax_n + F\,(\varphi_n, x_n),$$

$$\varphi_{n+1} = \varphi_n + \omega + f\,(\varphi_n, x_n)$$

(5.1)

where A is a constant $(s \times s)$-dimensional matrix; $F(\varphi, x)$ and $f(\varphi, x)$ are vector functions periodic in $\varphi = (\varphi^1, \varphi^2, \dots, \varphi^m)$ with period 2π, l_0 times continuously differentiable in the region

$$\| x \| = \left(\sum_{i=1}^{s} \left(x^i \right)^2 \right)^{\frac{1}{2}} \le \eta,$$

and such that

$$F(\varphi,0) = \left. \frac{\partial F(\varphi,x)}{\partial x} \right|_{x=0} = f(\varphi,0) = 0.$$

In this section, we establish conditions under which the system (5.1) can be reduced to the system of the form

$$y_{n+1} = A\,y_n, \quad \theta_{n+1} = \theta_n + \omega,$$

where the matrix A and the functions $F\,(\varphi,x)$ and $f\,(\varphi,x)$ possess the properties indicated in Lemma 6.5; they are variables of the first order of smallness. We set

$$N_k = N_{k-1}^{\kappa}, \quad \delta_k = \delta_{k-1}^{\kappa}, \quad \delta_k = N_k^{-\beta}, \quad k = 1, 2, 3, \dots$$

where the constants κ, β, k, and l_0 are such that

$$\kappa = \frac{3}{2}, \quad \beta = \frac{10}{9} + \frac{3}{2}\bigl(k_0 + s + 4\bigr),$$

$$l_0 = \frac{13}{2} + \frac{9}{2}(k_0 + s + 4), \quad k_0 = s + 8. \tag{5.2}$$

The following statement is valid for the parameters given above.

Theorem 6.6 (Martinyuk, 1975). *Suppose that the right-hand side of the system* (5.1) *is such that:*

(i) the eigenvalues of the matrix satisfy the inequalities

$$0 < |\lambda_j| < 1, \quad |\lambda_\alpha| \, |\lambda_\beta| \, |\lambda_j|^{-1} \le \gamma_1 < 1 \tag{5.3}$$

for all α, β and $j = 1, \ldots, s$;

(ii) the functions $F(\varphi, x)$ and $f(\varphi, x)$ are periodic in φ with period 2π and l_0 times continuously differentiable in the region $\|x\| \le \eta$; they satisfy the condition

$$F(\varphi, 0) = \frac{\partial F(\varphi, x)}{\partial x}\bigg|_{x=0} = f(\varphi, 0) = 0 \tag{5.4}$$

and the inequalities

$$|F(\varphi, x)| + |f(\varphi, x)| \le \delta_{k-1}, \tag{5.5}$$

$$\left| D_\varphi^\rho D_x^r f(\varphi, x) \right| + \left| D_\varphi^\rho D_x^r F(\varphi, x) \right| \le N_{k-1}^{l_0} \tag{5.6}$$

for $|\rho| + |r| = l_0$.

Then for sufficiently small δ_0 there exists a change of variables

$$x = y + Y(\theta, y), \quad \varphi = \theta + \Phi(\theta, y) \tag{5.7}$$

with the functions $Y(\theta, y)$ and $\Phi(\theta, y)$ periodic in θ with period 2π defined in the region

$$\|y\| \le \eta - N_{k-1}^{-1} \tag{5.8}$$

and satisfying the condition

$$Y(\theta, 0) = 0, \quad \frac{\partial Y(\theta, y)}{\partial y}\bigg|_{y=0} = 0, \quad \Phi(\theta, 0) = 0, \tag{5.9}$$

and the inequality

$$\left|Y(\theta, y)\right|_{k_0} + \left|\Phi(\theta, y)\right|_{k_0} \leq N_{k-1}^{-1} \tag{5.10}$$

such that in terms of θ *and* y *the system (5.1) takes the form*

$$y_{n+1} = A\, y_n + F^{(1)}(\theta, y),$$

$$\theta_{n+1} = \theta_n + \omega + f^{(1)}(\theta, y), \tag{5.11}$$

where the functions $F^{(1)}(\theta, y)$ *and* $f^{(1)}(\theta, y)$ *are periodic in* θ *with period* 2π
and l_0 *times continuously differentiable with respect to* θ *and* y *in the region*
(5.8). They satisfy the following condition

$$F^{(1)}(\theta, 0) = 0, \quad \left.\frac{\partial F^{(1)}(\theta, y)}{\partial y}\right|_{y=0} = 0, \quad f^{(1)}(\theta, 0) = 0 \tag{5.12}$$

and the inequalities

$$|F^{(1)}(\theta, y)| + |f^{(1)}(\theta, y)| \leq \delta_k, \quad (\|y\|) \leq \eta - 2N_{k-1}^{-1}; \tag{5.13}$$

$$\left|D_\theta^p D_y^r F^{(1)}(\theta, y)\right| + \left|D_\theta^p D_y^r f^{(1)}(\theta, y)\right| \leq N_k^{l_0}, \quad \textit{for } |p| + |r| = l_0. \tag{5.14}$$

Proof. Let

$$\Gamma(\theta, y) = L^{(1)} T_{N_k N_k}^1 F(\theta, y),$$

$$\Phi(\theta, y) = L^{(0)} T_{N_k N_k}^0 f(\theta, y), \tag{5.15}$$

where $T_{N_k N_k}^1$ *and* $T_{N_k N_k}^0$ *are smoothing operators* T_{NM}^1 *and* T_{NM}^0 *$(N = M = N_k)$,*
which were considered in §1; $L^{(1)}$ and $L^{(0)}$ are operators introduced in Lemma 6.5.
   It follows from the properties of the smoothing operators T_{NM}^1 and T_{NM}^0, and Lemma 6.5 that the functions $\Gamma(\theta, y)$ and $\Phi(\theta, y)$ are periodic in θ with period 2π and
continuously differentiable with respect to θ and y arbitrarily many times; furthermore,
they are defined in the region

$$\|y\| \leq \eta - N_{k-1}^{-1} \tag{5.16}$$

and satisfy conditions (5.12) and the inequality

$$\left|\Gamma(\theta,y)\right|_{k_0} + \left|\Phi(\theta,y)\right|_{k_0} \le 2a_0 c N_k^{\kappa(k_0+3)-\beta} = 2a_0 c N_k^{-\frac{1}{9}} N_{k-1}^{-1} \le N_{k-1}^{-1} \qquad (5.17)$$

Inserting (5.7) into (5.1), we get

$$y_{n+1} + \Gamma(\theta_{n+1}, y_{n+1}) = A\, y_n + \Gamma(\theta_n, y_n)$$

$$+ F\,(y_n + \Gamma(\theta_n, y_n), \theta_n + \Phi(\theta_n, y_n)),$$

$$\theta_{n+1} + \Phi(\theta_{n+1}, y_{n+1}) = \theta_n + \omega + \Phi(\theta_n, y_n) \qquad (5.18)$$

$$+ \Phi(y_n + \Gamma(\theta_n, y_n), \theta_n + \Phi(\theta_n, y_n)).$$

This system can be rewritten as

$$y_{n+1} - A\, y_n = -\Gamma(\theta_n + \omega, A\, y_n) + A\,\Gamma(\theta_n, y_n)$$

$$- [\Gamma(\theta_{n+1} - \theta_n - \omega + \theta_n + \omega, y_{n+1} - A\, y_n + A\, y_n) + \Gamma(\theta_n + \omega, A\, y_n)]$$

$$+ F\,(y_n + \Gamma(\theta_n, y_n), \theta_n + \Phi(\theta_n, y_n)),$$

$$\theta_{n+1} - \theta_n - \omega = -\Phi(\theta_n + \omega, A\, y_n) + \Phi(\theta_n, y_n) \qquad (5.19)$$

$$-[\Phi(\theta_{n+1} - \theta_n - \omega + \theta_n + \omega, y_{n+1} - A\, y_n + A\, y_n) - \Phi(\theta_n + \omega, A\, y_n)]$$

$$+ \Phi(y_n + \Gamma(\theta_n, y_n), \theta_n + \Phi(\theta_n, y_n)).$$

By choosing $\Gamma(\theta_n, y_n)$ and $\Phi(\theta_n, y_n)$ to be the solutions of the systems (2.1) and (2.2), respectively, we can represent (5.19) as follows:

$$y_{n+1} - A\, y_n = -[Y(\theta_{n+1} - \theta_n - \omega + \theta_n + \omega, y_{n+1} - A\, y_n + A\, y_n)$$

$$- \Gamma(\theta_n + \omega, A\, y_n)] + F(y_n + Y(\theta_n, y_n), y_n + \Phi(\theta_n, y_n)) - T^1 F(\theta_n, y_n),$$

$$\theta_{n+1} - \theta_n - \omega = -[\Phi(\theta_{n+1} - \theta_n - \omega + \theta_n + \omega, y_{n+1} - Ay_n + Ay_n)$$

$$- \Phi(\theta_n + \omega, A\, y_n)] + f(y_n + Y(\theta_n, y_n), \theta_n + \Phi(\theta_n, y_n)) - T^0 f(\theta_n, y_n). \qquad (5.20)$$

Solving the system (5.20) with respect to $y_{n+1} - A y_n$ and $\theta_{n+1} - \theta_n - \omega$ (this is always possible by virtue of (5.17)), we obtain the following system of equations

$$y_{n+1} = A\, y_n + F^{(1)}(\theta_n, y_n),$$

$$(5.21)$$

$$\theta_{n+1} = \theta_n + \omega + f^{(1)}(\theta_n, y_n).$$

The functions $F^{(1)}(\theta_n, y_n)$ and $f^{(1)}(\theta_n, y_n)$ defined in the region (5.16) are periodic in θ with period 2π and l_0 times continuously differentiable. They satisfy (5.12) and the following inequalities

$$| F^{(1)}(\theta, y)\,| + |f^{(1)}(\theta, y)\,| \leq c_1[|\, F(\theta + \Phi, y + \Upsilon) - T^1 F(\theta, y)\,|$$

$$+ |f(\theta + \Phi, y + \Gamma) - T^0 f(\theta, y)|] \leq c_1[|\, F(\theta + \Phi, y + \Gamma)$$

$$- T^1 F(\theta + \Phi, y + \Upsilon)| \; + |f(\theta + \Phi, y + \Gamma) - T^0 f(\theta + \Phi, y + \Gamma)\,|$$

$$+ |T^1 F(\theta + \Phi, y + \Upsilon) - T^1 F(\theta, y)| + |T^0 f(\theta + \Phi, y + \Upsilon) - T^0 f(\theta, y)\,|]. \quad (5.22)$$

If we take into account the properties of smoothing operators, then we get

$$|\, F(\varphi, x) - T^1 F(\varphi, x)\,| + |f(\varphi, x) - T^0 f(\varphi, x)|$$

$$\leq c N_s^{-l_0} \left[\sup_{\substack{|p|+|r|=l_0 \\ \|x\|\leq \eta}} \left(N_k \left| D_\varphi^p D_x^r F(\varphi, x)\right| + \left| D_\varphi^p D_x^r f(\varphi, x)\right| \right) \right]$$

$$\leq c N_k \left(\frac{N_{k-1}}{N_k} \right)^{-l_0} = c N_{k-1}^{\kappa + (1-\kappa)l_0} = c N_{k-1}^{\kappa(1+\beta)+(1-\kappa)l_0} \delta_k = c N_{k-1}^{-\frac{1}{12}} \delta_k \leq \frac{1}{2}\delta_k,$$

$$|T^1 F(\varphi, x) - T^1 F(\theta, y)| + |T^0 f(\varphi, x) - T^0 f(\theta, y)\,| \leq \frac{1}{2}\delta_k.$$

This yields

$$| F^{(1)}(\theta, y)|.+ |f^{(1)}(\theta, y)| \leq \delta_k.$$

The estimate (5.14) can be obtained just as in (Bogolyubov, Mitropolsky, and Samoilenko, 1969). This completes the proof of the inductive theorem.

By employing this theorem, we now prove the next theorem concerning reducibility in the vicinity of the toroidal set $x = 0$.

Theorem 6.7 (Martinyuk, 1975; Martinyuk and Samoilenko, 1974b).

Let the functions $F(\varphi, x)$ and $f(\varphi, x)$ be periodic in $\varphi = (\varphi^1, \dots, \varphi^m)$ with period 2π. Then for given $c_0, \gamma, \overline{\eta}$, and k_0, $k_0 \geq 2$, one can find $\delta_0 = \delta_0(c_0, \gamma, \overline{\eta})$ and an integer $l_0 = l_0(k_0)$ such that under the following conditions:

(i) the functions $F(\varphi, x)$ and $f(\varphi, x)$ are l_0 times continuously differentiable in the region

$$\| x \| \leq \eta, \quad x = (x^1, \dots, x^s) \tag{5.23}$$

and such that

$$F(\varphi, 0) = 0, \quad \left. \frac{\partial F(\varphi, x)}{\partial x} \right|_{x=0} = 0, \quad f(\varphi, 0) = 0; \tag{5.24}$$

they satisfy the inequalities

$$| F(\varphi, x) | + |f(\varphi, x) | \leq \delta_0, \tag{5.25}$$

$$| F(\varphi, x) |_{l_0} + |f(\varphi, x)|_{l_0} \leq c_0; \tag{5.26}$$

(ii) the eigenvalues of the matrix A satisfy the inequalities

$$0 < |\lambda_j| < 1, \quad |\lambda_\alpha| |\lambda_\beta| |\lambda_j|^{-1} \leq \gamma_1 < 1, \tag{5.27}$$

for every α, β and $j = 1, \dots, s$, there exist functions $\Gamma(\theta_n, y_n)$ and $\Phi(\theta_n, y_n)$ periodic in θ with period 2π and $k_0 - 1$ times continuously differentiable in the region

$$\| x \| = \left(\sum_{i=1}^{s} (x^i)^2 \right)^{\frac{1}{2}} \leq \eta, \tag{5.28}$$

for which

$$\Gamma(\theta, 0) = 0, \quad \left. \frac{\partial \Gamma(\theta, y)}{\partial y} \right|_{y=0} = 0, \quad \Phi(\theta, 0) = 0, \tag{5.29}$$

and

$$| \Gamma(\theta, y)|_{k_0 - 1} + |\Phi(\theta, y)|_{k_0 - 1} \leq \overline{\eta}. \tag{5.30}$$

These functions are such that by the change of variables

$$x_n = y_n + \Gamma(\theta_n, y_n),$$

$$\varphi_n = \theta_n + \Phi(\theta_n, y_n), \tag{5.31}$$

the system of equations

$$x_{n+1} = A x_n + F(\varphi_n, x_n),$$

$$\varphi_{n+1} = \varphi_n + \omega + f(\varphi_n, x_n), \tag{5.32}$$

is reduced to the form

$$y_{n+1} = A y_n,$$

$$\varphi_{n+1} = \varphi_n + \omega. \tag{5.33}$$

Proof. Theorem 6.7 implies that, for all $n \geq 0$, the general solution of the system (4.9) has the form

$$x_n = A^n y_n + \Gamma(n\omega + \theta_0, A^n y_0),$$

$$\varphi_n = n\omega + \theta_0 + \Phi(n\omega + \theta_0, A^n y_0),$$

where θ_0 and y_0 are arbitrary constants. The matrix A is real, and thus, by the real linear change of variables $x_n \to C x_n$, the system (5.32) can be transformed into the system of the same form but with the matrix A in the real canonical form with the parameter ε_1 satisfying the inequality $\varepsilon_1 + \gamma_1 < 1$. Having done this, one can choose δ_0 so small that Theorem 6.6 with $k = 1$ becomes valid for the system (5.32). This enables us to find the change of variables

$$x_n = x_n^{(1)} + \Upsilon^{(1)}\left(\varphi_n^{(1)}, x_n^{(1)}\right),$$

$$\varphi_n = \varphi_n^{(1)} + \Phi^{(1)}\left(\varphi_n^{(1)}, x_n^{(1)}\right), \tag{5.34}$$

which transforms (5.32) into

$$x_{n+1}^{(1)} = A^{(1)} x_n + F^{(1)}\left(\varphi_n^{(1)}, x_n^{(1)}\right),$$

$$\varphi_{n+1}^{(1)} = \varphi_n^{(1)} + \omega + f^{(1)}\left(\varphi_n^{(1)}, x_n^{(1)}\right),\tag{5.35}$$

etc. At the kth step the change of variables

$$x_{n+1}^{(k-1)} = x_n^{(k)} + Y^{(k)}\left(\varphi_n^{(k)}, x_n^{(k)}\right),$$

$$\varphi_{n+1}^{(k-1)} = \varphi_n^{(k)} + \Phi^{(k)}\left(\varphi_n^{(k)}, x_n^{(k)}\right),\tag{5.36}$$

transforms the system of equations for $x^{(k-1)}$ and $\varphi^{(k-1)}$ into

$$x_{n+1}^{(k)} = A^{(k)}x_n + F^{(k)}\left(\varphi_n^{(k)}, x_n^{(k)}\right),$$

$$\varphi_{n+1}^{(k-1)} = \varphi_n^{(k)} + \omega + f^{(k)}\left(\varphi_n^{(k)}, x_n^{(k)}\right).\tag{5.37}$$

By virtue of the previous theorems, all the functions $\Gamma^{(k)}$, $\Phi^{(k)}$ $F^{(k)}$, and $f^{(k)}$ in these successive changes are periodic in $\varphi^{(k)}$ with period 2π, l_0 times continuously differentiable in the region $\| x^{(k)} \| \le \eta - \left(N_0^{-1} + N_1^{-1} + \ldots + N_{k-1}^{-1}\right)$ and such that conditions of the type (5.29) hold. This means that the superposition of the changes (5.34)–(5.37), i.e., the change of variables

$$x_n = x_n^k + x_k\left(\varphi_n^{(k)}, x_n^{(k)}\right),$$

$$\varphi_n = \varphi_n^{(k)} + \Psi^{(k)}\left(\varphi_n^{(k)}, x_n^{(k)}\right),\tag{5.38}$$

converts the system (5.32) into (5.37).

Obviously, we have

$$X^{(k+1)}(\theta_n, y_n) = \Gamma^{(k+1)}(\theta_n, y_n) + X^{(k)}(\theta_n + \Phi^{(k+1)}(\theta_n, y_n), y_n + \Gamma^{(k+1)}(\theta_n, y_n)),$$

$$\Psi^{(k+1)}(\theta_n, y_n) = \Phi^{(k+1)}(\theta_n, y_n) + \Psi^{(k)}(\theta_n + \Phi^{(k+1)}(\theta_n, y_n), y_n + \Gamma^{(k+1)}(\theta_n, y_n)).\tag{5.39}$$

If we differentiate these relations and assume that $\| y \| \le \eta + \sum_{v=0}^{k} N_v^{-1}$, then we find that

$$\sum_{|p|+|r|=1} \left| D_\theta^p D_y^r X^{(k)}(\theta, y)\right|_0 \le b_1,$$

$$\sum_{|p|+|r|=1} \left| D_\theta^p D_y^r \Psi^{(k)}(\theta, y)\right|_0 \le b_1,\tag{5.40}$$

where $b_1 \to 0$ as N_0^{-1}.

We choose δ_0 so small that, in addition, the inequality $\sum\limits_{v=0}^{\infty} N_v^{-1} \le \overline{\eta}$ holds. Then for the variables θ and y in the region $\| y \| \le \eta - \overline{\eta}$ we obtain

$$
| X^{(k+1)}(\theta, y) - X^{(k)}(\theta, y) | + | \Psi^{(k+1)}(\theta, y) - \Psi^{(k)}(\theta, y)|
$$

$$
\le | \Gamma^{(k+1)}(\theta, y)| + | \Phi^{(k+1)}(\theta, y_n) | + | X^{(k)}(\theta + \Phi^{(k+1)}(\theta, y), y
$$

$$
+ \Gamma^{(k+1)}(\theta, y)) - X^{(k)}(\theta, y) | + | \Psi^{(k)}(\theta + \Phi^{(k+1)}(\theta, y), y
$$
(5.41)

$$
+ \Gamma^{(k+1)}(\theta, y)) - \Psi^{(k)}(\theta, y)| \le 2N_k^{-1} + 2b_1 N_k^{-1} = 2 (1 + b_1) N_k^{-1}.
$$

This inequality gives the criterion of uniform convergence of the sequences of functions $X^{(k)}(\theta, y)$ and $\Psi^{(k)}(\theta, y)$.

We set

$$
X^{(\infty)}(\theta, y) = \lim_{k \to \infty} X^{(k)}(\theta, y), \quad \Psi^{(\infty)}(\theta, y) = \lim_{k \to \infty} \Psi^{(k)}(\theta, y).
$$
(5.42)

The periodicity of $X^{(\infty)}(\theta, y)$ and $\Psi^{(\infty)}(\theta, y)$; moreover, the relations of type (5.29) and the inequality (5.41) yield

$$
X^{(\infty)}(\theta, 0) = \frac{\partial X^{(\infty)}(\theta, y)}{\partial y} \bigg|_{y=0} = \Psi^{(\infty)}(\theta, 0) = 0,
$$

and

$$
| X^{(\infty)}(\theta, y) | + | \Psi^{(\infty)}(\theta, y) | \le 2(1 + b_1) \sum_{v=0}^{\infty} N_v^{-1} \le \overline{\eta} ,
$$

respectively. Just as in Bogolyubov, Mitropolsky and Samoilenko (1969) one can show that the functions $X^{(\infty)}(\theta, y)$ and $\Psi^{(\infty)}(\theta, y)$ are $k_0 - 1$ times continuously differentiable. This means that, taking into account that

$$
| F^{(k)}(\theta, y)| + | f^{(k)}(\theta, y) | \to 0, \text{ as } k \to \infty,
$$

uniformly with respect to θ and y, we complete the proof of the theorem by setting

$$
\Gamma(\theta, y) = X^{(\infty)}(\theta, y), \quad \Phi(\theta, y) = \Psi^{(\infty)}(\theta, y).
$$

REFERENCES

All works *(except those marked by *)* are in Russian or Russian translation.

Alexandrov, P.S.

(1947) Combinatorial Topology. Gostekhizdat, Moscow, Leningrad, 1947.

Arnold, V.I.

(1961) *Small denominators. I. On a mapping of a circle onto itself.* Izv. Akad. Nauk SSSR, Ser. Math., **25** (1961), 21–36.

(1963a) *Small denominators. II. Proof of A.N.Kolmogorov's theorem on preservation of conditionally periodic motion under small changes of a Hamilton function.* Usp. Mat. Nauk, **18** (5), (1963), 13–40.

(1963b) *Small denominators and the problem of stability of motion in classical and celestial mechanics.* Usp. Mat. Nauk, **18** (6), (1963), 91–192.

Bellman, R.E., and Kalaba, R.E.

(1968) Quasilinearization and Nonlinear Boundary Value Problems. Mir, Moscow, 1968.
Translated from:
* Amer. Elsevier Publishing Company, Inc., New York, 1965.

Bers, L., John, F., and Schechter, M.

(1966) Partial Differential Equations. Mir, Moscow, 1966.
Translated from:
* Interscience Publishers, New York–London–Sydney, 1964.

Bogolyubov, N.N.

(1945) On Some Statistical Methods in Mathematical Physics. Izdatelstvo Akad. Nauk Ukrain.SSR, Kiev, 1945.

264 *References*

(1964) *On quasiperiodic solutions in the problems of nonlinear mechanics.*– In: Memoirs of the first summer math. school, **Part I**, 11–101. Naukova Dumka, Kiev, 1964.

Bogolyubov, N.N., and Mitropolsky, Yu.A.

(1963) Asymptotic Methods in the Theory of Nonlinear Oscillations. Fizmatgiz, Moscow, 1963.

Bogolyubov, N.N., Mitropolsky, Yu.A., and Samoilenko, A.M.

(1969) The Methods of Rapid Convergence in Nonlinear Mechanics. Naukova Dumka, Kiev, 1969.

Bol, R.G.

(1961) Selected Papers. Izdatelstvo Akad. Nauk Latv.SSR, Riga, 1961.

Boltyansky, V.G.

(1973) Optimal Control Over Discrete Systems. Nauka, Moscow, 1973.

Borisovich, Yu.G.

(1963) *On the Poincare-Andronov method in the problem of periodic solutions of differential equations with lag.* Dokl. Akad. Nauk SSSR, **152**, (1963), 779–782.

(1967) *On the Poincare-Andronov method in the problems on periodic and bounded solutions of differential equations with retarded argument.* – In: Proceedings of the Seminar in Theory of Differential Equations with Deviating argument. Univ. Druzhby Narodov im. P.Lumumby. **Vol. 5.**, 5–17, Moscow, 1967.

Borisovich, Yu.G., and Subbotin, V.F.

(1967) *Theorems on the existence of periodic semipositive solutions of differential equations with retarded argument.* – In: Proceedings of the Seminar on Functional Analysis, **Issue 9**, 111–115, Izdatelstvo Voronezh. Univ., Voronezh, 1967.

Bortei, M.S., and Fodchuk, V.I.

(1976) *Asymptotic reducibility of a nonlinear system of differential-functional equations to a linear system of ordinary differential equations.* Ukrain. Mat.Zhurn., **28**, (1976), 592–602.

(1979)   *On reducibility and construction of the solutions of systems of linear differential-functional equations with quasiperiodic coefficients.* Dif.Uravn., **15**, (1979), 771–783.

Cesari, L.

(1963)   * *Functional analysis and periodic solutions of nonlinear differential equations.* – In: Contributions to differential equations, **vol. 1**, 148–187, New York, 1963.

Dankanich, V.A.

(1984)   *Quasiperiodic solutions for systems with lag.* Ukrain.Mat.Zhurn., **36(1)**, (1984), 105–110.

Elsgoltz, L.E., and Norkin, S.B.

(1971)   Introduction to the Theory of Differential Equations With Deviating Argument. Nauka, Moscow, 1971.

Filatov, A.N., and Sharova, L.V.

(1975)   Integral Inequalities and the Theory of Nonlinear Oscillations. Nauka, Moscow, 1975.

Fodchuk, V.I.

(1965)   *Investigation of integral manifolds for a system of nonlinear differential equations with retarded argument.* Ukrain.Mat.Zhurn., **17(4)**, (1965), 92–102.

(1970)   *On integral manifolds for systems with lag.*– In: Proceedings of V Internat. Conf. on Nonlinear Oscillations, **vol.1**, 558–564, Inst. Mat. Acad. Nauk Ukrain. SSR, Kiev, 1970.

Grebennikov, E.A.

(1968)   *Some estimates of the averaging method for multi-frequency systems of ordinary differential equations.* Dif. Uravn., **4**, (1968), 459–473.

Grebennikov, E.A., and Ryabov, Yu.A.

(1971)   New Qualitative Methods in Celestial Mechanics. Nauka, Moscow, 1971.

(1978)   Resonances and Small Denominators in Celestial Mechanics. Nauka, Moscow, 1978.

(1979)   Constructive Methods for the Analysis of Nonlinear Systems. Nauka,
              Moscow, 1979.

Gurtovnik, A.S., Kogan, V.P., and Neimark, Yu.I.

(1975)   *Invariant toroidal manifolds and resonances on discrete dynamical sys-
              tems*. Ukrain. Mat. Zhurn., **27**, (1975), 167–182.

Halanay, A.

(1965)   *Periodic invariant manifolds for some class of systems with lag*. Rev. Ro-
              um. math. pures at appl., **10**, (1965), 251–261.

(1967)   * *Invariant manifolds for systems with time lag*. – In: Differential equati-
              ons and dynamical systems. Acad. Press, New York, 1967.

Halanay, A., and Veksler, D.

(1971)   Qualitative Theory of Pulse Systems. Mir, Moscow, 1971.

Hale, J.K.

(1966a)   Oscillations in Nonlinear Systems. Mir, Moscow, 1966.

(1966b)   * *Averaging methods for differential equations with retarded arguments
              and a small parameter*. J. Different. Equat., **2**, (1966), 57–79.

Kolmogorov, A.N., and Fomin, S.V.

(1976)   Elements of the Theory of Functions and Functional Analysis. Nauka,
              Moscow, 1976.

Krasnoselsky, M.A.

(1963)   *Alternative principle for the existence of periodic solutions of differential
              equations with retarded argument*. Dokl. Akad. Nauk SSSR, **152**, (1963),
              801–804.

(1965)   *On some new methods in the theory of the periodic solutions of ordinary
              differential equations*. – In: Proceedings of the Second Congress in Theo-
              retical and Appl. Mech., **vol. 2**, 81–96, Nauka, Moscow, 1965.

(1966)   Operator of Translation Along the Trajectories of Differential Equations.
              Nauka, Moscow, 1966.

Krasnoselsky, M.A., Burd, V.Sh., and Kolesov, Yu.V.

(1970)   Nonlinear Almost Periodic Oscillations. Nauka, Moscow, 1970.

Krasnoselsky, M.A., Lifchitz, E.A., and Strygin, V.V.

(1967)   *On a new method in the problem of periodic solutions of the equation
with deviating argument.* – In: Proceedings of the Seminar in Theory of
Different. Equat. with Deviat. Argument, **vol. 5**, 116–120, (Univ. Druzhby
Narodov im. P. Lumumba), Moscow, 1967.

Krasnoselcky, M.A., Perov, A.I., Povolotsky, A.M., and Zabreiko, P.P.

(1963)   Vector Fields on a Plane. Fizmargiz, Moscow, 1963.

Krasovsky, N.N.

(1959)   Some Problems in the Theory of Stability of Motion. Fizmatgiz, Moscow,
1959.

Krylov, N.M., and Bogolyubov, N.N.

(1937)   Introduction to Nonlinear Mechanics. Izdatelstvo Akad. Nauk Ukrain.SSR,
Kiev, 1937.

Levitan, B.M.

(1953)   Almost Periodic Functions. Gostekhizdat, Moscow, 1953.

Martinyuk, D.I.

(1967)   *Periodic solutions of nonlinear differential equations of the second order
with retarded argument.* Ukrain. Mat. Zhurn., **19(4)**, (1967), 125–132.

(1968a)   *On periodic solutions of countable systems of periodic differential equations with retarded argument.* Mat. Fizika, **Issue 4**, (1968), 84–89.

(1968b)   *Periodic solutions of countable systems of differential equations with lag.*
In: III Scientific Conf. of Young Ukrainian Mathematicians., 506–509,
Naukova Dumka, Kiev, 1968. (Ukrainian).

(1968c)   *Periodic solutions of nonlinear differential equations of the nth order
with retarded argument.* – In: Proceedings of the Second All-Union High
School Conference on the Theory and Appl. of Different. Eqns with Deviating Argument, 102–104, Izdat. Chernovitzkogo Univ., Chernovtzy,
1968.

(1971)   Lectures on the Theory of Stability of Solutions to the Systems with After-effect. Inst. Matematiki Akad. Nauk Ukrain.SSR, Kiev, 1971.

(1972)   Lectures on the Qualitative Theory of Difference Equations. Nauk. Dumka, Kiev, 1972.

(1975)   *Investigation of the vicinity of smooth invariant toroidal manifold for the system of difference equations.* Dif. Uravn., **11**, (1975), 1474–1484.

(1977)   *On periodic solutions of a functional equation.* – In: Analytic and Qualitative Methods for Investigation of Differential and Difference-Differential Equations, 121–127, Inst. Matematiki Akad. Nauk Ukrain. SSR, Kiev, 1977.

(1982a) *Bubnov-Galerkin's method for finding periodic and quasiperiodic solutions of systems with lag.* – In: Proceedings of the Second Conference "Different. Eqns and Appl.", **vol. 2**, 445–448, Ruse (Bulgaria), 1982.

(1982b) *Periodic and quasiperiodic solutions of difference-differential and difference equations:* Author's abstract of thesis ... doctor phys.-math. sci. Kiev, 1982.

Martinyuk, D.I., and Dankanich, V.A.

(1981)   *Galekrin's method for construction of quasiperiodic solutions of systems with lag.* – In: IX Internat. Conf. on Nonlinear Oscillations, 107–108, Nauk. Dumka, Kiev, 1981.

Martinyuk, D.I., and Fodchuk, V.I.

(1963)   *Periodic solutions of differential equation of the n-th order with lag.* Mat. Fizika, **Issue 4**, (1963), 90–92.

(1966)   *Asymptotic integration of quasi-linear autonomous systems with lag.* Uktrain. Mat. Zhurn., **18(3)**, (1966), 117–119.

Martinyuk, D.I., and Kharabovskaya, L.V.

(1970)   *On periodic solutions of systems of linear differential equations with lag.* Dokl. Akad. Nauk Ukrain.SSR, Ser. A, **1970(3)**, 217–220 (Ukrainian).

Martinyuk, D.I., and Kolomietz, V.G.

(1968)   *Periodic solutions of strongly non-linear systems with lag.* Mat. Fizika, **Issue 4**, (1968), 55–65.

Martinyuk, D.I., and Kozubovskaya, I.G.

(1968)   *On the problem of periodic solutions of quasilinear autonomous systems with lag.* Ukrain. Mat. Zhurn., **20(2)**, (1968), 263–265.

Martinyuk, D.I., Mironov, N.V., and Kharabovskaya, L.V.

(1971a)   *Stability of solutions of difference equations.* – In: Difference-Differential Equations, 45–58, Inst. Matematiki Akad. Nauk Ukrain. SSR, Kiev, 1971.

(1971b)   *Numerical-analytic method for investigation of periodic solutions of non-linear difference equations.* Ibid.: 58–66.

Martinyuk, D.I., and Perestyuk, N.A.

(1974)   *On reducibility of linear systems of differential equations with quasi-periodic coefficients.* Vychislit. i Prilk. Matem., **Issue 23**, (1974), 116 – 127.

(1975a)   *On reducibility of difference equations on torus.* Ibid.: **Issue 26**, (1975), 42–48.

(1975b)   *Reducibility of linear systems of difference equations with a smooth right-hand side.* Ibid., **Issue 27**, (1975), 34–40.

Martinyuk, D.I., and Samoilenko, A.M.

(1967a)   *On periodic solutions of non-linear systems with lag.* Mat.Fizika, Issue 3, (1967), 128–145.

(1967b)   *Periodic solutions of non-linear systems with lag.* – In: Abstracts of All-Union High-School Symposium on Appl. Math. and Cybernetics, 14–15, Izdat. Gorkovskogo univ., Gorky, 1967.

(1970)   *On a method for investigation of periodic solutions of nonlinear systems.* – In: Proceedings of V Internat. Conf. on Nonlinear Oscillations. Inst. Matematiki Akad. Nauk Ukrain. SSR, **vol. 1**, 423–428, Kiev, 1970.

(1974a)   *Existence of invariant manifolds for systems with lag.* Ukrain. Mat. Zhurn., **26**, (1974), 611–620.

(1974b)   *On reducing of the system of difference equations in the vicinity of toroidal manifold.* – In: Nonlinear Effects in Microelectronics and Their Application, p. 3, Izd. Ob-va Znanie Ukrain. SSR, Kiev, 1974.

270   *References*

(1975) *Invariant toroidal manifolds of systems with lag.* – In: IV All-Union Conf. in Theory and Appl. of Different. Equations with Deviating Arguments, 160–161, Nauk. Dumka, Kiev, 1975.

(1976) *Invariant tori for systems with aftereffect.* High School Annual. Appl. Math., **11, vol. 3,** (1976), Bulgaria, 47–53.

Martinyuk, D.I., and Tsyganovsky, N.S.

(1979a) *Invariant manifolds for systems with lag under pulse influence.* Dif. Uravn., **15,** (1979), 1783–1795.

(1979b) *Existence of toroidal manifolds for pulse systems with lag.* – In: V All-Union Conf. in Qualitative Theory of Differential Equations, 116–117, Shtiinza, Kishinev, 1979.

Mitropolsky, Yu.A.

(1964) Problems in the Asymptotic Theory of Nonstationary Oscillations. Nauka, Moscow, 1964.

(1968) *On construction of solutions and reducibility of differential equations with quasiperiodic coefficients.* – In: III All-Union Congress in Theoret. and Appl. Mechanics, 212–213, Nauka, Moscow, 1968.

Mitropolsky, Yu.A., and Lykova, O.B.

(1974) Integral Manifolds in Nonlinear Mechanics. Nauka, Moscow, 1974.

Mitropolsky, Yu.A., and Martinyuk, D.I.

(1969) Lectures on the Theory of Oscillations for Systems with Lag. Inst. Matematiki Akad. Nauk Ukrain. SSR, Kiev, 1969.

(1979) Periodic and Quasiperiodic Oscillations for Systems with Lag. Vyshcha Shkola, Kiev, 1979.

Mitropolsky, Yu.A., Martinyuk, D.I., and Dankanich, V.A.

(1980) *Galerkin's method in the theory of quasiperiodic solutions of nonlinear differential equations with lag.* Ukrain. Mat. Zhurn., **32,** (1980), 553–537.

Mitropolsky, Yu.A., and Mikhajlovskaya, N.A.

(1972) *Periodic solutions of discrete difference equations of the second order.* Ukrain.Mat.Zhurn., **24,** (1972), 543–547.

Mitropolsky, Yu.A., and Moseenkov, B.I.

(1976) Asymptotic Solutions of Partial Differential Equations. Vyshcha Shkola, Kiev, 1976.

Mitropolsky, Yu.A. and Samoilenko, A.M.

(1964) *To the problem of structure of trajectories on toroidal manifolds.* Dokl. Akad. Nauk Ukrain.SSR. Ser.A, **1964** (8), 984–985 (Ukrainian).

(1965) *On construction of solutions of linear differential equations with quasiperiodic coefficients by using the method of rapid convergence.* Ukrain. Mat. Zhurn., **17**(6), (1965), 42–59.

(1972a) *On quasiperiodic oscillations in nonlinear systems.* Ukrain. Mat. Zhurn., **24**, (1972), 179–194.

(1972b) *Quasiperiodic oscillations in the systems of nonlinear mechanics.* Mat. Fizika, Issue 12, (1972), 86–105.

(1976a) *Asymptotic investigation of weakly non-inear system.* Preprint: Akad. Nauk Ukrain.SSR. Inst. Matematiki; 76.5, Kiev, 1976.

(1976) *Investigation of oscillating systems of the second order.* Preprint: Akad. Nauk Ukrain. SSR, Inst. Matematiki; 76.6, Kiev, 1976.

Mitropolsky, Yu.A., Samoilenko, A.M., and Perestyuk, N.A.

(1977) *On the problem of justification of the averaging method for equations of the second order under pulse influence.* Ukrain. Mat. Zhurn., **29**, (1977), 750–762.

Mitropolsky, Yu.A., Samoilenko, A.M., and Tsydilo, K.V.

(1977) *On invariant toroidal manifolds for nonlinear systems with lag.* – In: Differential Equations with Deviating Arguments, 207–214, Naukova Dumka, Kiev, 1977.

Mozer, Yu.

(1968) *Rapidly convergent method of iterations of non-linear differential equations.* Usp. Mat. Nauk, **23**(4), (1968), 179–238.

Myshkis, A.D.

(1972) Linear Differential Equations With Retarded Argument. Nauka, Moscow, 1972.

Neimark, Yu.I.

(1972)   Method of Point Mappings in the Theory of Nonlinear Oscillations. Nau-
              ka, Moscow, 1972.

(1975)   *Invariant manifolds of differential equations with variating arguments.* –
              In: IV All-Union Conference in Theory and Appl. of Differential Equati-
              ons with Deviating Arguments, 180–191, Naukova Dumka, Kiev, 1975.

(1978)   Dynamical Systems and Controlled Processes. Nauka, Moscow, 1978.

Nurzanov, O.D.

(1977a)   *Numerical-analytic method for investigation of a class of nonlinear peri-
              odic systems of integro-differential equations.* Mat. Fizika, **Issue 22**,
              (1977), 22–30.

(1977b)   *On periodic solutions of nonlinear integro-differential equations.* Dokl.
              Akad. Nauk Ukrain.SSR, Ser. A, **1977** (7), 595–599.

Ordynskaya, Z.P.

(1976)   *On the behavior of solutions of a systems of differential equations with
              lag near integral manifold.* Dif. Uravn., **12**, (1976), 1446–1454.

(1977)   Invariant Toroidal Manifolds for Systems with Lag.: Author's abstract of
              the thesis ... cand, phys.-math. sci. Kiev, 1977.

(1979)   *On the existence of the invariant manifold for a system of equations with
              lag.* Mat. Fizika, **Issue 19**, (1979), 49–57.

Parasyuk, I.O.

(1978)   Construction and Investigation of Quasiperiodic Solutions of Some Clas-
              ses of Differential Equations.: Author's abstract of the thesis ... cand.
              phys.-math. sci., Kiev, 1978.

Pelyukh, G.P., and Sharkovsky, A.N.

(1974)   Introduction to the Theory of Functional Equations. Nauk. Dumka, Kiev,
              1974.

Petrovsky, N.G.

(1964)   Lectures on the Theory of Ordinary Differential Equations. Nauka, Mos-
              cow, 1964.

Pinney, E.

(1961)   Ordinary Difference-Differential Equations. Izdat. Inostr. Liter., Moscow, 1961.

Translated From:
            * Univ. of California Press, Berkeley & Los Angeles, 1958.

Pliss, V.A.

(1964)   Non-Local Problems in the Theory of Oscillations. Nauka, Moscow-Leningrad, 1964.

Rozhkov, V.I.

(1968)   *Asymptotics of the periodic solution of the equation of neutral type with small lag.* Dokl. Akad. Nauk SSSR, **180**, (1968), 1041–1044.

(1970)   *Asymptotic properties of the periodic solutions of systems of equations of neutral type with small lag.* – In: Proceedings of V Internat. Conf. on Non-Linear Oscillations, **vol. 1**, 475–482, Inst. Matematiki Akad. Nauk Ukrain. SSR, Kiev, 1970.

Rubanik, V.P.

(1969)   Oscillations of Quasilinear Systems with Lag. Nauka, Moscow, 1969.

Ryabov, Yu.A.

(1960a)   *Application of the small parameter method to construction of solutions of differential equations with retarded argument.* Dokl. Akad. Nauk SSSR, **133**, (1960), 288–291.

(1960b)   *Application of the small parameter method to investigation of automatic regulation with lag.* Avtomatika i Telemekhanika, **21**, (1960), 729–739.

(1961)   *Application of the Lyapunov-Poincare small parameter method in the theory of systems with lag.* Inzhenern. Zhurn., **1**(2), (1961), 3–15.

(1962)   *Small parameter method in the theory of periodic solutions of differential equations with lag.* – In: Proceedings of Seminar in Theory of Differential Equations with Deviating Arguments, **Issue 1**, 103–113, Univ. Druzhby Narodov im. P. Lumumba, Moscow, 1962.

(1964)   *Analysis of the nonlinear oscillations in systems with small lag.* – In: The Second All-Union Congress in the Theoret. and. Appl. Mechanics, 188–189, Nauka, Moscow, 1964.

Ryabov, Yu.A., and Tolmachev, I.L.

(1969)   *Construction of quasiperiodic solutions in the problems of the theory of nonlinear oscillations by computer.* – In: Abstracts of Papers of V Internat. Conf. on Nonlinear Oscillations, p.189, Inst. Matematiki Akad. Nauk Ukrain. SSR, Kiev, 1969.

Sacker, R.

(1965)   * *A new approach to the perturbation theory of invariant surfaces.* Communs Pure and Appl. Math., **18**, (1965), 717–732.

(1969)   *A perturbation theorem for invariant manifold and Holder continuity.* Math. and Mech., **18**, (1969), 705–762.

Samoilenko, A.M.

(1964)   *To the problem on the structure of trajectories on the torus.* Ukrain. Mat. Zhurn., **16**, (1964), 769–782.

(1965)   *Numerical-analytic method for investigation of periodic systems of ordinary differential equations.* I. Ibid., **17(4)**, (1965), 82–83.

(1966a)   *Numerical-analytic method for investigation of periodic systems of ordinary differential equations.* II. Ibid., **18(2)**, (1966), 50–59.

(1966b)   *On the reducibility of a system of ordinary differential equations in the vicinity of a smooth integral manifold.* Ukrain. Mat. Zhurn., **18(6)**, (1966), 41–64.

(1966c)   *On reducibility of a system of ordinary differential equations in the vicinity of a smooth toroidal manifold.* Izv. Akad. Nauk SSSR, Ser. Math., **30**, (1966), 1047–1072.

(1968)   *On reducibility of systems of linear differential equations with quasiperiodic coefficients.* Ukrain. Mat. Zhurn., **20**, (1968), 279–281.

(1970a)   *On the preservation of invariant torus under perturbation.* Izv. Akad. Nauk SSSR, Ser. Math., **34**, (1970), 1219–1240.

(1970b)   *To the theory of perturbations of invariant manifolds for dynamical systems.* – In: Proceeding of V Internati. Conf. of Nonlinear Oscillations, **vol. 1**, 495–499, Inst. Matematiki Akad. Nauk Ukrain. SSR, Kiev, 1970.

(1975)   *On exponential stability of the invariant torus of a dynamical system.* Dif. Uravn., **11**, (1975), 820–834.

Samoilenko, A.M., Martinyuk, D.I., and Perestyuk, N.A.

(1973) *Invariant tori of difference equations.* Dif. Uravn., **9**, (1973), 1904–1910.

Samoilenko, A.M., and Nurzhanov, O.D.

(1979) *Bubnov-Galerkin's method for construction of periodic solutions of integro-differential equations of the Volterra type.* Dif. Uravn., **15**, (1979), 1503–1517.

Samoilenko, A.M., and Parasyuk, I.O.

(1977) *On Galerkin's method in the theory of perturbations of invariant tori.* Dokl. Akad. Nauk Ukrain.SSR. Ser. A, **1977** (2), 112–115 (Ukrainian).

Samoilenko, A.M., and Ronto, N.I.

(1976) Numerical-Analytic Methods for Investigation of Periodic Solutions. Vyshcha Shkola, Kiev, 1976.

Samarsky, A.A.

(1977) Theory of Difference Schemes. Nauka, Moscow, 1977.

Samarsky, A.A., and Gulin, A.V.

(1973) Stability of Difference Schemes. Nauka, Moscow, 1973.

Samarsky, A.A., and Karamzin, Yu.N.

(1978) Difference Equations. Znanie, Moscow, 1978.

Shimanov, S.N.

(1957) *On a method for finding under what condition periodic solutions of nonlinear systems exist.* Prikl. Matem. i Mekhan., **19**, Issue 2, (1957), 225–228.

(1959) *To the theory of oscillations of quasi-linear systems with lag.* Ibid.: **23**, Issue 5, (1959), 836–844.

(1960) *Oscillations of quasi-linear autonomous systems with lag.* Izv. Vuzov. Radiofizika, **3**, (1960), 456–466.

(1965) *To the theory of differential equations with aftereffect.* Dif. Uravn., **1**, (1965), 102–116.

(1970) *To the theory of periodic oscillations of quasi-linear non-autonomous periodic systems with periodic lags.* – In: Proceedings of V Internat. Conf. on Nonlinear Oscillations, **vol. 1**, 617–622, Inst. Matematiki Akad. Nauk Ukrain. SSR, Kiev, 1970.

Sobolev, S.L.

(1950) Some Applications of the Functional Analysis in Mathematical Physics. Izdat. Leningrad. Univ., Leningrad, 1950.

(1966) Equations of Mathematical Physics. Nauka, Moscow, 1966.

Tkach, B.P.

(1969) *On periodic solutions of a countable system of differential equations with deviating argument of neutral type.* Ukrain. Math. Zhurn., **21(1)**, (1969), 73 – 85.

Tsydilo, K.V.

(1973) Oscillations of Nonlinear Systems with Lag.: Author's abstract of the thesis ... cand. phys.-math. sci., Kiev, 1973.

Urabe, M.

(1965) * *Galerkin's procedure for nonlinear periodic systems.* Arch. Ration. Mech. and Anal., **20(2)**, (1965), 120–152.

(1966) *Galerkin's method for nonlinear systems.* Mekhanika, **97(3)**, (1966), 3–34.

(1972) * *Existence theorems of quasiperiodic solutions to nonlinear differential systems.* Funkc. Ekvacioj, **15(1)**, (1972), 65–100.

(1973) * *On a modified Galerkin's procedure for nonlinear quasiperiodic differential systems.* – In: Equations Differentielles et Fonctionnelles Non Lineaires, 223–258, Paris, 1973.

Vakhabov, G.

(1968) *On periodic solutions of nonlinear systems of integro-differential equations.* – In: Proceedings of the Seminar in Math. Physics and Nonlinear Oscillations, **vol.1**, issue 2, 95–107, Naukova Dumka, Kiev, 1968.

(1969) *Numerical-analytic method for the investigation of periodic systems of integro-differential equations.* Ukrain. Mat. Zhurn., **21**, (1969), 675–683.

Volosov, V.M. and Morgunov, B.I.

(1971) Averaging Method in the Theory of Nonlinear Oscillating Systems. Izdat. Mosk. Univ., Moscow, 1971.

Vuitovich, B.

(1982) *On numerical-analytic method for investigation of integro-differential equations.* Vestnik Kievsk. Univ. Matematika i Mekhanika, **Issue 24,** 1982, 14–21. (Ukrainian).

Vuitovich, B. and Nurzhanov, O.D.

(1982) *Bubnov-Galerkin's method for nonlinear periodic systems of integro-differential equations with infinite aftereffect.* Preprint: Akad. Nauk Ukrain. SSR, Inst. Matematiki; 82.50, Kiev, 1982.

Zavakykut, G.D.

(1983) *Numerical-analytic method for the investigation of periodic solutions of a class of operator-differential equations.* Dif. Uravn., **19,** (1983), 569–575.

Zverkin, A.M.

(1970) *On integral manifolds for systems with lag.* – In: Proceedings of V Internat. Conf. on Nonlinear Oscillations, vol. **1,** 275–282. Inst. Matematiki Akad. Nauk Ukrain. SSR, Kiev, 1970.

Zverkin, A.M., Kamensky, G.A., Norkin, S.B., and Elsgoltz, L.E.

(1962) *Differential equations with deviating argument.* Usp. Mat. Nauk, **17(2),** (1962), 77–164.

(1963) *Differential equations with deviating argument.* – In: Proceedings of the Seminar in Theory of Differential Equations with Deviating Argument, vol. **2,** 3–49, Univ. Druzhby Narodov im.P.Lumumba. Moscow, 1963.

Subject Index